STATISTICS

ABOUT THE AUTHOR Since receiving his Ph.D. from the University of Michigan in 1972, Allan G. Johnson has taught at Wesleyan University, Harvard University, Dartmouth College, Hartford College for Women, and the University of Connecticut, Hartford. His professional interests include social stratification, the sociology of gender, research methods and statistics, and demography.

STATISTICS

Allan G. Johnson

Under the General Editorship of
Robert K. Merton
Columbia University

HARCOURT BRACE JOVANOVICH, PUBLISHERS

San Diego   New York   Chicago   Austin   Washington, D.C.
London   Sydney   Tokyo   Toronto

Copyright © 1988 by Harcourt Brace Jovanovich, Inc.

All rights reserved. No part of this publication may be reproduced or transmitted in any form or by any means, electronic or mechanical, including photocopy, recording, or any information storage and retrieval system, without permission in writing from the publisher.

Requests for permission to make copies of any part of the work should be mailed to: Permissions, Harcourt Brace Jovanovich, Publishers, Orlando, Florida 32887.

ISBN: 0-15-583542-4

Library of Congress Catalog Card Number: 87-81141

Printed in the United States of America

Copyrights and Acknowledgments and Illustration Credits appear on pages 424–25, which constitute a continuation of the copyright page.

PREFACE

To Students

The use of statistical evidence and reasoning has expanded enormously since World War II and can now be found in virtually all of the social and behavioral sciences and related areas of research and practice, including sociology, psychology, political science, economics, anthropology, history, education, public health, and business. I have written *Statistics* to help you make sense of it all, to understand statistical language, techniques, and principles well enough to evaluate critically and learn from the vast and growing literature that makes use of them.

I have not set out to make researchers of you. Instead, I have written for those who may or may not consider themselves mathematically sophisticated (who may, in fact, dislike math), but who nonetheless want to understand what science has to offer—who, as they skip over all the figures and tables, have an uneasy feeling that they are not equipped to decide for themselves whether an author's arguments and conclusions and the evidence on which they are based are worth believing, much less to draw their *own* conclusions from a set of data.

To learn statistics, you have to be willing to think hard about what you read, to go back and try again if you feel uneasy after the first few attempts. In cases where students have experienced difficulty in the past, I have explained concepts in more than one way. If as you read you sense some redundancy from time to time, try to resist the temptation to skip ahead, for each example, each approach to a concept, procedure, or interpretation, offers its own nuances that can make the difference between superficial learning and a deeper understanding.

It would be a mistake to get the idea that, because I am not trying to make statisticians of you, this book makes statistics easy. I have tried to omit those formal mathematical aspects of statistics that do not add enough to most people's understanding to justify the effort it takes to understand them. In this way I have tried to humanize statistics in order to make it more understandable, more useful, and less intimidating for the ever-widening audience of students in many different fields who must depend on it for some portion of their academic work. If you find yourself wanting to probe more deeply, consider the "One Step Further" boxed inserts within the chapters and the references listed at the back of the book.

Try to be patient with yourself as you learn new concepts. Acquire them at your own pace and try to learn them so that you can explain them in your own words to someone who does not already understand. As a first-year sociology graduate student at the University of Michigan, I found statistics a challenging subject but was able to earn A's in most of my courses. When I was asked to tutor other students, however, and found myself trying to explain the same concepts I had understood so well on exams, I had some rude shocks and had to wonder just what it was that I *did* understand. I discovered that being able to give back an answer on an exam reflects only one level of learning (and not a very deep one at that). The deeper understanding comes when you can teach others, when you understand something well enough to put it in your own words and illustrate it with examples from your own experience.

Each chapter concludes with a summary, a list of key terms, and a set of problems. Please note that the answers to most end-of-chapter problems are found at the end of the text. The only answers you will not find there are those to questions that clearly have many different acceptable answers, such as those that ask you to think of an example of something. Also note that there is an extensive glossary complete with appropriate symbols and page references to initial discussions of each concept.

To Instructors

I have tried to write *Statistics* in a way that will allow instructors and their students to have their cake and eat it, too. For instructors who stress interpretation over computation, the text isolates much of the mathematical formulations that are unnecessary for a basic understanding of statistics and, by intimidating many students, only interfere with their learning. For those who want their students to have a deeper grounding, however, the boxed inserts, "One Step Further," contain additional major materials you will need for an introductory course in statistics.

This organization allows students to acquire a basic understanding of statistics presented in conventional language and with the option of pursuing deeper, more mathematical subtleties at any time and without disrupting the flow of the text. Students can digest new materials at a more gradual pace, rather than having to deal with new ideas and mathematical formulations at the same time. This slows the pace and lets students concentrate on understanding where statistics come from and what they mean without cluttering the discussion unnecessarily with computations and theoretical subtleties. And yet, the important materials are still readily available, carefully placed at the appropriate points in the text.

The book is divided into two parts, descriptive statistics and inferential statistics. I have tried to order the topics in Part I as they occur in the research process, beginning with measurement and working toward increasingly sophisticated techniques. The progression from chapter to chapter is cumulative, with the possible exception of Chapter 5 (probability), which can be presented earlier if you wish (although I do not recommend discussing it much later).

The topics in Part II are closely interconnected, although it would be possible to reverse the order of Chapters 12 and 13, presenting hypothesis testing before estimation. I do not recommend this because I think that students find it easier to tackle estimation first and that they grasp the inference process as a whole more firmly if they begin with confidence intervals. The reverse logic of hypothesis testing can be confusing to many students, who may be still struggling with sampling distributions and the normal curve when they begin their study of inference techniques. Beginning with estimation and using this as a foundation for hypothesis testing (with the potential for some interesting comparisons) has worked very well with my students, and I strongly recommend it. Reversing the order, however, can be done with the chapters in their present form without too much groundwork on your part.

You may notice the absence of some relatively advanced topics. I do not discuss analysis of variance or log-linear analysis, for example, because I think they are too advanced for most one-semester undergraduate courses. For those who miss a discussion of nonparametric statistics, please note that the relevant tables (in addition to random number and F tables) are nonetheless available in the Appendix for those who want to teach these topics on their own without the bother of reproducing copies of tables for their students.

To accompany *Statistics*, I have written a Study Guide for students, which includes a large number of additional problems (with answers) similar to those found at the end of each chapter, as well as exercises designed to help students organize the materials and solidify their understanding of key concepts. In addition, I have prepared an Instructor's Manual that contains a large number of problems (also with answers), almost all of which rely on real data and which can be used either as classroom examples or on exams. In writing it, I have tried to provide the kinds of resources that instructors find most difficult to develop on their own.

Acknowledgments

I want to thank several people for their contributions to all that it took to make *Statistics* a reality. James A. Davis, Edmund Meyers, and David Goldberg are three sociologists from whom I acquired an early appreciation for the rigors and pleasures of serious empirical research. More recently, Harcourt Brace Jovanovich's advisory editor in sociology, Robert K. Merton, has brought to this project not only his extraordinary knowledge of the social sciences, but his sharp, critical insistence on utter clarity coupled with his eye for enriching detail—both of which, I hope, have helped to make this a book that students not only will learn from , but will actually enjoy reading in ways rarely if ever associated with statistics texts.

I am indebted to Edmund Meyers for his encouragement and detailed suggestions that contributed immeasurably to the spirit and quality of an earlier published version of this book, and to Candace Clark for her particularly thoughtful, penetrating review of the manuscript for this one. I also want to express my appreciation for the many useful suggestions of Steven Green, Michigan State University, Lawrence L. Lapin, San Jose State University, Richard Rubinson, Florida State University, and Bruce C. Straits, University of California, Santa Barbara.

I am grateful to the Kiewit Computation Center at Dartmouth College and its director, Eugene A. Fucci, for their generosity in allowing me to use their facilities to analyze data from the General Social Surveys. My thanks also go to Joel Levine, professor of mathematical social science, whose development of a PC version of Dartmouth's *IMPRESS* mainframe data analysis software enabled me to generate additional examples from the General Social Surveys; and to James A. Davis, whose innovative *Pathfinder* software gave me access to a treasure trove of ratio-scale data and the wherewithal to analyze them using multiple regression and path analysis on my humble PC.

I am grateful to the literary executor of the late Sir Ronald A. Fisher, F.R.S., to Dr. Frank Yates, F.R.S., and to Longman Group Ltd., London, for permission to reprint tables from their book, *Statistical Tables for Biological, Agricultural, and Medical Research*, sixth edition, 1974.

I have now had the privilege of seeing two books published with HBJ, and this latest experience has shown the professionalism and sensitivity with which the editorial and production staff goes about its work on projects both large and small. My special thanks go to Marcus Boggs for his continuing support for my work, to Johanna Schmid for her timely efforts at times of crucial decisions, to Karen Davidson for her expert editing of the manuscript, to Ruth Cornell for overseeing a smooth transition into print, to Jane Carey for designing the book and creating a cover that I knew was just right the instant I saw it, to Robin Risque for her coordination of the art program, and to Lesley Lenox for managing the production process.

Finally, I want to express my special appreciation to George Zepko for his friendship and for being one of the rare people in my life who both understands and supports those quantitative, computative aspects of me that made this book possible; and to Stuart Alpert, Naomi Bressette, Brent Harold, Susan Weegar, Ellen Allen, Annalee, Valdemar, Geraldine, Emily, and Paul Johnson, and, especially, Nora Jamieson go my love and my deep and abiding appreciation for the richness they bring to this writer's life.

Allan G. Johnson

Brief Contents

I Description and Explanation

1	An Introduction to Statistics and the Problem of Knowing	3
2	Measurement	13
3	Distributions, Comparisons, and Tables	30
4	Summarizing Distributions	65
5	Elementary Probability Theory	101
6	Relationships between Two Variables: Nominal and Ordinal Scales	123
7	Relationships with Interval and Ratio Scales: Regression and Correlation	164
8	Relationships of More Than Two Variables: Controlling	191
9	Multivariate Analysis	226

II Statistical Inference

10	Sampling	248
11	Statistical Inference	268
12	Estimating Population Means and Proportions	295
13	Hypothesis Testing	325

Detailed Contents

Preface	v

I Description and Explanation

1 An Introduction to Statistics and the Problem of Knowing — 3

Statistics and Science	6
Asking the Right Questions	8
Learning a New Language	9
Where We Go from Here	10
Summary	12
Key Terms	12

2 Measurement — 13

From Concepts to Variables	13
Validity and Reliability	16
Measurement Error	19
Types of Variables	21
Qualitative (Nominal) Variables	22
Quantitative Variables	22
Ordinal-Scale Variables	23
Interval-Scale Variables	24
Ratio-Scale Variables	25
Why Measurement Scales Are Important	25
The Dichotomy: A Special Case	26
Summary	27
Key Terms	28
Problems	28

3 Distributions, Comparisons, and Tables — 30

Frequency Distributions	30
Tables and How to Read Them	30
Grouped Data	33

Comparing Distributions	34
Marginals and Cells	35
Percentages and Proportions	36
Cross-tabulations	39
Which Way to Percentage?	39
More Complicated Tables	42
Ratios	45
Graphics: Describing Distributions with Pictures	48
Bar Graphs	49
Line Graphs	52
Histograms and Frequency Polygons	55
Stem and Leaf Displays	56
Ogives	58
Scatterplots	59
Statistics and Truth: Some Perspective	61
Summary	61
Key Terms	62
Problems	62

4 Summarizing Distributions   65

Central Tendency: The Average	66
The Mean	67
The Balance-Beam Analogy	68
Box 4–1 One Step Further: The Mathematics of Means	69
Using Means to Compare Groups	70
Box 4–2 One Step Further: The Mathematics of Means with Grouped Data	72
Race and Fertility: An Example	73
The Median	75
The Balance Beam Revisited	76
Box 4–3 One Step Further: The Mathematics of Medians	77
The Median with Ordinal Variables	77
Income, Gender, and Education: An Example	79
Some Cautions about Means and Medians	80
Percentiles	80

The Mode	81
Choosing a Measure of Central Tendency	82
Measuring Trends: Panels, Cohorts, and Longitudinal Effects	82
Measuring Variation	86
The Variance and Standard Deviation	86
Interpreting the Variance and Standard Deviation	90
Box 4–4 One Step Further: The Mathematics of Variance and the Standard Deviation	91
The Range	92
Central Tendency, Variation, and the Shape of Distributions	92
Summary	96
Key Terms	97
Problems	97

5 Elementary Probability Theory 101

The Importance of Probability in Statistics	101
Some Basic Ideas	102
Probabilities of Simple Events	105
Box 5–1 One Step Further: Venn Diagrams and Simple Events	106
Conditional and Unconditional Probabilities	106
Probability Distributions	107
Probabilities of Joint Events	110
Box 5–2 One Step Further: Venn Diagrams and Joint Events	111
Independence between Events	111
Independence Is Symmetrical	113
Some Examples of Independence	114
More Complicated Joint Events	115
Compound Events: Adding Probabilities	116
Mutually Exclusive Events	117
Box 5–3 One Step Further: Venn Diagrams and Compound Events	117
Expected Values	118
Summary	119

Key Terms	120
Problems	120

6 Relationships between Two Variables: Nominal and Ordinal Scales — 123

Independence and Dependence	123
A Probability Approach	125
How to Describe Relationships between Variables	128
The Direction of Relationships	128
Linear Relationships	128
Nonlinear Relationships	129
The Strength of Relationships	133
Relationships between Nominal-Scale Variables	134
Direction in Nominal-Scale Relationships	134
Nominal-Scale Measures of Association	137
Yule's Q and the Logic of Paired Comparisons	138
The Phi Coefficient	143
Goodman and Kruskal's Tau	146
Relationships between Ordinal-Scale Variables	149
Ordinal-Scale Measures of Association	149
Goodman and Kruskal's Gamma	149
Somers' d	153
Rank-Order Measures of Association	154
Spearman's r_s	154
Kendall's Tau	156
Summary	158
Key Terms	160
Problems	160

7 Relationships with Interval and Ratio Scales: Regression and Correlation — 164

The Basics of Linear Regression	165
The Regression Line	166
Calculating the Slope	167
Calculating the Regression Constant	169
Box 7–1 One Step Further: Computing the Slope	170

Correlation: How Strong Is the Relationship?	171
Box 7-2 One Step Further: Computing the Correlation Coefficient	174
Explained Variance	177
Unexplained Variance and the Standard Error of Estimate	180
Interpreting Slopes and Correlations	182
Box 7-3 One Step Further: Variance and the Relative Size of Slopes and Correlations	183
Stochastic and Mean Independence	184
A Note on Curvilinear Regression	186
Summary	187
Key Terms	188
Problems	188

8 Relationships of More Than Two Variables: Controlling   191

The Logic of Controls	192
The Order of Relationships	193
Spuriousness: Is the Relationship Real?	194
Testing for Spuriousness	196
The Importance of Testing for Spuriousness	198
Box 8-1 One Step Further: Understanding Spuriousness with Concordant and Discordant Pairs	199
Partials and Extraneous Variables	200
Understanding Causal Relationships: Intervening Variables	200
Intervening Variables: An Example with Data	203
Component Variables	206
Antecedent Variables	208
Suppressor Variables	209
Suppressor Variables: An Example with Data	211
Distorter Variables	213
Specification and Interaction Effects	214
Controls and Small Samples: Running Out of Cases	218

	Summary	219
	Key Terms	222
	Problems	222
9	**Multivariate Analysis**	**226**
	Multiple Regression and Correlation	227
	Standarized Slopes	229
	An Example of Multiple Regression Analysis in Action	231
	Using Duncan's Equations	235
	Inside the Equations	236
	Path Analysis	238
	Interpreting Path Diagrams	240
	Summary	242
	Key Terms	242
	Problems	243
II	**Statistical Inference**	
10	**Sampling**	**248**
	Sampling the Assembly Line: A "Simple" Sample	249
	Sample Size	249
	Making a Frame	249
	The Three Rules of Scientific Sampling	250
	Random Sampling	252
	Systematic Sampling	253
	Sampling Error	254
	Stratification: Minimizing Random Error	255
	Response Rates: Minimizing Bias	256
	Oversampling	258
	Complex Samples	260
	Multistage Samples	260
	Cluster Samples	261
	Let the Buyer Beware	264
	Summary	265
	Key Terms	266
	Problems	266

11 Statistical Inference — 268

- Some New Ways of Looking at Distributions — 268
  - Some New Graphics: Probability Density Functions — 269
- Theoretical Probability Distributions: Sampling Distributions — 272
- The Characteristics of Sampling Distributions — 275
  - The Standard Error — 276
    - Calculating the Standard Error — 278
  - The Central-Limit Theorem and the Shape of Sampling Distributions — 279
- Getting Some Perspective — 284
- The Normal Distribution — 285
  - Using the Normal Distribution — 286
  - The Normal Curve and Sampling Distributions — 289
  - Translating between Standard Errors and Unstandardized Scores — 290
- Summary — 292
- Key Terms — 293
- Problems — 293

12 Estimating Population Means and Proportions — 295

- Point Estimates — 296
- Confidence Intervals: Estimating Population Means — 297
- How Confidence Intervals are Actually Constructed — 304
  - Student's t-Distribution: Z for Small Samples — 307
- Estimation and Sample Size — 309
- Estimating Population Proportions — 310
- Box 12–1 One Step Further: PQ as a Variance — 311
  - Some Instant Confidence Intervals for Proportions — 312
- Estimating Differences between Sample Proportions — 314
  - Open-ended Confidence Intervals — 316
- Estimating Differences between Sample Means — 317
- The Perpetual Trade-off: Confidence, Precision, and Sample Size — 319

Interpreting Confidence Intervals: Some Cautions	320
Summary	320
Key Terms	321
Problems	322

13 Hypothesis Testing 325

Hypothesis Testing as an Everyday Activity	326
Testing the Null Hypothesis	326
Rejecting the Null Hypothesis: The Probability of Error	328
Rolling Dice: A Second Example	328
Single-Sample Hypothesis Tests: Means	329
Interpreting Hypothesis Test Results	332
Single-Sample Hypothesis Tests: Proportions	334
Two-Sample Tests: Testing for Differences between Populations	336
The Confidence Interval Alternative	339
About "Significant Differences"	340
Chi-Square: Hypotheses about Relationships between Variables	341
Testing the Null Hypothesis	342
Chi-Square and Goodness-of-Fit Tests	349
The Index of Dissimilarity	352
Hypothesis Tests for Relationships between Variables	353
On the Uses and Limitations of Hypothesis Testing	355
Summary	356
Key Terms	357
Problems	357

Epilogue	361
Appendix Tables	365
Table 1 Areas under the Normal Curve (the Distribution of Z)	365
Table 2 Student's t-Distribution	366
Table 3 The Distribution of Chi-Square	367
Table 4 Some Random Numbers	368

Table 5 The Distribution of F — 370
Table 6 Critical Values for the Wilcoxon Signed-Rank Test — 376
Table 7 Critical Values for the Mann-Whitney Test — 377
Table 8 Critical Values for the Wald-Wolfowitz Runs Test — 379

Glossary (with symbols and page references) — 380

References — 395

Answers to Selected Problems — 401

Index — 427

STATISTICS

I

Description
and Explanation

1

An Introduction to Statistics and the Problem of Knowing

Statistics is a word that can be used to describe a set of techniques for organizing, summarizing, and interpreting quantitative information. By "quantitative information," I mean information that lends itself to being counted and sorted and represented by numbers, such as the number of suicides that occur each year, the number of people who vote in an election, or the number of artifacts found in an archeological dig. Thus, the various techniques that can be used to describe how suicide rates rise with increasing age among men but not among older women are included in the idea of statistics as a set of techniques generated by this field of study.

Statistics also refers to the specific results of statistical practice. A percentage, for example, can be referred to as a **statistic** (notice that it is singular here, rather than plural as above), as can a *mean* or *median* or *correlation coefficient* (all of which you will understand by the end of this book).

If you are interested in what is going on in the world around you, it is likely that you find yourself increasingly reading articles and books that use statistics. Whether they are found in newspapers and magazines, or assigned readings in scores of college courses in many different fields, statistics pose particular problems for the untrained reader. If we do not understand where statistics come from and how they are used, we are often caught between two unattractive alternatives. On the one hand, we may accept conclusions that are not supported by the evidence, especially when they happen to agree with our own views of how the world works. On the other hand, we may feel tempted to reject a conclusion, because of the commonly held position that "since you can prove anything with statistics there is no more reason to accept a conclusion than to reject it." In short, a lack of training in statistics encourages us to reject valid conclusions that we disagree with and to accept false ones that are more to our liking. With the increasing use of statistics in so many diverse fields, a poor understanding of statistics is a prescription for ignorance and misunderstanding.

In a most general way, then, statistics represents a set of tools used to help make a particular kind of sense of the world, a kind of sense rooted in the scientific method. All of us, every day of our lives, are involved on a continuing basis in the general problem of understanding what is going on around us, of gathering, organizing, and interpreting observations. When you get up in the morning, you may wonder about something as simple as what the weather is going to be like today, and whether you should take a raincoat or wear a heavy jacket. You might look out of the window to see what the sky looks like, or you might stick your head out to see how cold it is. The way the sky looks and the way the air feels might give you some clues, but not enough on which to base a firm conclusion, so you turn on the radio or scan the morning newspaper for a forecast. Regardless of what the local meteorologist tells you, you have been around long enough to know that weather forecasts are often wrong, so you have to include some consideration of the probability of error in your decision. In your informal gathering of weather information, you inevitably select some bits of information and ignore others. You might look out of an east-facing window and see sunny skies, but had you looked to the west—the direction from which weather changes typically come—you might have seen clouds moving in.

Something as simple as predicting the weather is full of the same problems confronted by scientists in their work. When researchers gather information—whether on people, events, or entire societies in different historical eras—they face the problem of selecting a set of representative observations from what is often an infinite number of possibilities, and organizing and interpreting them in a way that both makes sense and represents some identifiable portion of reality. They want to find the story their observations tell, and they want the story to tell something about the way things really are (or, in the case of history, about the way things really were). No one wants to be misled. We want information to actually mean what we think it means. In short, we want facts that are gathered, organized, summarized, analyzed, and presented in ways that allow us to draw intelligent conclusions about ourselves and the world.

In many cases, the things we want to understand are far more consequential than the weather, and the answers we arrive at affect not only the theories we adopt to explain the world, but the lives and well-being of the people who live in it. Consider, for example, the general problem of race relations in the United States. Anyone who is reasonably well informed knows that the U.S. has racial problems, from cultural prejudice based on race to the large underclass of poor blacks who are excluded from what is otherwise one of the most prosperous societies in the world.

It is one thing to have a general perception of social conditions in a society, of how those conditions have changed historically, or of how they compare with those found in other societies. It is quite another thing, however, to understand the causes and consequences of something like racism in enough detail to be able to bring about meaningful change, or to be aware of it as it occurs. Many Americans share the general perception, for example, that race is a factor in legal proceedings, that black and white defendants are treated unequally, or that the legal system defends the interests of white victims more vigorously than it does those of blacks. There are many others, however, who perceive the legal system as being essentially color blind, and who deny that race is any longer a factor in court proceedings.

How do we find the truth? Some are content to cling to their own perceptions of how the world works, perceptions gathered unsystematically from diverse and often unreliable sources, from television shows to informal conversations with friends. Scientists, however, measure their understanding of the world by a more rigorous standard, one that requires them to gather and interpret information in ways that would lead others who use the same methods to arrive at the same conclusions. The scientific method demands that rather than settle for the first potential explanation that presents itself, we test a variety of alternatives, and thereby remain open to the possibility that even our most cherished ideas may turn out to be wrong.

The problem of documenting the influences of race in social life is a classic case in the use of statistics and scientific methods in social and behavioral research. Consider, for example, a case that was decided by the U.S. Supreme Court in 1987. Statistical evidence that courts were more likely to use the death penalty against blacks than against whites convicted of similar offenses played a prominent part in the 1968 Supreme Court decision that banned use of the death penalty. In the most recent case, however, the issue is not racial discrimination based on the race of the offender, but discrimination against the offender based on the race of the *victim*. As was the case in 1968, the argument relies heavily on statistical evidence and scientific reasoning.

The case centers on a black Georgia man sentenced to death for killing a white policeman during a robbery. The defense argues that Georgia discriminates against those who kill whites, and uses as evidence the fact that those who kill whites are 11 times more likely to be sentenced to death than are those who kill blacks. Georgia argues that the defense is incorrect, not in its statistical evidence but in its interpretation of the evidence. Georgia argues that the defense ignores a crucial third factor in addition to race and the use of the death penalty, and that is the nature of the crime in which the killing occurs. Specifically, Georgia maintains that crime statistics show that whites are more likely than blacks to be killed in crimes such as robbery that tend to provoke "the moral outrage of the community," and therefore, of juries drawn from those communities. Blacks, on the other hand, are more likely to be killed in disputes in family quarrels and other informal settings that do not involve criminal activity.

Aside from its outcome (see Chapter 8), the Georgia case shows the important ways in which systematic statistical evidence contributes to our understanding of how the world works and its consequences for groups and individuals. It also shows, however, that things are rarely as simple as they may seem—a problem that is no less serious for the scientist than it is for us as we go about the business of making sense of our own personal lives.

Statistics and Science

The uses of research can be broken down into three major goals. The first is **description**, which focuses on gathering evidence that tells us how things are or were. How does the average income of women in the U.S. compare with that of men? Who is the most likely to feel stress on the job, high-level executives or their secretaries? Are cancer death rates going up, going down, or staying the same? How does the amount of political freedom found in most capitalist societies compare with that found in most socialist societies? Did the level of social inequality in societies such as Britain and the United States change during the Industrial Revolution? Whether they focus on individuals or entire societies, the answers to such questions simply describe what is going on.

A second major goal of scientific research is **explanation**. Why, for example, do full-time employed women still make only 60 cents for every dollar earned by comparable men? Why do secretaries tend to experience more stress than their bosses? Why are death rates for some cancers going up while they decline for others? Why are cancer death rates higher in some communities than others? How do we explain the fact that democratic institutions tend to be stronger in some societies than in others?

Any explanation is likely to involve a variety of factors that are linked together. Consider, for example, the problem of explaining the income gap between full-time male and female employees. What factors might be involved?

Women are crowded into poorly paid occupations (they are).

Women have less education (they don't).

Men control salaries and discriminate against women even when they perform the same jobs as men (they do).

Women leave the job market to have children, thereby interrupting their career development (they do).

Both wives and husbands tend to value the husband's career more than the wife's (they do, although less than they used to).

Employers assume that families depend more on the earnings of working men than on the earnings of working women (they used to, but they don't anymore, especially in single-parent families).

The factors, in turn, generate questions:

Why are women crowded into poorly paid jobs?

Why doesn't college education pay off for women as much as it does for men?

Why are women discriminated against?

Why don't fathers share child-care responsibilities with mothers?

Why is a wife's work considered of lower value than a husband's?

The statistical fact that women receive less income than men is called a **finding**, as are the answers to each question that we pursue in search of an explanation. Each finding contributes to an increasingly complex explanation of how the different pieces of social life fit together to produce income differentials between men and women. The ultimate goal of scientific research is to formulate possible explanations that can be tested with systematic observations of the world, and then to gather appropriate information and see how well our ideas fit the information. When we find, for example, that women are paid less than men who have much less education, we must reject the idea that income differences are based on inferior education for women.

A third major goal of research is **prediction**, the ability to use what we know about how things are in order to make statements about how things will be in the future. Will U.S. investment in Third World countries improve conditions for the average citizen in those societies or will it make things worse? What kinds of jobs will be most in demand in the next decade and in what occupations are layoffs going to be most likely? How large will the school age population be ten years from now, and how will this affect the job market for teachers at each level of schooling? How will the age structure of the population change, and what kinds of products and services will be most in demand? Which candidates will the voters elect in the next election? Will students who get good grades generally have more successful careers once they get out of school than those who get mediocre grades?

In some cases, the ability to predict the future is based on statistical explanations of cause and effect relationships. We already know, for example, roughly how many high school freshmen there will be 12 years from now because we already know how many babies were born two years ago. In other cases, however, the ability to predict the future is based on statistics that do not necessarily have anything to do with cause and effect. Students who get good grades in high school, for example, tend to get good grades in college; but although grades in one school tend to predict grades in another with some accuracy, this does not mean that we understand why. It could be that grades reflect differences in intelligence, but they could also reflect the degree to which students please their teachers or how hard they work or what kind of cultural background they came from. In short, the ability to predict something does not necessarily mean that we understand it.

Regardless of which of these major goals we focus on, either alone or in combination, science involves the systematic gathering and interpretation of information, a process that involves decisions at each step along the way. Each of these decisions, in turn, affects the kind of information that is gathered and the kinds of conclusions that are drawn from it. Anyone who wants to be in a position to evaluate scientific information has to understand what those decisions are and how they affect the research process.

ASKING THE RIGHT QUESTIONS

Consider, for example, the annual General Social (GENSOC) Survey conducted by the National Opinion Research Center (NORC) at the University of Chicago. This survey gathers information on a wide variety of topics of interest to social scientists in general and sociologists in particular, and we will be relying heavily on the NORC's findings throughout the book. For the moment, consider just one of the questions it asked in 1985: "Please tell me whether or not *you* think it should be possible for a pregnant women to obtain a *legal* abortion if the family has a very low income and cannot afford any more children."

NORC drew a sample of 1,948 dwellings in the continental United States, from which it obtained interviews from 1,534 adults. Trained interviewers went to the selected houses and asked this question among many others and recorded the answers on a printed interview schedule. Each type of response (in this case, yes or no) was assigned a number (in this case, 1 or 2) called a code. The codes are recorded on large sheets of paper (called "code sheets") and then entered into computers. The result is one set of codes for each respondent, including a code of 1 or 2 telling us what his or her answer was to this question. Once in the computer, the responses can be "read" quickly, and accurately counted and sorted.

A consumer of this kind of information should begin by asking several important questions:

1. What does it *mean* when someone says yes or no to this question when the interviewer asks it? How do we know the answer means what we think it means?
2. What is the best way to handle 1,534 bits of information in order to describe what U.S. adults think about this question?
3. To what degree do the answers given by 1,534 adults represent the opinions of the entire adult population that numbers more than 170 million people?
4. Given our information on a sample of U.S. adults, what kinds of things can we say about all U.S. adults? How precise can we be? What are the odds that we are wrong?
5. How do we explain the fact that some adults think abortion under the circumstances stated in the question should be legal while others do not?

These are the kinds of questions that arise with all kinds of research, whether the information is gathered by interviewers asking questions of people in their homes, by experimenters observing subjects, or by historians combing through government archives. Regardless of the field or the topic, whenever statistical information is involved it is important to know what kinds of critical questions to ask about the process through which information is gathered and interpreted. How, for example, were the respondents selected, who did the interviewing, and where? How were the questions worded? Did all of the people selected in the sample agree to answer the questions? Does the information agree with the author's conclusions?

Having asked such questions, you then need to know enough about the sources and uses of statistics to evaluate the answers intelligently. To prepare you for that task is the purpose of this book.

LEARNING A NEW LANGUAGE

As you read on in the chapters that follow, keep in mind that learning statistics—as with learning any subject—is partly a matter of learning to speak and think in a new language. You will run into new terms, and it is important that you make them a part of your active vocabulary. One of the most common words, for example, is **data**. To some it sounds very scientific, which may give it an undeservedly intimidating air of mystery. *Data* is the plural of **datum**. A datum is simply a bit of information, such as in 1985, 42 percent of U.S. adults agreed that pregnant women with very low income who cannot afford any more children should be able to obtain legal abortions. "I am over six feet tall" is also a datum, as are "the United States is a representative democracy"; "the Industrial Revolution began in the 17th century"; and "12 percent of whites, 36 percent of blacks, and 30 percent of people of Spanish origin were below the poverty level in 1982." Calling such statements *data* instead of "information" or "facts" does not make them any different, but *data* is the word used most often by those who use statistics, so it is worth your while to get comfortable with it and its correct usage (the data *are* interesting, not the data *is* interesting).

A **population** is any precisely defined group of anything, such as all women who are currently married and living in the continental United States, or all personal computers sold in the United States during 1987. A **sample** is any subgroup selected from a population. When we gather data on all members of a population, we have a **census**.

You will also run into the terms **descriptive statistics** and **inferential statistics**. When we gather data on any group, whether it is a census or a sample, and use them to describe how the group looks on a particular characteristic, then we are using *descriptive* statistics. On the other hand, when we use the information contained in a sample in order to make statements about the entire population from which the sample was drawn, then we are using *inferential* statistics (sometimes called **inductive** statistics). When we draw samples from populations, we must contend with the fact that a perfectly drawn sample is inherently less accurate than a perfectly conducted census, and that there are new sources of error that must be estimated and considered in the interpretation of our findings. With inferential statistics we are trying to accomplish the same descriptive, explanatory, and predictive tasks that form the core of scientific work, but using data based on samples. The only difference between inferential and descriptive statistics lies in the consideration of error inherent in the sampling process.

Where We Go from Here

This book is organized around the major uses of statistical techniques. Part I focuses on descriptive statistics, beginning with problems of measurement. We then move on in Chapters 3 and 4 to explore the different ways of organizing, summarizing, and presenting data, including the uses of percentages, averages, and standard deviations. Chapter 5 introduces elementary probability theory, which is crucial for understanding everything that follows. The remaining three chapters of Part I focus on increasingly complex (and interesting) questions having to do with the uses of statistical relationships to better understand cause and effect, as well as differences between groups, communities, and societies.

Part II is devoted to inferential statistics. Its purpose is to enable us to accomplish everything described in Part I using data drawn from samples.

In this short introductory chapter I have tried to give you some feel for what is to come. If this book has a theme, it is that the work of empirical researchers closely parallels what all of us try to do in our daily lives, but with important differences in method—differences such as precision, systematic observation, a known chance of error, and perhaps a higher degree of self-consciousness than most of us are used to. If we are successful in the chapters that follow, you will be better prepared to protect yourself from the occasional abuses of statistics that you will encounter. More important, however, will be your increasing ability to consume intelligently the enormous amounts of high-quality data being made available in what can only be described as an information revolution based on both statistical knowledge and advances in computer science.

It is important to keep in mind that science in general and statistics in particular represent only one of the many ways of knowing about the world. Systematic, quantitative information is very useful for certain kinds of problems that lend themselves to relatively objective, direct observation; but it is not very useful for many of the most important areas of human life. Statistics can help us document some of the consequences of different forms of government, for example, but they cannot tell us which consequences are good and which are bad. Nor can they tell us much about the vital and yet unobservable spiritual and emotional aspects of human existence. They can document the incidence of violence against children or the torture of political prisoners, for example, but they can do little to convey the horror of those events.

It is also important to be aware of the influence of two contrary tendencies in the use of statistical information. On the one hand is the tendency to discount everything statistical, especially when we disagree with it. There is a common belief that it is possible to "prove anything with statistics," but in fact it is far easier to lie with words than with scientific results, for the requirements of science are far more strict that those of ordinary conversation. If it is easier for people to misrepresent the world with statistics, it is only because relatively few people have enough training in statistics to know which critical questions to ask and how to interpret the answers.

On the other hand, the same power of numbers that makes us wary of them can also lead us to give them more weight than they deserve. Numbers are often used to sanctify assertions of fact, as if attaching a number to a statement makes it more believable. In a society as fixated with science as ours is, it is understandable that we often accord more credibility to numbers than to words. Listen to debaters, for example, and you will hear how much they depend on numbers to convey an impression of precision and certainty.

Somewhere between these extremes there lies an intelligent approach to statistics, an approach that appreciates the many uses to which they can be put while also keeping in mind their limitations and the critical questions that have to be asked about where they come from and how they are interpreted. It is toward that intelligent, informed middle ground that I hope the experience of reading and studying this book will lead you.

Summary

1. Statistics is a concept that applies both to a set of techniques for analyzing quantitative information and to the specific numerical results of those techniques. Statistics are used in the scientific process of systematically testing ideas about the world.

2. The use of statistics in scientific research aids the three major goals of science: description, explanation, and prediction.

3. The critical consumption of statistical information requires that readers ask important questions about how data are gathered, whom they represent, how they are interpreted, and how they are used to make inferences about populations.

4. As with most areas of study, learning statistics involves the acquisition of a new vocabulary that draws attention to otherwise overlooked aspects of the world. This chapter has already introduced a number of new concepts. Go over the list of key terms that follows, and see if you can explain them to someone else.

Key Terms

census 10	finding 7
code 8–9	inferential (inductive) statistics 10
data 10	population 10
datum 10	prediction 7
description 6	sample 10
descriptive statistics 10	statistic 3
explanation 6	statistics 3

2

Measurement

Whether we want to use scientific methods to describe the world or explain it, we must begin by observing it. If we want to describe something, we must decide in what *terms* to describe it. If we want to compare two groups of people, we must decide how to determine who is in which group and how we are going to observe and classify their differences. In short, all research begins with the problem of measurement.

From Concepts to Variables

Consider, for example, something as straightforward as measuring income differences between males and females. In order to do this, we have to determine a way of categorizing people according to sex. At first glance, this might seem to be a relatively straightforward problem. Theoretically, sex is defined in terms of genetic characteristics: men have one X and one Y chromosome, and women have two X chromosomes. A biologist could examine a sample of cells from a person and determine the person's sex. There are, of course, exceptions, such as those few men who have an extra Y chromosome. Even conceptually, then, we run into some odd cases that have to be dealt with at least formally, such as by defining all such people as "male" in spite of the fact that they deviate from the pure male genetic model.

We can also measure sex on the basis of secondary characteristics such as genitals, internal reproductive organs, breasts, body hair; but here there is even more room for ambiguity and cases that do not fit easily into a precisely defined sex category. The possession of what are defined as male and female secondary sex characteristics is not a case of either/or; rather, it is a continuum. There are women, for example, who have more body hair than some men, and there are men with more developed breast tissue than some women. In more extreme cases, there are people who are born with a combination of male and female genitals (see Money and Ehrhardt, 1972).

If we are going to compare males and females as part of our research, we are going to have to devise a rule, a set of procedures, that allows us to unambiguously classify people as either male or female. In doing so, we need a rule that results in as little error as possible: in other words, we do not want to classify someone as "female," for example, who should be classified as "male" according to our concept of what a male human being is. Such a set of procedures is known as a **measurement instrument**, and the process of constructing one is known as **operationalization**. In other words, when you sit down and decide just how you are going to measure a concept, you are operationalizing it, and the resulting set of procedures is a measurement instrument.

Regardless of what measurement instrument we use to categorize people in terms of sex, we will find that not everyone falls in the same category. The fact that we find variation when we observe this characteristic means that sex is a **variable**. If, however, we measure a characteristic for which everyone falls in one and only one category, then we are measuring a **constant**. If we measure whether or not people have any chromosomes at all, we will have a constant, since everyone will fall in the same category ("yes, I have chromosomes"). But if we measure what *kinds* of chromosomes people have, we will find that people fall in different categories, a fact that makes this characteristic a variable. Note that a characteristic may be a constant for one set of observations and a variable for another. The characteristic, "having some formal schooling," for example, would be a constant for U.S. business executives; but in many Third World countries it would be a variable, since although many have had some formal schooling, there are many others who have not.

There are several ways we could operationalize the concept of sex. We could classify people according to their genetic composition. Most people would be clearly either male or female, but some people would have something other than an XX or XY combination, and we would have to figure out what to do with them. We could create a third category such as "in between" or "mixed," or assign them based on some other set of criteria such as their genital configuration. The important point is that any system of classification must unambiguously assign each observation to one and only one category of the variable.

A second operationalization of the concept of sex would be to have an observer note the observable secondary sex characteristics of each person and decide on the basis of these whether someone is male or female. This is what is done, for example, when physicians write death certificates and when most survey interviewers decide the sex of their respondents. This is the instrument that most of us use to determine the sex of people we interact with: we look at them and make a decision based on what we see. For the most part, we have to rely on a very narrow range of clues, such as facial hair and body shape, a limitation that forces us often to depend on less reliable indicators, such as the way people style their hair or the clothes they wear. Of course, we are hardly ever able to observe such characteristics as genitals and virtually never privy to the intimate details of chromosome configurations. This means that we are more likely to make mistakes with this second operationalization than with the first, since people can—and often do—present an appearance that renders their sex at least somewhat ambiguous.

This leads us to a third operationalization of the concept of sex, and that is to simply ask people what sex they consider themselves to be. In almost all societies, the pressure is so great to be of one sex or another that virtually everyone other than small children will place themselves in just one of the two categories, male or female (Garfinkel, 1971).

These three measurement instruments represent the concept of sex in three ways. In some ways, the results of using these different instruments will overlap—in other words, people categorized as female using the first instrument also will be categorized as female using the other two. But in other ways the results will not overlap. The first instrument, for example, by defining sex in terms of genetic characteristics, ignores the fact that some people have secondary characteristics that suggest a sex other than that indicated by their chromosomes. It also ignores the fact that people identify their own gender without any knowledge of their chromosomes, and this can lead to contradictory results.

The second instrument relies on what we can observe of people's external appearance, while the third focuses on how people observe and classify themselves. With these approaches, there is considerably more room for ambiguity and contradiction. A person with male chromosomes, for example, who undergoes a sex-change operation in order to look like a female, might be classified as female by an attending physician (or by a prospective husband). Or a person with female chromosomes and subdued female secondary sex characteristics (such as undeveloped hips and breasts) might feel and act like a man but be clinically classified as a female.

In choosing a measurement instrument, researchers have to decide which aspect of a concept they want to focus on, and this depends on what it is their research is intended to find out. A psychologist studying sex differences in the incidence of color blindness (a condition known to be tied to genetic characteristics that are more likely to be found in males than in females) would focus primarily on "male" and "female" as genetically defined categories. A sociologist interested in dating behavior, on the other hand, would be more interested in how people think of themselves and how they present themselves to others in terms of gender. From this perspective we would be more interested in how people define their own genders or in how they are perceived by others.

VALIDITY AND RELIABILITY

Regardless of which aspect of a concept we focus on, it is important to use measurement instruments that are both *reliable* and *valid*. An instrument with **reliability** is one that if applied to the same observation at two different times yields the same result when the characteristic being measured has not changed. This should also be true when a measurement instrument is used at the same time by two different observers. A ruler, for example, should not expand and contract with changes in temperature to the extent that readings differ from one place to another; the gas gauge in a car should not move up and down erratically; an intelligence test administered at short intervals should give the same results each time; crime rates should not change unless the actual incidence of crime has changed.

Without such constancy, we have a hard time knowing whether *measured* differences reflect *real* differences or some process that has nothing to do with the concept we are trying to measure. Crime statistics gathered by the U.S. Federal Bureau of Investigation, for example, have long been criticized because they rely on victims to report crimes to the police. The fact that most crimes go unreported means that FBI crime statistics reflect both the incidence of crime *and* the tendency of victims to report crimes to the police. This means that the crime rate can go up if people become more likely to report crimes, even if the actual frequency of crime goes down in any given year. Thus, the FBI crime reporting system is an unreliable measure of crime because changes in the measurements often do not correspond to changes in what is being measured. A partial solution to this problem has been the Department of Justice's Victimization Surveys in which large national samples are interviewed to determine the number of crimes they have experienced, but even here under-reporting is a problem (U.S. Department of Justice, 1976).

Reliability problems are common both in science and in our everyday lives. The categories that psychiatrists use to diagnose patients, for example, are so generally defined that several different psychiatrists will typically diagnose the same patient in several different ways. A great deal of research has been done on the grading process in schools, and one consistent finding is that if several teachers are asked to evaluate the same piece of work, they may assign a wide range of different grades (Kirschenbaum et al., 1971). We all run into the same kinds of problems in taking our own measurements of what goes on around us. Our sense of time, for example, is highly unreliable. Depending on how we are feeling, five minutes can seem to pass as slowly as twenty or as quickly as one. As a measurement instrument, our subjective estimates of elapsed time are extremely unreliable.

While reliability is an important quality in all measurement instruments, it does us little good if we do not know what the results mean. **Validity** is a concept that refers to the degree to which instruments measure what we intend them to measure, and it is, therefore, one of the most important concepts in all of science. The most precise and reliable measurement instrument is useless if we do not know with some confidence just *what* it is measuring with such great precision and reliability.

Consider, for example, the long-standing problem of trying to understand the persistence of racial segregation in U.S. neighborhoods. Residential segregation, especially between whites and blacks, is so extreme that more than 80 percent of Americans would have to change their neighborhoods in order to bring about racial balance in housing. One explanation has been that whites do not want to live with blacks, but most survey results have found that whites are willing to live in integrated neighborhoods. In their 1976 study, however, Farley and his associates suspected that the questions used to measure people's positions on segregation failed to get at people's true racial values.

Typical questions in previous studies, for example, focused on general propositions such as "Would you be upset if a black family of your same economic status moved into your neighborhood?" to which large majorities of whites said they would not be upset. In search of a more valid measure of values about integration, Farley presented respondents with diagrams showing actual neighborhood possibilities (Figure 2-1). Each diagram showed 15 houses, with the respondent's house shown at the middle. Houses were either white or black, thus showing levels of integration ranging from 7 percent to 53 percent. With this new measurement instrument, Farley found that the more integrated the hypothetical neighborhood, the more likely whites were to say they would feel uncomfortable in such a neighborhood, would try to move out, and would not move in (Farley, Bianchi, and Colasanto, 1979). Undoubtedly, this instrument will also benefit from further research validation.

Unlike reliability, there are several different types of validity, all of which focus on the meaning of what is being measured. **Construct validity**, for example, refers to the appropriateness of a measurement instrument for the population under study. If we use familiarity with classical ballet and opera

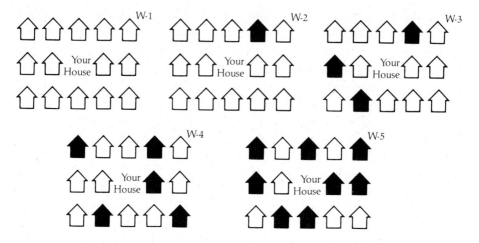

Figure 2-1 Diagrams for measuring white resistance to neighborhood integration.

SOURCE: Farley, Bianchi, and Colasanto, 1979, Figure 6.

music to measure the musical development of black ghetto children, for example, we might draw invalid conclusions since their musical experience rests on a cultural background that generally does not include that kind of music (unlike, for example, the background of many educated social scientists).

The same criticism has often been leveled at intelligence tests that are based on the experiences of white middle-class people and yet are applied to working- and lower-class nonwhites. Tests depend on language and on examples from everyday life that inevitably reflect cultural backgrounds as much as, if not more than, intelligence. Consider, for example, the following questions:

1. A "handkerchief head" is:
   (A) A bad dude; (B) A porter; (C) An "Uncle Tom"; (D) A preacher
2. Which word is most out of place here?
   (A) Bagel; (B) Lox; (C) Gefilte

Blacks would be more likely than whites of equal intelligence to answer the first question correctly (C), while Jews would have the edge with the second question (A is out of place since B and C are seafoods). Knowing the correct answer to the second question depends on cultural background as well as the general cognitive ability to categorize things on the basis of similarities and differences. To the extent that such questions are intended to measure intelligence alone, they lack construct validity.

Predictive validity refers to the degree to which the results obtained from using a measurement instrument allow us to accurately predict some other variable *regardless* of what the first variable actually means. There is a considerable body of evidence, for example, indicating that school grades at one level do predict how well a student will perform at higher levels. Regardless of what high school grades actually measure, then— conformity, intelligence, motivation—they have predictive validity as measures indicating how well students will do later on. There is another body of research, however, that strongly suggests that grades do not predict performance in settings other than school—being a lawyer, for example, as distinct from being a law student (Wallach, 1972). Thus, grades have predictive validity *within* school situations but lack predictive validity as measures of how well students will do beyond school.

Predictive validity is also found in epidemiological studies of the distribution of various kinds of disease. Cancer deaths, for example, vary sharply from one place to another, and these geographical differences are specific for particular kinds of cancers. Thus, living in some communities brings with it a statistically higher chance of contracting cancer of the colon, while in other communities lung cancer is more common. As an indicator of the chances of getting cancer, place of residence is a measurement instrument with considerable predictive validity, in spite of the fact that researchers are not yet able to explain why there should be such distinct geographical patterns.

Validity is the most serious problem in any effort to gather information. Our lives are full of decisions and choices based on our observations of others and ourselves, observations that serve as measurements, whether of someone's suitability as an employee or as a spouse. When you read a research report, you have to decide for yourself whether the measurement instruments used are valid. You should be careful to look for statements describing the actual instruments used, since these are not always reported, especially in the mass media. They may refer to concepts such as "support for a political candidate" or "confidence in the economy" instead of indicating exactly what questions were used to operationalize them. Even when the actual instruments are reported, you still must use your judgment. The greater the number of interpretations for a question and its answers you can think of, for example, the more tenuous does any single interpretation become.

MEASUREMENT ERROR

All measurement involves some degree of error. Consider, for example, the problem of determining the cause of death when someone dies, a task that involves considerable measurement problems. First, when a person dies, it may be from several causes. Suppose Roberta has a heart attack on Tuesday, enters the hospital that same day, seems to be stabilizing, then contracts pneumonia on Friday. Saturday, she weakens steadily, and on Monday she finally dies. What was the cause of death? Coronary heart disease? Pneumonia? Some other condition that escaped the attending physician's attention? When physicians fill out death certificates, they must determine the primary cause of death as well as secondary causes. Since no one really knows what the "real" cause was, or whether it was primarily one cause, we must depend on the physician's judgment.

In another case, suppose that a coroner responds to a call from Clare, who has just come home from work to find her husband dead on the living room floor. How does the coroner figure out the cause of death? An autopsy is perhaps the surest method, but it may not be performed, and even if performed may not lead to a clear result. The physician must rely on certain observable signs as well as expectations based on experience, such as the fact that some causes of death are statistically more likely than others, depending on the victim's age, race, gender, and social status. In arriving at a decision, the coroner may use such expectations, perhaps unconsciously.

There are thus both reliability and validity problems in the measurement of cause of death, each of which contributes to **measurement error**. Error usually takes two forms. The first is **random error**; the second—and by far the more serious—is systematic error that is called **bias**.

The word *random* has a specific and often misunderstood scientific meaning. If you want to select someone at random from your community and you walk down your street and stop the first person you meet, this is not a random selection. If you take people in an area, put their names on a list, assign a number to each name, and draw numbers out of a well-mixed hat, then you have a random selection. The difference is that random means that each possible alternative has *an equal chance* of happening. The first case is not a random selection, since there is no way to demonstrate that everyone has an equal chance to walk down your street.

In our death diagnosis example, if attending physicians are as likely to make one kind of diagnosis error as another, then the errors would be random *and they would tend to cancel one another out in the long run*. For every heart attack that was misdiagnosed as a stroke, there would be a stroke that would be misdiagnosed as something else, and something else misdiagnosed as a heart attack. If, however, doctors tend to make certain errors more than others, then the errors will accumulate and not cancel one another out. If, for example, doctors expect older men to die of heart attacks and therefore tend to report heart attacks too often, then their measurements will be *biased*. Bias is error that tends to go in one direction more than others.

There are many examples of bias in both science and everyday life. People tend to exaggerate their educational attainment and to report ages that end in

9, 0, or 1 more than other ages. People tend to overestimate the height of prominent people. There is some evidence that highly educated people are more likely than others to report socially desirable responses to survey interviewers, regardless of their actual views (Crowne and Marlowe, 1964). Whites are less likely to admit racial bias in the presence of blacks, even when the attitudes are recorded on anonymous written questionnaires (Summers and Hammonds, 1966). Blacks are more likely to say that whites can be trusted to white interviewers than they are to black interviewers (Schuman and Converse, 1971).

Good researchers always try to avoid bias in their measurement procedures. If it is unavoidable, the next best thing is to detect it and try to estimate its direction and magnitude. Random error, however, is far less problematic. To illustrate, consider a national study of fertility patterns known as the Princeton Study. In an effort to predict future national birth rates, researchers asked a national sample of women to predict the number of children they would have in the next five years. On an individual basis, women were generally poor predictors of their own fertility, but as a *group*, their individual predictions yielded an average fertility level that was extremely accurate. Since the study was intended to predict birth rates rather than births for individuals, the random error was not a problem. Had the women *consistently* underestimated or overestimated their fertility, then the findings would have contained bias, which would have distorted the results.

The most important thing to keep in mind is that any measurement instrument—from government unemployment figures to subjective estimates of how tall someone is—is an attempt to operationalize an underlying concept, and there are usually several, if not many, different ways to operationalize the same concept, each of which carries with it some chance of bias or random error. There is simply no such thing as an error-free measurement, even in the most carefully controlled laboratory. What is possible is some fairly precise estimate of the amount of error most likely involved in a particular measurement process. As we will see in later chapters, the estimation of error is one of the most important of all scientific activities and one of the key uses of statistics.

Types of Variables

Variables are used to classify observations of anything—from people to things and events—into categories; but variables themselves can be divided into different types, or *scales* of measurement, according to the kinds of distinctions they allow us to make among observations. It is very important that you learn to distinguish between different scales, because they determine which statistical techniques can be used.

QUALITATIVE (NOMINAL) VARIABLES

Qualitative variables only allow us to make relatively crude distinctions in terms of similarities and differences. They are more often known as **nominal** (or **categorical**) **variables**, because they allow us to do no more than *name* observations or sort them into categories such as "single" and "married" without saying anything more about the differences between them. Social scientists focus on many nominal-scale variables, including race, gender, political party, ethnicity, type of political system, religion, place of birth, type of classroom situation, type of school grading system (letter grade or pass/fail), marital status, occupation, given names, opposition to or support for abortion or the death penalty, and many other positions on social issues.

A useful way to distinguish among different types of variables is to consider how we might go about using a number to represent each category in such a way that the numbers reflect the actual differences among the categories. With nominal variables, we can assign virtually *any* numbers to the categories so long as no number is used for more than one category. We have the freedom to use any numbers we want because with nominal variables the only requirement is that different categories have different numbers. Your individual social security number, for example, is a unique category of a nominal scale that uses nine-digit numbers to distinguish among each and every U.S. citizen. Long, complex numbers are also used as nominal scales in many other applications, from bank credit card numbers to drivers' licenses. It does not matter which number you have so long as someone else does not have it too.

QUANTITATIVE VARIABLES

Quantitative variables allow us to make distinctions based on the magnitude of some characteristic such as height, income, gross national product, or risk of job-related injury. Quantitative variables allow us to speak in terms of "high" and "low" or "more" and "less." Among quantitative variables there are several further distinctions we can make. **Discrete variables**, for example, have categories that can take on only certain values. If we measure the number of times people have been unemployed, we can get results such as 0, 1, 2, or 3, but we cannot get 0.5 or 2.164 unemployment experiences. Variables such as age, however, are **continuous** because their categories can take on any one of an infinite number of values within a given range. We can measure age, for example, in decades, years, months, weeks, days, hours, minutes, seconds, milli-seconds—theoretically, the degree of detail possible is infinite.

In general, quantitative variables differ from qualitative variables in that they allow us to make distinctions based on the relative magnitude of some characteristic. There are, however, three different types of quantitative variables—ordinal, interval, and ratio—that differ from one another in how precisely they allow us to describe relative magnitudes.

Ordinal-Scale Variables An **ordinal-scale variable**, such as social class, consists of categories that can be ranked from high to low according to the characteristic you are measuring. As such, social class has nominal properties (upper-class George's class position differs from lower-class Pat's). But it also has an ordinal property which allows us to compare categories in terms of rank (George's position is higher than Pat's). In comparison with nominal-scale variables, ordinal scales enable us to make more precise statements about differences between categories, and are thus referred to as a "higher" scale of measurement. Ordinal scales are also common in the social sciences, including measures of social class, the amount of political freedom found in different societies, age groups (child, young adult, adult, middle-aged, etc.), highest educational degree (grade school, high school, B.A., M.A., Ph.D.), level of economic development, relative complexity of kinship systems, and the degree to which people agree or disagree with different positions on social issues.

Notice particularly with the last example that the same concept can be measured with different scales of measurement. We could, for example, ask people if they agree with the statement that "governments should never negotiate with terrorists" and record answers as merely "yes" or "no," giving us a nominal variable. We could also ask respondents to give us additional information by telling us how *strongly* they agree or disagree (such as "very" or "mildly"), giving us an ordinal measure of the same concept.

Since ordinal variables allow us to make more precise distinctions than we can make with nominal variables, we have less freedom to use numbers to represent categories. With nominal variables, we could assign any numbers so long as different categories had different numbers. But since ordinal variables rest on the idea of order—of "greater than" and "less than"—any numbers we assign must reflect the rank position of different categories. With the five educational attainment categories of grade school, high school, B.A., M.A., and Ph.D., for example, we could assign any numbers we wanted so long as they maintain the proper order of the categories. "Grade school" can have any number so long as it is lower than those assigned to the other four categories; "high school" can have any number so long as it is higher than the one assigned to grade school and lower than those assigned to the other three categories; and so on up to Ph.D., which must have the highest number of all.

Interval-Scale Variables With **interval-scale variables** we are able to make all of the distinctions allowed by nominal and ordinal scales, which is to say, we can say that one category is different from another and we can rank categories from high to low. Interval scales, however, have an additional property, and that is that we can also compare the actual magnitude of the *quantitative distance between categories.*

Consider, for example, temperature measured in degrees centigrade. Suppose we compare the temperatures of three objects, the first at 45 degrees, the second at 25 degrees, and the third at 15 degrees. We can see that this measurement scale has nominal properties (the temperatures are different) and ordinal properties (the first is warmer than the second, which is warmer than the third). It has the additional property of allowing us to compare the *distances* between categories—the difference between the first and the second (45 − 25 = 20 degrees) is exactly twice the difference between the second and the third (25 − 15 = 10 degrees). Since this kind of scale allows us to directly compare the intervals that separate categories, it is called an *interval* scale.

Notice that although we can precisely compare the distance between categories of interval-scale variables, the most we can say about the categories themselves is that one is greater or less than another. To be more specific, we can compare the relative magnitude of differences between scores by dividing one by another (20 is twice as large as 10). This is something we could not do with ordinal scales. But we cannot do the same thing with the scores themselves. Sixty degrees centigrade does *not* represent twice as much heat as 30 degrees. Why not? Because on the centigrade scale, zero degrees does not represent the complete absence of heat. The "zero" point on this scale (as with the Fahrenheit scale) is not a "true" zero; rather, it is an arbitrary zero point. If we used the Kelvin scale, however, whose zero point is absolute zero, we could say that 60 degrees Kelvin is twice as hot as 30 degrees Kelvin.

Any scale that allows you to compare distances between intervals but which lacks a true zero is an interval scale. With college board scores, the lowest possible score is 200, which means that a score of 600 is not twice as good as a score of 300; nor is a score of 400 twice as good as a score of 200. Another example is the result of a horse race when expressed in terms of the distance between horses rather than the time it takes each to complete the course (Sea Biscuit wins by a nose, followed at one length by Lucky Larry, with Blue Beauty two lengths behind, and so on).

Notice that with interval scales we are still more restricted in our freedom to use numbers to represent categories of the variable. With interval scales, we must assign numbers in such a way that the *relative distance between categories* is correctly expressed. With the college board scores mentioned above—scores of 600, 400, and 300—we could assign any other numbers we wanted to those categories so long as the distances between them are of the same relative magnitude. In other words, the distance between the first two must be twice the distance between the second and third. Scores of 6, 4, and 3, for example, would do just as well—as would scores of 55, 35, and 25. (Before going on, be sure you understand why the preceding sentence is true.)

Ratio-Scale Variables Although interval scales are uncommon, their properties are also found in the third and most powerful type of quantitative variable, the **ratio scale**.

A ratio scale, such as age, has nominal properties (45-year-old Pat is the same age as 45-year-old Caesar; 30-year-old Bob differs in age from his 60-year-old father; 50-year-old Martha and her 40-year-old husband are of different ages); it has ordinal properties (Bob is younger than his father; Martha is older than her husband); and it has interval properties (the age difference between Bob and his father [30 years] is three times as great as the difference between Martha and her husband [10 years]). Since there is a "true" zero in the age scale, however, we can also compare the magnitude of the scores themselves—not just the intervals between them—by dividing one by the other. Thus, we can say that Bob's father is twice as old as Bob is, a kind of comparison that can only be made with variables that have ratio properties.

Any scale that has nominal, ordinal, and interval properties as well as a "true" zero constitutes a ratio scale. Age in years, income in dollars, education in years, height in inches, the size of national defense budgets measured in dollars, the number of years it takes for technological innovations to reach Third World countries, the number of condemned prisoners awaiting execution on death row, death rates, the number of seconds it takes to memorize a list of words—all of these are examples of ratio scales that allow us to make the most precise comparisons between relative magnitudes of scores.

Notice again that a single concept can be measured using more than one scale. Income, for example, can be measured as an ordinal scale (high, medium, and low) or as a ratio scale (in actual dollar amounts). As you will see later on, which scale of measurement we use often depends on the kinds of precision we need in answering specific research questions and the kinds of analytical techniques we want to use. The fact that ratio scales allow the greatest precision does not mean that they are always the preferred form of measurement.

WHY MEASUREMENT SCALES ARE IMPORTANT

For our purposes, the most important differences between scales of measurement are those that distinguish nominal, ordinal, interval, and ratio

Table 2-1 The Kinds of Comparisons between Scores
That Can Be Made with Different Scales of Measurement

	Scale of Measurement			
Comparisons That Can Be Made	*Nominal*	*Ordinal*	*Interval*	*Ratio*
Similarity or Difference	Yes	Yes	Yes	Yes
Greater Than/Less Than	No	Yes	Yes	Yes
Magnitude of Differences	No	No	Yes	Yes
Magnitude of Scores	No	No	No	Yes

scales (see Table 2-1). Notice that in moving from lower to higher scales each higher scale has all of the properties of the scales below it plus an additional property that sets it apart. Thus, all ordinal scales have nominal properties in addition to ordinal properties; all interval scales have nominal, ordinal, and interval properties, and all ratio scales have nominal, ordinal, interval, and ratio properties. Thus, while there are relatively few variables that are "pure" interval scales, there are many variables that have interval properties (all ratio variables).

Why do we make these distinctions between different types of variables? It is important that you be able to tell one scale from another *because the choice of statistical tools depends on the scale of measurement of the variables involved.* Techniques that are appropriate for ratio-scale variables, for example, cannot be used with nominal and ordinal scales, and techniques for ordinal scales cannot be used with nominal scales. You cannot take a statistic such as an average, for example, and use it on a nominal-scale variable. Otherwise you might think you could "add" an orange, a pear, an apple, and a banana, and somehow arrive at an "average fruit."

While the silliness of this case is apparent, there are many cases in statistics where the inappropriateness of particular techniques is less obvious. If you cannot tell which scale of measurement you are dealing with, then you cannot know which techniques are appropriate to use. Nor will you be in a position to evaluate someone else's use of statistical techniques.

The Dichotomy: A Special Case A **dichotomy** is a variable that has only two categories, and it represents a special case in statistical analysis because it can be used as if it were a much "higher" level of measurement than it may appear to be. Since there are only two categories in a dichotomy, it can be thought of as having ordinal properties even when it is otherwise nominal. The variable "gender" is a nominal-scale dichotomy, for example, but we could think of the category "female" as representing people who have more femaleness than "males" (or "male" as representing people who have more maleness than "females").

Even more important is the fact that dichotomies involve only a single interval, such as that separating the categories "male" and "female." Since there is only one interval, we can treat a dichotomy in the same way that we treat interval scales, because *with only one interval, the requirement that we be able to compare the relative magnitude of intervals is irrelevant.*

Because interval scales allow us to use more sophisticated techniques than either nominal or ordinal scales, researchers are sometimes tempted to convert an ordinal- or nominal-scale variable that has more than two categories into a dichotomy. If, for example, we have a variable such as religious affiliation that consists of several categories (Christian, Jew, Moslem, Hindu, Buddhist, etc.), we might "collapse" it into a dichotomy such as "Christian" and "non-Christian." In some cases this is quite justified, but in others it has the effect of distorting our results. If, for example, Moslems and Buddhists differ substantially from each other, lumping them together for the sake of allowing the use of more sophisticated statistical techniques may do more harm than good.

Summary

This chapter has not focused on the numbers and equations that will occupy us from here on, but it raises important questions about the procedures through which statistics are produced. There is a great deal of judgment involved in the measurement process, and it is important that you learn to ask critical questions about that process and its results.

1. All measurement begins with the operationalization through which concepts are transformed into one or more measurement instruments, each of which may reflect a different aspect of what we are trying to measure.

2. Measurement instruments are judged in terms of how reliable and valid they are. Even the most reliable and valid instruments, however, involve some degree of random error or bias.

3. Variables are either qualitative (nominal) or quantitative (ordinal, interval, or ratio) scales. Quantitative variables can be further divided into discrete and continuous scales. Each higher scale of measurement has all of the properties of the scales below it, plus an additional property that sets it above the others.

4. Understanding the differences between the various measurement scales is enormously important, because the appropriateness of specific statistical

techniques depends directly on the scale of the variables involved. Techniques that are appropriate for higher levels of measurement cannot be used on lower level variables.

Key Terms

categorical variable 22
constant 14
continuous variable 22
dichotomy 26
discrete variable 22
interval-scale variable 24
measurement error 20
  bias 20
  random error 20
measurement instrument 14
nominal-scale variable 22

ordinal-scale variable 23
operationalization 14
qualitative variable 22
quantitative variable 22
ratio-scale variable 25
reliability 16
validity 17
  construct 17
  predictive 19
variable 14

Problems

1. Define and give an example of each of the key terms listed at the end of the chapter.

2. Suppose we want to measure the degree to which someone can be considered to be "educated." How would you define the concept of "educated"? How might you go about operationalizing it? Can you think of instruments that would measure this concept on a nominal scale? An ordinal scale? An interval or ratio scale? What problems of validity and reliability might you encounter in using your measures?

3. Identify the following variables in terms of the scale of measurement (nominal, ordinal, interval, ratio) that each represents.
   a. left- or right-handedness
   b. size of car (subcompact, compact, midsize, luxury size)
   c. temperature in degrees Fahrenheit
   d. school class (freshman, sophomore, junior, senior)
   e. the year (1986, 1987, 1988, etc.)
   f. the number of these questions that you get right the first time around

g. approval of a president's performance in office (strongly approve, approve, disapprove, strongly disapprove)
h. ethnicity
i. educational attainment measured in years
j. social class (upper class, upper-middle class, middle class, working class, lower class)
k. the amount of rainfall last year (measured in inches)
l. gross national product (measured in dollars)
m. class rank in school (first, second, third, etc.)
n. degree of religiosity (high, medium, low)
o. military rank (lieutenant, captain, major, lieutenant colonel, colonel, etc.)
p. type of society (hunter-gatherer, industrial, agrarian, horticultural, herding, fishing)
q. economic system (e.g., capitalism, socialism, feudalism)
r. number of abortions a woman has experienced during her entire lifetime

4. For each of the variables in problem 3, determine whether it is continuous or discrete.

5. For each of the variables in problem 3 that is lower than a ratio scale, can you think of a way of operationalizing the same concept that will result in a higher scale of measurement?

6. Why is it so important to be able to distinguish one scale of measurement from another? Be specific and give examples that show what you mean.

7. Why is the dichotomy considered to be a special case in measurement? Be specific and give examples that show what you mean.

༺ 3 ༻

Distributions, Comparisons, and Tables

The most immediate challenge facing anyone who wants to analyze a set of data is to take a large number of separate observations—sometimes numbering in the hundreds of thousands—and organize and summarize them in a way that allows them to tell a meaningful story that gets us as close to the truth as possible. As a reader of someone else's research reports, the most immediate challenge facing you is to understand and critically evaluate the ways in which researchers have gone about meeting the challenge facing them. In this chapter we will look at some of the most fundamental problems involved in this process of organizing and summarizing statistical data, with the goal of improving both your critical and your applied skills.

Frequency Distributions

We begin with our raw materials, which may consist of thousands of individual pieces of data, the accumulated results of all our measurements. As an example, consider one of the questions asked of 1,470 respondents in the 1986 NORC General Social Survey: "If you were asked to use one of four names for your social class, which would you say you belong in: the lower class, the working class, the middle class, or the upper class?" We have 1,470 bits of information such as those in Table 3-1, one corresponding to each respondent. What can we do with them?

TABLES AND HOW TO READ THEM

The simplest way of presenting what we have is to count up the number of people who gave each of the possible responses. This is called a **frequency**

Table 3-1 Social Class Self-identification for 20 Hypothetical Respondents in the 1986 NORC Survey

Respondent	Social Class
1	middle
2	middle
3	lower
4	upper
5	lower
6	middle
7	working
8	working
9	working
10	lower
11	middle
12	upper
13	middle
14	working
15	lower
16	don't know
17	working
18	middle
19	working
20	middle

distribution. Any arrangement of numbers in columns and rows is called a **table**, and Table 3-2 shows how people answered this question in 1986. There are several principles about reading tables that you can learn from this one.

When you look at a table, start with the description. Read it carefully. Next, examine the column and row headings so that you are sure of the meaning of each number in the body of the table. In this case, the first column lists the possible answers the respondents in the sample could have given. The second column lists the number of people who gave each response. So, 48 people placed themselves in the upper class, 683 in the middle class, 631 in the working class, and 95 in the lower class, for a total of 1,457 people. The letter N (sometimes capitalized, sometimes not) usually stands for the number of **cases** (a frequently used synonym for "observations"). You should also be aware that the letter f is often used as an abbreviation for "frequency."

If you have been paying sharp attention, you may have already noticed a slight but significant discrepancy in our reported findings: although there

Table 3-2 Social Class Self-identification, U.S. Adults, 1986

Social Class	frequency
Upper	48
Middle	683
Working	631
Lower	95
TOTAL	1,457

SOURCE: Davis, 1986, p. 218.

were 1,470 respondents in the NORC study, only 1,457 are represented in Table 3-2. Where are the missing 13 people? The answer is that for any given variable there are usually some observations with missing data. There should *always* be an explanation for missing cases, either in the table itself or in a footnote. Without this, the table loses some of its meaning, not to mention its credibility. The greater the number of missing cases, the more doubts there are about the respresentativeness of the data. Sometimes the level of nonresponse to specific questions can be very large, and as we will see in Chapter 10, this raises serious questions about the representativeness of the data and the validity of their interpretation. So, *always* make sure that the frequencies in the table add up to the total number of observations in the study.

Properly presented, Table 3-2 should look like Table 3-3. This tells us that there were in fact 1,470 respondents, of whom 5 said they did not know in which class they would place themselves and 8 gave no answer (which can also mean that the interviewer neglected to record the answer or that it was lost in some other way). The main body of the table shows us how those who did answer the question were distributed on the class variable.

Table 3-3 shows that the largest class category with which people identify themselves is the middle class, followed closely by the working class. The smallest class category is the upper class. The table also tells us that virtually everyone in the sample gave a response to the question. As tables go, this one is about as simple as you will find. Even with such simple distributions, however, there is no substitute for starting at the top and carefully working your way through, making sure that you know what each and every number means. It may sound too elementary, but even professionals have to take great care with tables if they want to be sure they understand what is in them. Hurrying through data just does not work in most cases.

Table 3-3 Social Class Self-identification, U.S. Adults, 1986

Social Class	frequency
Upper	48
Middle	683
Working	631
Lower	95
Don't Know	5
No Answer	8
TOTAL	1,470 = N

SOURCE: Davis, 1986, p. 218.

Grouped Data We often find that making a separate category for each possible response results in enormous and unwieldy tables. If we asked people to report their exact annual income to the nearest dollar, for example, and made each unique dollar amount into a separate category, we would wind up with thousands of rows in a simple frequency distribution. To avoid this, we often abbreviate the table by lumping similar responses together. The resulting data are called **grouped data** and the procedure is called **grouping**. A grouped income distribution for the NORC sample is found in Table 3-4 (before going on, examine this table and make sure nothing is wrong with it).

You will notice two things about the effects of grouping in this table. First, when we group data, we lose detail. Instead of knowing someone's exact income, we now know only that the person's income fell somewhere, say, between $15,000 and $19,999. This is most problematic with open-ended categories ("$25,000 or over") that include everyone from $25,000-a-year managers to billionaires.

In many cases, this loss of detail may not make much of a difference in the conclusions we draw about the world, but in others it can matter a great deal. If, for example, we made a single category of lower- and working-class people in Table 3-3 and made a second category by lumping middle- and upper-class people together, we would lose some very important details, such as the fact that the extreme class positions are much smaller than those toward the middle.

The second thing I want you to notice about the grouped data in Table 3-4 is that there is no magic formula that tells us how to define the groups. Often, it is done rather arbitrarily (there is no particularly compelling reason, for

Table 3-4 Distribution of Pretax Family Income, United States, 1985

Total Family Income	frequency
Under $1,000	15
$1,000 to $2,999	22
$3,000 to $4,999	100
$5,000 to $6,999	70
$7,000 to $9,999	92
$10,000 to $14,999	169
$15,000 to $19,999	145
$20,000 to $24,999	147
$25,000 or over	586
Refused	58
Don't Know	65
No Answer	1
TOTAL	1,470 = N

SOURCE: Davis, 1986, p. 62.

example, to make a category of $10,000 to $14,999, rather than, say, $11,221 to $16,108). Sometimes lines are drawn at socially meaningful points, such as official poverty levels in the case of income.

A variable operationalizes a concept and allows us to assign observations to categories. A frequency distribution is the simplest kind of summary that shows how many observations fall in each category. It is the most basic technique for reducing unmanageable masses of information on individual observations to a manageable description of a set of data.

Comparing Distributions

It may have occurred to you that although frequency distributions are useful, they do not take us very far toward answering the kinds of questions we are most likely to ask. We would rarely be content to find out what the distribution on a particular variable looks like and stop there. We usually want to go on to make comparisons between different populations or of the same population at different times. How, for example, do the social class distributions of whites, blacks, Hispanics, and other racial/ethnic groups differ? How do religious beliefs vary from one type of society to another? How does the distribution of population between urban and rural areas in 1986 compare with corresponding distributions in 1886 and 1786?

Table 3-5 Existence of Social Class Systems among
Different Types of Nonindustrial Societies

Existence of a Class System	Type of Society			
	Hunting & Gathering	Horticultural*	Agrarian**	All
Yes	3	133	63	199
No	140	160	26	326
TOTAL	143	293	89	525

SOURCE: Adapted from Lenski and Lenski, 1978, p. 101, Table 4.5.
* Horticultural societies are those in which food is grown in small gardens without the benefit of plows.
** Agrarian societies are those in which food is grown in large fields cultivated with the use of plows.

To answer the most interesting kinds of questions about distributions, we need to place them in some kind of meaningful comparative context. The first step in comparing groups would be to compare their frequency distributions. For example, take a moment to study Table 3-5, which shows the incidence of social class systems in societies that differ in the ecological arrangements through which they meet their material needs. Hunting and gathering societies depend on what they can find on the land, produce none of their own food, and therefore do not generate a surplus. Horticultural societies use digging sticks to cultivate small gardens and are able to produce a small surplus; agrarian societies use plows to cultivate large fields and can produce huge surpluses that support cities.

Since the ability to produce a surplus will, in turn, make social inequality possible by allowing some to accumulate wealth at the expense of others, we would expect that as we move from hunting and gathering societies to agrarian societies that class systems will become more common. To test this expectation, we could first compare the frequency of class systems in the three different types of societies. Table 3-5 shows these three frequency distributions as well as that for the entire sample of societies (all of the societies lumped together).

Marginals and Cells The first thing to notice about this table is that it has two different kinds of numbers. The bottom row (TOTAL) shows the frequency of each type of society (143 hunting and gathering, 293 horticultural, and 89 agrarian) and the last column (All) shows the overall frequency of societies with class systems (199) and societies without them (326). These numbers are called *marginal frequencies* (or **marginals**, for short) because they appear in the margins of the table. They show the number of people with each characteristic represented in the table.

Each number in the body of the table (in other words, inside the marginals) shows how many societies have each *combination* of ecological and class characteristics. There are 3 societies that both rely on hunting and gathering and have class systems, and there are 140 societies that both rely on hunting and gathering and do *not* have class systems. There are 133 societies that are horticultural and have class systems, and there are 160 that both are horticultural and do *not* have class systems. There are 63 societies that both are agrarian and have class systems, and there are 26 societies that are agrarian and without class systems. These numbers that show the incidence of combinations of characteristics are called **cells**.

Finally, there are 525 societies in the entire sample, with no cases of missing data. Be sure you can identify all of these numbers in the table before you go on.

Percentages and Proportions We can make a couple of comparisons with Table 3-5. In general, societies without class systems are more common than societies with class systems. This holds true in both hunting and gathering and horticultural societies. In agrarian societies, however, there are more societies with class systems than without.

What about our original expectation, that as we move from hunting and gathering societies to horticultural and agrarian societies class systems will become more common? If we compare the frequencies, we find that there are more horticultural class societies (133) than there are agrarian class societies (63), but is this a fair comparison? Since there are only 89 agrarian societies to begin with, it would be impossible for the number of agrarian class societies to match the number of horticultural class societies even if *every* agrarian society had a class system.

The problem we have run into here is that our sample does not include equal numbers of each type of society, and this is making it difficult to compare the different types. If we had equal numbers of each ecological type, then we could make comparisons quickly and directly. It would, of course, be very convenient if data always came in such neat packages, in which each category had the same number of cases, but that is the exception, not the rule. Fortunately, there is a much easier solution to the problem of comparing categories of unequal sizes, and the key lies in the idea of likelihood, which is central to most of the questions that comparisons are about.

When we talk about the likelihood of agrarian societies having class systems, for example, we are not talking about the absolute number of agrarian class societies, but the number of agrarian class societies *relative* to the total number of agrarian societies. To get one number relative to a second, we divide it by the second. The result is a **proportion** which, when multiplied by 100, is a **percentage**. Thus, proportions and percentages start out as fractions. The denominator is always whatever directly follows the word "of," as in "the percentage of *all agrarian societies* that have class systems."

If we convert the cell frequencies in Table 3-5 to proportions, then we can make comparisons without having equal numbers of each type of society, because we have, in effect, forced each type to have the same total (1.00). For each type, we divide both the number with class systems and the number without class systems by the total number of societies of that type (see Table 3-6). Before you continue, be sure you understand this paragraph and be sure you see how Table 3-6 represents it.

If we carry out the divisions in Table 3-6, the result is the proportions of each type of society that have class systems and do not have class systems (Table 3-7). If we multiply each of these proportions by 100, then we get the same results expressed in terms of percentages (Table 3-8). Percentages and proportions are equivalent; which one we use is a matter of personal preference.

Before drawing any conclusions from these tables, we should note a few things about percentage and proportional distributions. First, percentages must add to 100.0 and proportions must add to 1.00. If percentage totals are off by more than .1 (above 100.1 percent or below 99.9 percent), or if proportion totals are off by more than .001 (above 1.001 or below .999), then an error other than a rounding error has been made somewhere. You should check it by adding it yourself.

Second, if a percentage or proportional distribution is based on a small number of cases (under 25 or 30), the researcher is risking considerable

Table 3-6   Existence of Social Class Systems by Type of Society (Calculating Proportions down the Columns)

Existence of a Class System	Type of Society			
	Hunting & Gathering	*Horticultural*	*Agrarian*	*All*
Yes	3/143	133/293	63/89	199/525
No	140/143	160/293	26/89	326/525
TOTAL	143/143	293/293	89/89	525/525

SOURCE: Adapted from Lenski and Lenski, 1978, p. 101, Table 4.5.

Table 3-7 Existence of Social Class Systems by Type of Society (Proportions)

	Type of Society			
Existence of a Class System	Hunting & Gathering	Horticultural	Agrarian	All
Yes	.02	.45	.71	.38
No	.98	.55	.29	.62
TOTAL	1.00	1.00	1.00	1.00
(N)	(143)	(293)	(89)	(525)

SOURCE: Adapted from Lenski and Lenski, 1978, p. 101, Table 4.5.

Table 3-8 Existence of Social Class Systems by Type of Society (Percentages)

	Type of Society			
Existence of a Class System	Hunting & Gathering	Horticultural	Agrarian	All
Yes	2%	45%	71%	38%
No	98	55	29	62
TOTAL	100%	100%	100%	100%
(N)	(143)	(293)	(89)	(525)

SOURCE: Adapted from Lenski and Lenski, 1978, p. 101, Table 4.5.

inaccuracy. For example, suppose we have a set of data with only 5 cases, each case representing 20 percent (1/5 × 100) of the total. If a single error has been made at any point in the gathering or processing of the data, then the distribution has shifted by 20 full percentage points. Thus, one error can change a distribution from

60%		40%
40%	to	60%
100%		100%

which reverses the majority. A single error makes too great a difference in the findings. Therefore, you should not rely on interpretations of such distributions when the number gets down around 20 cases or so (at which point a single error makes a difference of 5 percent).

Third, when computing percentages and proportions, it is important to avoid presenting the results with too many significant digits, a practice that fosters the impression that the data are more precise than they really are. Table 3-8, for example, shows that 45 percent of horticultural societies have class systems. This percentage has been rounded to the nearest whole percentage point, although the actual calculation yielded a percentage of 45.0512. The four significant digits (.0512) were dropped because with these kinds of data—as with most of those encountered outside strict laboratory conditions—the process of gathering and processing information contains enough error that it is at best pretentious to claim the kind of precision implied by a percentage carried out to the nearest one ten-thousandth of a percent. For most of the data you will encounter, it is rarely appropriate to present percentages that are accurate to any more than the nearest tenth (45.1 percent) or one hundredth (45.01 percent) of one percent.

Cross-tabulations We use tables in order to compare the ways in which groups are distributed on a single variable, such as the presence or absence of a class system. When the categories we are comparing—such as the different ecological arrangements found in societies—themselves constitute a second variable, we have a **cross-tabulation** (often referred to as a "cross-tab"). From this perspective, we can look at Table 3-8 as a cross-tabulation of two nominal scale variables, type of society on the one hand and the presence of a class system on the other.

Which Way to Percentage? The purpose of making a cross-tabulation is to compare the categories of one variable (such as different social classes) in terms of a second variable. Percentages and proportions allow us to compare categories even though they differ greatly in size. If you look at a table, however, you can readily see that it is possible to percentage in two different directions—across each row or down each column—and it is very important to understand that the direction in which we percentage must be appropriate for the kind of question we are trying to answer.

In Table 3-9, for example, we have the incidence (in millions of people) of poverty among whites and blacks. Going down the first column, we can see that in 1982, there were 23.5 million poor whites and 172.3 million whites who were not poor. Similarly, there were 9.7 million poor blacks and 17.5 million blacks who were not poor. If we want to know which category is most likely to be poor, we cannot simply compare the frequency of poverty for each race because there are more than seven times as many whites as blacks to begin

Table 3-9 Frequency Distribution of Poverty by Race, United States, 1982 (in millions of people)

Poverty Status	Race		All
	White	Black	
Poor	23.5	9.7	33.2
Not Poor	172.3	17.5	189.8
TOTAL	195.8	27.2	223.0

SOURCE: U.S. Bureau of the Census, 1983, p. 471, Table 777.

with. So, we must use percentages, but it is here that things can get tricky if we are not careful.

If we want to compare whites and blacks, we percentage down each column, giving the results in Table 3-10. Fifteen percent of the population is poor, but blacks are three times as likely as whites to be—36 percent versus 12 percent. This is straightforward enough, but what if we had percentaged across the rows instead of down the columns? What would these results have told us?

To see what a difference a simple change in the direction of percentaging can make, look at Table 3-11. We have rearranged nothing in this table; the variables and their categories are in the same position. What we have done is to divide each cell by the total of its row instead of the total for its column. Yet, if you tried to determine from this table which racial category is most likely to be poor, you might be tempted to choose whites rather than blacks because 71 percent is larger than 29 percent. You would be wrong, however,

Table 3-10 Poverty by Race, United States, 1982

Poverty Status	Race		All
	White	Black	
Poor	12%	36%	15%
Not Poor	88	64	85
TOTAL	100%	100%	100%

SOURCE: U.S. Bureau of the Census, 1983, p. 471, Table 777.

Table 3-11 Race by Poverty, United States, 1982

Poverty Status	Race		All
	White	Black	
Poor	71%	29	100%
Not Poor	91%	9	100%
TOTAL	88%	12	100%

SOURCE: U.S. Bureau of the Census, 1983, p. 471, Table 777.

because Table 3-11 does not describe whites and blacks and therefore cannot be used to compare them. Instead, Table 3-11 describes poor people and nonpoor people as categories in terms of race. It tells us that most poor people are white (71 percent) as are most nonpoor people (91 percent) and most people in general (88 percent).

Tables 3-10 and 3-11 are set up to answer very different kinds of questions, and it is vital that you learn to distinguish between them. Table 3-10 is percentaged to answer questions that compare whites and blacks: Which category is most likely to be poor? Which is most likely to be nonpoor? Notice that the percentaging is done within each group that is being compared—among whites and among blacks—which, in this case, means down the columns. In general, *percentaging should always be done within each of the categories that is being compared.* This means that if the 100 percent totals are at the bottoms of columns, then it is the columns that are being compared (as in Table 3-10).

If, however, the 100 percent totals are at the ends of rows, then it is the row categories that are being compared, as in Table 3-11. This table is set up to answer questions very different from those appropriate for Table 3-10: Are poor people more likely to be white than black? Are nonpoor people more likely to be white than black? Because the percentaging is done *among* poor people and *among* nonpoor people, then it is those categories that we can compare, *not* whites and blacks.

Whenever you examine a table that is in the form of percentages or proportions, first check the direction of percentaging. Establish firmly in your mind which categories are being compared and in what terms they are being compared. Do this by first looking for the columns or rows with nothing but 1.0 (proportional distributions) or 100 percent (percentages) in them. If the columns have been percentaged, then the categories of the column variable are the ones being compared and the row-variable categories are the terms in which you are comparing them (as in Table 3-10). If the row variable categories are the ones that are percentaged, then everything holds in reverse (Table 3-11).

If you misinterpret the direction of percentaging, you may arrive at the correct conclusion only by luck. Even the most seasoned professionals can make this simple mistake, often resulting in interpretations of findings that are the opposite of what they should be (see Hajda, 1961; Bordua and Somers, 1962; and Hajda, 1962). To give yourself some immediate practice before going on, return to Table 3-5, percentage it first across the rows and then down the columns, and then see if you can explain to someone else the difference between the two. What kinds of questions can you answer by percentaging across the rows? What kinds of questions can you answer by percentaging down the columns? What are the answers?

MORE COMPLICATED TABLES

You are likely to run into more than simple two-variable cross-tabs before long. These more complicated tables can look so imposing that you are tempted to pass them by; but just remember to start at the top and carefully work your way through them, making sure as you go along that you understand what each number represents.

The most fundamental way in which tables are made more complicated is by including more than just two variables. Consider, for example, Table 3-12 which shows distributions of educational attainment in years by two other variables, race and gender. The table has seven columns of figures—three for each race (men, women, and all) and one for both races combined. It has nine rows—seven for the different categories of educational attainment, ranging from zero to four years of elementary school to four or more years of college; one for the totals of each column; and the last for the number of people on which each percentage is based (N).

Each column totals 100 percent (except for the last column which totals to 99.9 percent because of rounding error). This should tell you immediately that with the table percentaged in this way, we can compare *columns only*. We can compare the educational distributions of different racial and gender categories, such as white men (column I) compared with black men (column IV). We *cannot*, however, compare rows, which is to say, we cannot compare categories of people with different educational attainments in terms of their race or gender composition. To do this we would have to percentage across the rows. (Be sure not to just nod your head in agreement here. Do you understand why the statements in this paragraph are true? Could you explain it to someone else?)

Table 3-12 Years of School Completed by Race and Gender, United States, 1982

Years of School Completed	Whites			Blacks			(VII) All Races
	(I) Men	(II) Women	(III) All	(IV) Men	(V) Women	(VI) All	
Elementary 0-4	2.6%	2.3%	2.4%	9.0%	6.0%	7.3%	3.0%
5-7	5.1	4.9	5.0	10.3	10.9	10.6	5.6
8	7.1	7.4	7.2	6.5	7.0	6.8	7.1
High School 1-3	11.9	13.1	12.6	18.5	21.9	20.4	13.3
4	34.5	42.7	38.8	32.1	32.8	32.5	37.9
College 1-3	15.8	15.2	15.5	14.5	12.9	13.6	15.3
4+	23.0	14.4	18.5	9.1	8.5	8.8	17.7
TOTAL	100.0%	100.0%	100.0%	100.0%	100.0%	100.0%	99.9%
N(1,000s)	(56,253)	(62,539)	(118,792)	(5,984)	(7,615)	(13,599)	(135,526)

SOURCE: U.S. Bureau of the Census, 1983, p. 114, Table 223.

There is much to be learned from a table such as this. Blacks, for example, are much more heavily concentrated in the lower levels (compare columns III and VI) with percentages that are more than double those for whites. These race differences hold for both men and women (compare columns I and IV and columns II and V).

Among whites, men and women have similar educational distributions at the lower levels, but beginning with high school diplomas, differences emerge. Women are considerably more likely to stop at high school graduation (42.7 percent versus 34.5 percent) and considerably less likely to have four or more years of college (14.4 percent versus 23.0 percent). The pattern among blacks is different, however, for men and women have quite similar distributions, with men having only a slight edge at the highest educational level (9.1 percent versus 8.5 percent).

Thus, one general finding is that there are numerous gender differences in educational attainment, but their magnitude depends on which race and which level of schooling you are looking at. If you examined comparable data for several different years—such as 1950, 1960, and 1970—in comparison with those in Table 3-12, you could see a great deal about the ways in which gender and race differences in education are changing in the United States. With a little patience and care, you can learn a great deal from statistical tools as humble as a percentage.

One possible source of confusion in percentage tables is that in some tables percentages do not add up to 100 percent—a fact that makes it all the more important to read carefully the title and footnotes to make sure you understand what the numbers mean. Table 3-13, for example, shows the percentages of people in different social classes who agree with the propositions listed down the left-hand column. As you read across each row, you can see that the higher people's class positions are, the more likely they are to agree with perceptions and explanations of inequality that support the idea that the class system in the United States functions fairly. The percentages do not add to 100 percent anywhere because those who disagree have been omitted from the table in order to save space. (If you know what percentage agrees, then by subtracting from 100 you will get the percentage who disagree.)

The most reliable way to avoid misinterpreting the data in any table is to take numbers from the table and see if you can put them in a sentence that makes sense. "Forty-five percent of lower-class people agree that all young people of high talent have a fairly equal chance of going to college, but 67 percent agree in the middle class and almost everyone in the upper class agrees." The higher your class position, the more likely you are to believe that there is equal educational opportunity.

Table 3-13 Perceptions and Explanations of Inequality by Social Class (Income), Muskegan, Michigan, 1970

	Percentage Who Agree		
Perceptions	*Lower Class*	*Middle Class*	*Upper Class*
All young people of high ability have a fairly equal chance of going to college.	45	67	96
Young people whose parents are poor are just as likely to be in college as anyone else.	29	35	43
If they both work hard, a rich man's son and a poor man's son have equal chances of making a given amount of money.	34	42	57
Poor people generally don't work as hard as rich people.	10	24	39
Poor people basically don't care much about getting ahead.	12	23	46
Personal attributes account for wealth.	28	34	72

SOURCE: Computed from Rytina, Form, and Pease, 1970, Tables 1, 3, and 4.

Ratios The purpose of a **ratio** is to compare two numbers directly by dividing one by the other. As of 1987, for example, the U.S. Senate had 98 male and 2 female members. The ratio of male to female members, therefore, was 98/2 = 49. This ratio—which would be written as 49:1—means that there were 49 times as many male senators as females.

Ratios are usually used for one of two purposes, both of which are subject to the same kind of misinterpretation. First, as in the case of the male and female senators, we compare the relative magnitudes of two numbers. Average income for male workers is twice that for the average female worker; for blacks, the percentage unemployed typically averages twice that for whites; life expectancy in the United States is 1.8 times longer than it is in Afghanistan; suicide rates for U.S. males are 2.8 times higher than they are for U.S. females; and among those 65 years old and older, the male:female ratio climbs to 5.7. In the first case, we divided the male average income by the female average income; in the second, we divided the black unemployment rate by the white rate; in the third, we divided life expectancy in the U.S. by that for Afghanistan; and in the fourth example, we divided the male suicide rates by the female rates.

The second major use of ratios is to indicate change, and there are two conventional ways of doing it. Suppose, for example, that we look at changes in people's values about men's and women's careers. In 1977, for example, 55 percent of U.S. adults agreed with the proposition that "It is more important for a wife to help her husband's career than to have one herself." In 1986, agreement had dropped to just 36 percent (Davis, 1986). One way of using a ratio to represent this change is to subtract the later figure (36) from the earlier (55) and divide that difference by the earlier figure (55). This gives us the ratio of the amount of change relative to the starting point:

$$\frac{(1977 - 1986)}{(1977)} = \frac{55 - 36}{55} = 35\%$$

In this case, we would say that agreement dropped by 35 percent between 1977 and 1986.

A second way to represent this change with a ratio is to simply divide the later number by the earlier: 36/55 = .65 or 65 percent. In this case we would say that agreement in 1986 was only 65 percent of what it was in 1977.

All of these uses of ratios suffer from the same potential for misinterpretation: a ratio tells us nothing about the *absolute* difference between the two numbers involved. If Grace earns $20,000 a year and Margaret earns $10,000 the ratio of Grace's to Margaret's incomes is 2:1; but incomes of only $20 and $10 a year would produce the same ratio. In the first case, the absolute difference in incomes is substantial, but in the second it is trivial.

A political candidate winning an election by 800,000 to 400,000 votes wins by a 2:1 ratio, as does a candidate winning 4 votes to 2 votes. A 35 percent drop in support for the idea that wives should value their husbands' careers above their own is significant when it involves an absolute decline of 29 percentage points (from 55 percent to 36 percent), but it would be far less important if the change consisted of a 35 percent drop from, say, 2 percent to 1.3 percent.

The problem raised by ratios is not so much one of misstatement as of impact: ratios can create an impression whose effect is way out of line with the absolute size of differences between groups or changes over time. As we have seen above, they can sound very impressive even when they represent a trivial absolute difference. Or, they can sound insignificant when in fact they represent substantial differences. A ten percent increase in the poverty rate, for example, may not sound like much until we realize that it means that roughly three *million* people are living in poverty who were not living in poverty before.

The solution for the reader who encounters ratios is direct and relatively simple: whenever possible, convert ratios back to the original numbers that are used in the numerator and denominator of the ratio. If the author does not provide the original numbers, and if you have no idea of their magnitude, then you must remember that the ratio could represent differences ranging from major to miniscule. It might help to keep in mind that, like everyone else, authors and speakers like to feel that what they have to say is important. To say "Support tripled in three years" is more impressive than "Support rose from 1 percent to 3 percent." Such uses of ratios are not misstatements of fact; rather, they are a more dramatic way of presenting a piece of information. By the same token, users of statistics may have an interest in making differences seem as small as possible and may use ratios to create that impression.

These cautions notwithstanding, it is important to point out that ratios can be very useful and revealing. Suppose we wanted to compare the incidence of coronary heart disease deaths in 1950 and 1980. In 1950, 535,705 people died of heart disease; but in 1980, 761,195 people died from heart disease. Does the fact that more people died from heart disease in 1980 than in 1950 mean that heart disease has become more of a health problem?

The problem here is the same one we encountered earlier in this chapter when we tried to compare frequencies: if we want to measure the likelihood of dying from heart disease in 1950 and compare it with the likelihood for 1980, we must take into account the different numbers of people who were eligible to die in each of those years. What we want, then, is the number of people who died in 1950 *relative* to the number of people alive during 1950, and a comparable ratio for 1980. Such ratios are known as **rates**.

In 1950, there were 152,378,000 people in the United States compared with 226,546,000 in 1980. In measuring death rates, it is conventional to take the ratio of the number of deaths in the numerator to the size of the population measured in 100,000s in the denominator. The resulting ratios for the two years look like this:

1950: 535,705/1523.78 = 351.6 deaths:100,000 population
1980: 761,195/2265.46 = 336.0 deaths:100,000 population

The heart disease death rates for 1950 and 1980 are 351.6 and 336.0, respectively. These ratios reveal that although the absolute number of people who died from heart disease was greater in 1980 than in 1950, the *relative* number of people who died—as measured by the death rates—was smaller.

An equally useful application of the ratio is found in Table 3-14. Here we have taken the ratio of male and female death rates for all causes of death and for different age categories in order to see how the gender gap in death rates has changed over time. Among 15 to 24-year-olds in 1950, for example, men

Table 3-14 Ratios of Male Death Rates to Female Death Rates, All Causes, United States, 1950–1980

Age	1950	1960	1970	1980
Under 1	1.31	1.32	1.29	1.25
1–4	1.21	1.22	1.24	1.33
5–14	1.45	1.51	1.59	1.54
15–24	1.89	2.49	2.78	2.97
25–34	1.52	1.76	2.11	2.58
35–44	1.48	1.63	1.74	1.88
45–54	1.66	1.88	1.85	1.86
55–64	1.70	1.93	2.07	1.94
65–74	1.48	1.71	1.89	1.91
75–84	1.24	1.33	1.50	1.62
85+	1.13	1.11	1.15	1.27

SOURCE: U.S. Bureau of the Census, 1983, p. 75, Table 105.

were 1.89 times more likely than women to die, but in 1980, the ratio had grown so that men were 2.97 times more likely than women to die (check the table for yourself to verify that this statement is true). If you look at each age category and go across its row of figures, you can see how the gender gap has generally grown over the last 30 years (largely because death rates have fallen more rapidly for women than for men). This simple use of ratios allows us to quickly and directly compare the death rates of men and women and reveal an important story about epidemiological trends.

Graphics: Describing Distributions with Pictures

Sometimes, in order to present a set of data in the clearest, most readily understandable form, we resort to a large set of techniques generally known as **graphics**. While they may add drama, interest, or clarity to a distribution, they contain the same information as the less glamorous tables on which they are based.

To begin, take a look at Figure 3-1. Stop for a moment and read out loud what is printed within the triangles in (a), (b), and (c). Then answer this question: Which line segment is longer in (d), *w-z* or *x-y*? Don't measure, just look and trust your eyes.

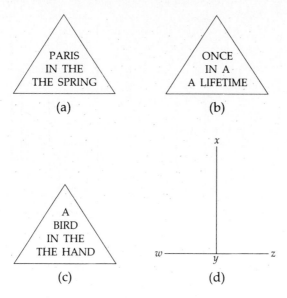

Figure 3-1

In reading the triangles, did you notice that in each one a word is repeated? Did you read "Paris in the *the* spring" out loud? Or did you omit the repeated word as if it was not there? If you did not notice the repetition, you have a lot of company, for most people tend to skip over the extra words simply because they do not expect them to be there, and this is an important element in the reading of graphic presentations. We tend to see what is familiar to us, what we expect to see, and block out the unexpected.

In part (d) of Figure 3-1, if you answered that both segments are of the same length, you were right, but many people will answer that the vertical segment (x-y) is the longer. Once again, our eyes and minds are not perfect instruments of perception. We are subject to many different kinds of systematic misperceptions of our environments, a fact that makes graphic presentations of statistical information a source of misunderstanding as well as clarity. They make data so easy to look at that they encourage us to be less critical than we should be.

BAR GRAPHS

Consider, for example, one of the most commonly used graphic techniques, the **bar graph**. In many bar graphs, the left-hand side (or *axis*, as it is called) shows the number (or proportion or percentage) of observations that have the particular characteristic described on the horizontal axis.

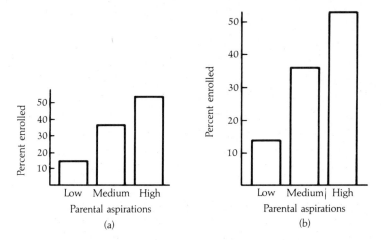

Figure 3-2 Percent enrolled in college preparatory classes by level of parental aspirations, U.S. high school students, 1969.

Figure 3-2 shows a pair of graphs, (a) and (b), both of which represent the same set of data, but in different ways that produce different effects. In both, the vertical axis represents the percentage of students enrolled in college preparatory programs, while the horizontal axis represents three categories of an ordinal-scale variable, the level of parental aspirations. As in most bar graphs, the relative height of the bar indicates the number or percentage of observations having a particular characteristic, and the bars are drawn with equal widths. Occasionally, the *area* of the bar is used to indicate percentages rather than the height of the bar, but this practice is much less common.

What do the graphs tell us? In both cases, you can see that the percentage of students enrolled in college preparatory programs climbs steadily as parental aspirations increase. Although the pattern of differences is the same in both cases, I think you will agree that the differences seem more pronounced in (b) than they do in (a). The actual differences are the same, but the combination of width and height of the bars creates different impressions. A general rule of thumb is that the horizontal axis should be approximately three-quarters of the length of the vertical axis.

As a reader, the only way to avoid such misimpressions is to look carefully at the scales on the vertical and horizontal axes to determine independently of the graph just what it means. If you rely solely on the visual effect you run the risk of drawing conclusions more extreme than those warranted by the data. Remember that graphics do not create new information; they simply express a

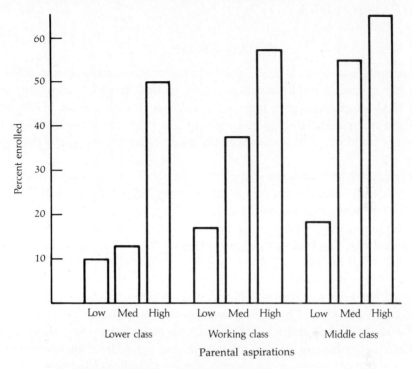

Figure 3-3   Percent enrolled in college preparatory classes by level of parental aspirations and by social class, U.S. high school students, 1969.

SOURCE: Boocock, 1972, p. 63, Table 4-2.

table in pictorial form. Since it is a lot harder to be misled by a table than by a picture, when in doubt translate the picture back into tabular form, if only in your mind.

Bar graphs can be considerably more complex than the ones in Figure 3-2. In Figure 3-3, for example, we have added a third variable—social class—to the pair of variables represented in Figure 3-2. In this bar graph, there are three sets of bars, one for each level of social class. Each of these sets has one bar for each of three levels of parental aspiration (low, medium, and high). This more complex graph allows the reader to see quickly not only the effect of parental aspirations on enrollment in college preparatory programs (by comparing the bars *within* each set), but also how those effects are themselves affected by social class (by comparing corresponding bars in different sets). Stop for a moment and see if you can summarize how the effect of parental aspirations varies from one class to another.

LINE GRAPHS

The problems of scale and misinterpretation that make it important to read bar graphs carefully are also found in **line graphs**, which are often used to show changes over time, such as those depicting stock averages or demographic trends. As you can see in Figure 3-4, the ways in which the numbers are marked off on the axes affect our interpretation of the data they represent.

This figure shows how the size of the male nobility in Venice fell during the 16th and 17th centuries, in part because of the strict rules of entry into the nobility that were applied. Notice first that the vertical axes on graphs (a) and (b) are broken by double slashes just above the point where they join the horizontal axis. The broken lines are necessary because the distance between zero and 1,400 on the vertical axis has been compressed so that it is only as long as the distance between 1,400 and 1,500. In other words, a distance representing zero to 1,400 nobles has been allotted only one-fourteenth as much room as it should have. This was done because the lowest point on the line is quite high—around 1,500—and to include the big empty space between that point and zero on the vertical axis would be a waste of room on the printed page. Without the double slashes to indicate that this has been done, however, readers might get the false impression that the number of nobles in Venice approached zero in the mid-17th century.

Although both graphs represent the same historical trend, the drop in (b) is more impressive than in (a) unless we look at each vertical axis and see that the *actual* drops are identical in both graphs. Why the different appearances? In (b) the axes have been drawn to conform to the "three-quarters" rule; but in (a), the horizontal axis has been stretched out and the vertical scale shortened to create the impression of much less change.

When properly displayed, line graphs are a particularly useful method for showing trends, for they allow readers to grasp quickly what is going on in a set of data. Figure 3-5, for example, shows almost a century of unemployment rates for the United States, highlighting the close correspondence between low unemployment and war.

As Figure 3-6 shows, line graphs can also be used to compare two distributions. Here we can see how male and female death rates from suicide vary over the life course. The graphs show clearly not only that male death rates are higher than those for females at every age, but that male suicide rates rise steadily with advancing age, while those for females peak at middle-age and then decline. Such strikingly different patterns raise fundamental questions about the different social and psychological effects of aging on men and women.

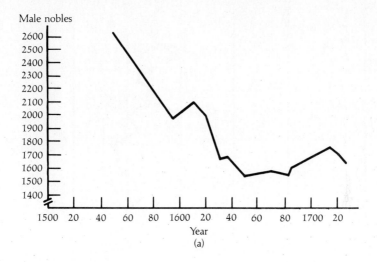

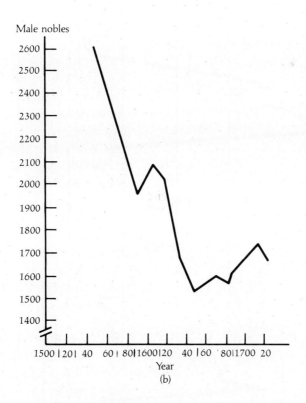

Figure 3-4 Number of male nobles in Venice, 1500–1720.

SOURCE: Braudel, 1982, p. 472.

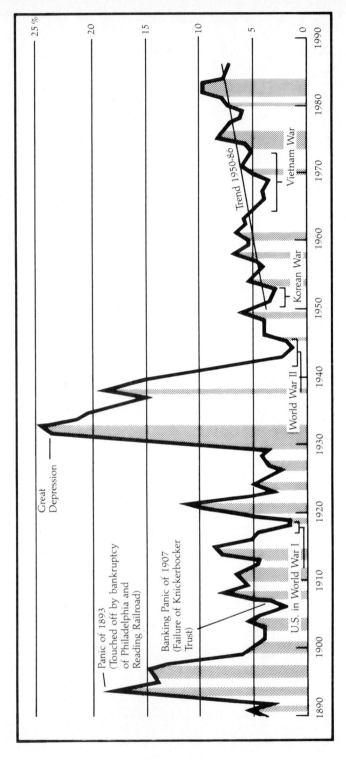

Figure 3-5 Percent unemployed, 1890–1986, United States.

SOURCE: Bureau of Labor Statistics, 1986.

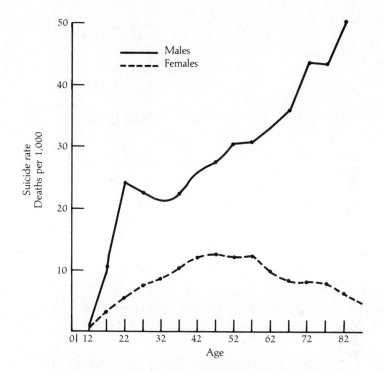

Figure 3-6   Male and female suicide rates by age, United States, 1973.

SOURCE: National Center for Health Statistics, 1975.

HISTOGRAMS AND FREQUENCY POLYGONS

Another common graphic technique is similar to the bar graph, but rather than representing discrete variables (such as the ordinal categories of parental aspiration levels in Figure 3-2) the **histogram** is used with continuous variables such as age and income. Figure 3-7, for example, shows the percentage distribution of age at menopause for U.S. women. The horizontal axis shows the age at menopause; the vertical axis shows the percentage of women experiencing menopause in a given age interval.

We can see immediately from this figure that most women (54 percent) experience menopause between their 45th and 55th birthdays, and that a small percentage (2 percent) experience menopause before their 25th birthdays. The histogram tells the story quickly and clearly.

An alternative way of presenting the information in Figure 3-7 is to fit a line graph to a histogram by connecting the midpoints of the tops of each bar, as

56 / CHAPTER 3 DISTRIBUTIONS, COMPARISONS, AND TABLES

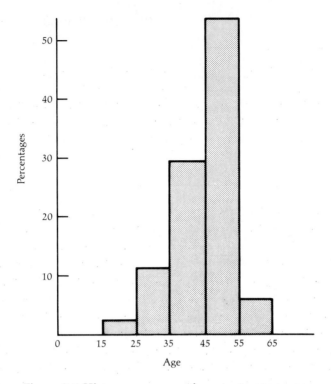

Figure 3-7 Histogram representing age at menopause, U.S. women, 1959–1961.

SOURCE: U.S. Department of Health, Education, and Welfare, 1961.

in Figure 3-8. This is called a **frequency polygon**. Which of this pair of techniques is used is largely a matter of personal preference. Both convey the same information.

STEM AND LEAF DISPLAYS

An ingenious method for displaying the distribution of an interval or ratio scale variable is Tukey's **stem and leaf display**. Suppose, for example, we have data on the age at menopause for 100 women. To summarize this distribution of ages with a stem and leaf display, we show the first digit of each age along the left-hand side. This column of numbers is called the *stem* (see Figure 3-9). We then place each woman's score in the appropriate row, showing the second digit of her age at menopause.

Thus, reading down from the top, there were no women who experienced menopause before the age of 10, because the first row (ages 01 through 09) is

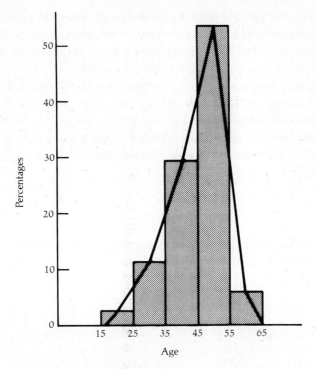

Figure 3-8 Frequency polygon representing age at menopause, U.S. women, 1959–1961.

SOURCE: U.S. Department of Health, Education, and Welfare, 1961.

```
0
1  9
2  56799
3  3444555678888899999
4  011112222333444455556667777788889999
5  000011111122222233333334444445555556789
6  01
7
8
9
```

Figure 3-9 Stem and leaf display representing age at menopause for 100 women.

empty, but one woman experienced menopause at age 19 (second row). The third row contains five women with ages at menopause of 25, 26, 27, 29, and 29. At the other end of the display, two women experienced menopause in their 60s, one at age 60 and the other at age 61.

Stem and leaf displays are useful because they show a great deal of detailed data in a relatively small space. We can easily see the relative frequency of different ranges of scores (most women experience menopause in their 50s, for example) as well as which scores are most common. Stem and leaf displays have all the advantages of a histogram with the added advantage of greater detail.

OGIVES

Figure 3-8 shows the percentage of women who experience menopause in each age category. Since menopause is a once-in-a-lifetime experience,

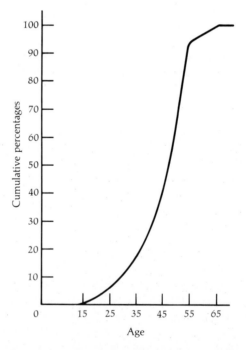

Figure 3-10 Ogive representing the cumulative percentage of U.S. women experiencing menopause by age, 1959–1961.

SOURCE: U.S. Department of Health, Education, and Welfare, 1961.

however, we could also think of this variable in terms of the percentage of women who have this experience *by* each age. In other words, since 2 percent of women experience menopause between the ages of 15 and 25, and an additional 12 percent between the ages of 25 and 35, we can say that 2 percent have this experience *by* the age of 25, and 14 percent (2 + 12) have gone through it *by* the age of 35. This is called a *cumulative* distribution because it is formed by adding the percentage for one age group to the percentages of all the groups below it. The graphic form of a cumulative distribution is called an **ogive**. We can quickly see from Figure 3-10 that by age 25 very few women have experienced menopause, but that after age 25 the percentages increase sharply. By age 55, 96 percent of all women have experienced menopause.

SCATTERPLOTS

In all of the graphics we have discussed thus far, each point on the horizontal axis has corresponded with only one point on the vertical axis. In Figure 3-5, for example, there was only one unemployment rate for each year. This enabled us to use just one line to connect all of the points. Much social research data, however, involves multiple observations, with the result that for each value on the horizontal axis there are many different scores on the vertical axis. In this kind of situation, a **scatterplot** is the most effective way to display the data.

In Figure 3-11, for example, each point on the graph represents the combination of two characteristics for a given country. The variable on the horizontal axis is per capita income (the amount of income each person would receive if all income were distributed evenly). The variable on the vertical axis is the ratio of children enrolled in secondary school to the number of children under the age of 15, expressed as a percentage. West Germany, for example (in the upper right corner), has a per capita income of $9,000, and the number of students enrolled in secondary school is equal to around 45 percent of all those under the age of 15.

The scatterplot allows us to see how differences in per capita income tend to go along with a certain pattern of differences in secondary school enrollment: the wealthier a society is, the greater is its secondary school enrollment. (Notice also that almost three-quarters [72 percent] of all the world's people are crowded into countries that are both very poor and have very low levels of school enrollment.) If we had presented these data in the form of a table—a long list of countries with their income and school enrollment figures alongside—it would have been much more difficult to see this pattern of differences.

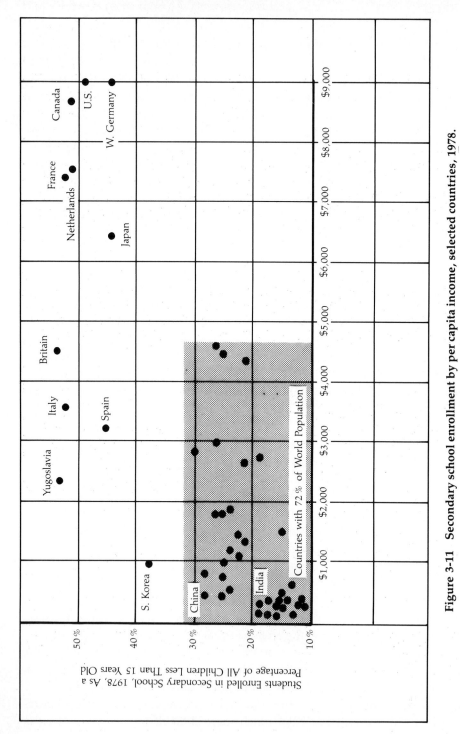

Figure 3-11 Secondary school enrollment by per capita income, selected countries, 1978.

SOURCE: Computed from U.S. Bureau of the Census, 1981, Tables 1544, 1546, 1549.

Statistics and Truth: Some Perspective

Scientists and artists have shown great ingenuity in devising ways to present data, from the most humble frequency distribution to the most complex and colorful graphic display. Although space does not allow me to familiarize you with any more than some of those most commonly used, what is most important is that you appreciate the difficulties that scientists face in trying to distill vast quantities of information into a clear story. Because there are many ways of presenting the same set of data, your task as a reader is to pay careful attention to the underlying set of facts on which a table or figure is based. There is no magic formula that tells us which kind of table or graph to rely on to represent a set of data; there is no "naturally" best or most objective way (see Tufte, 1982).

The ambiguities and opportunities for misinterpretation that arise with graphic techniques in particular may bring to your mind the old reference to "lying with statistics." There are those who suggest rather strongly that it is possible to prove anything with statistics, but in truth it is no easier to lie with statistics than it is to lie with words. In fact, the precise language and careful definitions of statistics make it *more* difficult to fool an informed and careful reader. And that is the key—to be a reader who appreciates the sources and limitations of statistical techniques, who does more than read the text and glance at the tables and graphs. If we are, as a society, more easily deceived by statistics, I think it is because we are badly trained.

The only way to be sure that data are presented and interpreted correctly is to have a combination of patience, tenacity, and critical attention. A teacher can help you become more aware of the technical aspects of statistics and their uses, but in the end what results from your encounters with them will depend on how much you care about truth and accuracy.

Do not get discouraged if you find it difficult to make your way through tables and graphs; we *all* do. I know of nothing that makes it easy, although practice certainly makes it much easier. What I am certain of is that enormous amounts of important information about the world are available to anyone who takes the time to learn how to evaluate statistical information.

Summary

1. Having gathered data, researchers must organize and summarize large numbers of individual observations, often in the form of statistical tables. In reading tables, it is important to pay attention to each of their parts, from headings to footnotes.

2. Frequency distributions are the simplest way of presenting a set of statistical results. Grouping categories enables us to reduce a large number of categories to a relative few.

3. Proportions and percentages allow us to compare categories that differ in size. In interpreting cross-tabulations, it is important to pay attention to the direction—down columns or across rows—in which the percentaging is done. In general, percentaging is done within each category that is being compared with other categories.

4. Ratios are another technique for comparing categories. Although they do indicate the relative magnitude of scores, they do not tell us anything about absolute differences between scores.

5. Graphic techniques are useful for displaying a set of data in a form that is relatively easy to grasp. Bar graphs are used with nominal- and ordinal-scale variables; line graphs, histograms, frequency polygons, stem and leaf displays, ogives, and scatterplots are used with interval- and ratio-scale variables. When interpreting graphics, it is particularly important to be aware of ways in which visual displays can give misleading impressions of the underlying data.

Key Terms

bar graph 49
case 31
cell 36
cross-tabulation 39
frequency distribution 30
frequency polygon 56
graphics 48
grouped data 33
grouping 33
histogram 55

line graph 52
marginal 36
ogive 59
percentage 37
proportion 37
rate 47
ratio 45
scatterplot 59
stem and leaf display 56
table 31

Problems

1. Define and give an example of each of the key terms listed at the end of the chapter. See if you can explain each of them to someone else.

2. Using Table 3-15, respond to parts a–j on p. 63. In your answers, state the direction in which you must percentage the table and calculate the appropriate percentages.

Table 3-15 Answers to the Question, "Can you imagine circumstances in which you would approve of an adult male hitting a male stranger?" by Education, U.S. Adults, 1979

Approve of Hitting?	Educational Attainment			All
	Less Than High School	High School Only	More Than High School	
Yes	244	563	199	1,006
No	226	241	56	523
TOTAL	470	804	255	1529

a. Who is most likely to approve of hitting, someone with less than high school or someone with more than high school?
b. Who is most likely to disapprove of hitting, someone with high school only or someone with more than high school?
c. Who is most likely to have more than a high school education, someone who approves of hitting or someone who disapproves?
d. Do most people approve of hitting or do most disapprove?
e. Which educational attainment level are people most likely to have?
f. Rank the three educational attainment categories from high to low in terms of the likelihood of support for hitting.
g. Which educational attainment category is *least* likely to approve of hitting?
h. Which educational attainment category is *most* likely to approve of hitting?
i. Is there any important information missing from Table 3-15? If so, what? What difference is its absence likely to make?
j. How many variables are there in this table?

3. In a sentence or two, summarize the general principle that governs the direction of percentaging in cross-tabulations.

4. The overall crime rate in the United States was 4,154 per 100,000 inhabitants in 1973; in 1982 it was 5,553. Use ratios to express the degree of change between the two years. Use both approaches described in the text.
   a. Are these two pieces of data sufficient to describe the trends in crime rates between 1973 and 1982? Why or why not?

5. Why is the use and interpretation of ratios potentially misleading? Give an example that shows what you mean.

6. Use a bar graph to display the data in Table 3-15.
7. The homicide rates (per 100,000 population) for the United States between 1973 and 1982 were as follows:

1973	9.4
1974	9.8
1975	9.6
1976	8.8
1977	8.8
1978	9.0
1979	9.7
1980	10.2
1981	9.8
1982	9.1

Display these data using:
a. a line graph
b. a histogram

4

Summarizing Distributions

In Chapter 3 we introduced some relatively straightforward ways of comparing distributions through the use of frequencies, percentages, proportions, ratios, and various graphic techniques. As useful as such methods are, there are many instances in which distributions are so complex or detailed that it is difficult to make comparisons that are clear and efficient. In other words, in spite of the fact that something like a percentage distribution can reduce an enormous set of individual observations into a relatively compact, easily grasped form, there is often a need to further reduce or simplify a set of data. In the ideal case, we are able to summarize important characteristics of distributions with a single number, which is what this chapter is primarily about.

You may have noticed that the process of data reduction moves us farther and farther away from individuals. In the last chapter, in fact, we never mentioned individuals, who got lost somewhere in the tables and graphs. It is important to remember that by its nature, statistical information never focuses on individuals as anything other than what goes into constituting a set of data. Statistics is a set of techniques used to study and categorize sets of observations, from molecules to societies. It is relevant to individual members of those sets only insofar as they take on or are affected by the characteristics of the categories they belong to. As any social scientist knows, of course, this connection between groups and individuals is a vitally important one.

You may still have some feelings about the idea that statistics can be used to bury the individual in the obscurity of the "social mass." Although it is true that statistics cannot be used to make precise statements about individuals, they can tell us a great deal about the environments in which individuals exist and the effects those environments tend to exert. If a woman applies for a job, for example, and is turned down, statistics cannot by themselves explain why *she* was not hired. They can, however, tell us a great deal about how people in a similar social status (female) tend to fare in job interviews, and this can tell us much about the ways in which our positions in social environments affect what happens to us.

Questions about individuals and questions about their environments, although related, are often different and call for different techniques of gathering and interpreting information. Sociologically, for example, questions about an individual's experience can make use of statistics only insofar as those statistics provide an understanding of the environments in which the individual acts. Questions about those environments, however, including questions about the extent of discrimination and the ways in which economic arrangements sort men and women into different occupational and family statuses, must rely heavily on systematic statistical information.

Are women discriminated against in hiring? We could go out and compare a man and woman, equally qualified, who apply for the same job, but how would we know that whatever we observed was not peculiar to those individuals? If we want to talk about individuals, it is enough to observe them, but if we want to talk about male and female as social categories and how membership in one or the other affects people's lives, then we must gather data on representative samples of men and women.

In this chapter we will continue to work with the problem of using data in order to tell a story that both makes sense and faithfully reflects the evidence. In each case, the techniques we are able to use will depend on the scale of measurement of the variables we are dealing with.

In this chapter you will encounter for the first time in this book some mathematical formulas, some of which your instructors may suggest that you ignore. Regardless of how you pay attention to them, keep in mind that what is most important is an understanding of what the numbers mean and how their interpretation can inform us or mislead us.

Central Tendency: The Average

There are times when it is enough to present a single frequency or percentage distribution in order to tell a meaningful story, but most of the time we are interested in making comparisons between groups of observations, which is to say, between the distributions that describe them. We saw in Chapter 3 how difficult it is to compare categories of unequal size, a problem we solved through the use of percentages and proportions. Most of our examples, however, used variables with only a handful of categories.

Comparing distributions with many categories can be very cumbersome. If the variables are nominal scales, all we can do to simplify things is to combine categories, a practice that often sacrifices detail in exchange for simplicity. As we will see in the rest of this chapter, however, with higher levels of measurement, we can summarize distributions—and thus simplify our comparisons—in more sophisticated and efficient ways.

We begin with measures of **central tendency**, all of which allow us to describe a distribution in terms of its "center." The goal here is to find a way to represent an entire distribution of scores with only a single score. Of course, in reducing a set of data to this bare minimum we inevitably lose detail; but at the same time, we gain in our ability to make comparisons.

THE MEAN

As a measure of central tendency the **mean** is the statistic most often associated with the idea of an "average." Computation of the mean is quite straightforward: *for a set of scores, the mean is simply the sum of all the scores divided by the number of scores.*

The mean is appropriate *only* for interval- or ratio-scale variables, such as per capita income for countries. Per capita income is a measure of the amount of income a country has, relative to the size of its population—in other words, the amount of money each person would get if income were distributed evenly. Table 4-1 shows per capita income figures for seven countries. To calculate the mean income for this group, we add these seven numbers together for a total of $31,312 and then divide by 7 for a mean income of $4,473. The symbol for the mean of a sample is usually a letter with a bar over it, such as $\bar{X}$. Population means are usually represented by the Greek letter *mu* (μ).

Table 4-1 Per Capita Income, Selected Countries, 1980

Country	Per Capita Income
Switzerland	$16,210
United States	$10,408
Mexico	$1,901
Syria	$1,254
Ghana	$846
Egypt	$485
India	$208
TOTAL	$31,312

SOURCE: U.S. Bureau of the Census, 1983, p. 865, Table 1509.

As a measure of central tendency for quantitative variables the mean adds together the "quantities" for all observations (in this case, dollars) and then distributes them evenly among all the cases. Notice that no country has an income equal to the mean of $4,473. In fact, most amount to less than half the mean. Although this may strike you as strange, it will make more sense if you remember that means are used to describe groups of observations, not their individual members. To say that "the average country in this group had a per capita income of $4,473 in 1980" is a misstatement because it implies that the mean says something precise about individual countries. The correct interpretation is that this *group* of countries had an average income of $4,473 per capita.

While the mean cannot be used to draw conclusions about individuals, it is very useful for describing groups of observations. The mean locates a distribution along a continuum going from the lowest possible score to the highest. It allows us to compare groups on the basis of how much of a characteristic the groups possess relative to their size.

Notice also that the mean can be affected by extremely large or small scores. In Table 4-1, for example, the inclusion of two very high income countries—the United States and Switzerland—produces a mean that is far higher than the income of most countries in the distribution and considerably lower than that for the high-income countries. If we omitted these two extreme scores, the mean for the remaining five would fall drastically to $939, a figure that is much closer on the average to the majority of countries in the original distribution.

The Balance-Beam Analogy It might help you to understand the mean better if we use a physical analogy. Imagine that we have a long wooden beam laid out on the ground, whose length has been marked off in equal sections each of which represents $1,000 of per capita income. We use blocks of equal weight to represent each country and we place them at points on the beam that correspond to their income (forcing some crowding at the lower end; if we had many countries we would have to put some blocks almost on top of each other). The result might look something like this:

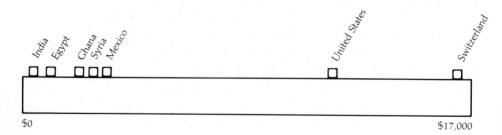

Now we put a fulcrum under the beam in such a way that it exactly balances (with all seven blocks laid out on top of it). If each block has the same weight, the result will look like this:

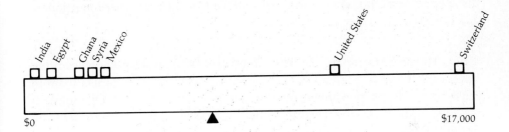

The point at which we would have to place the fulcrum in order to balance the beam is the point that corresponds exactly to the mean.

Box 4-1   One Step Further: The Mathematics of Means

To understand the formula for the mean (as well as many others you will encounter in other boxes like this one) you must become familiar with a special symbol called a *summation sign*, which looks like this:

$$\Sigma$$

As a symbol it represents a mathematical operation. It tells you to add up all of the scores for the variable that appears immediately after it. If you take as an example the seven countries in Table 4-1, in order to calculate the mean we must add up their per capita incomes (X). Mathematically, it would look like this:

$$\Sigma X_i$$

The subscript i is a variable that is used to keep count of each case in the distribution. In our example, each country is identified from 1 to 7. Thus, for Switzerland, $i = 1$, and for India $i = 7$. To calculate the mean, we then divide this sum by N, which stands for the number of observations. Mathematically, then, the formula that *defines* the mean looks like this:

$$\bar{X} = \frac{\Sigma X_i}{N}$$

This analogy shows both how the mean is used to represent the "center" of a distribution and how extreme scores affect it. Just one country with an income of, say, $50,000 would force us to move the fulcrum considerably to the right. Even though it is just one country among several others, its great distance from all the others forces us to move the fulcrum a disproportionate amount.

Using Means to Compare Groups Interval- and ratio-scale variables often have a relatively large number of categories, and if we want to use them to make comparisons, percentage distributions are not very useful. One solution is to use a measure of central tendency such as the mean to represent the entire distribution.

Suppose, for example, that we ask members of two groups—Group 1 with 50 people and Group 2 with 200—to tell us how many children they have and get the results shown in Table 4-2. How might we compare them? One possibility would be simply to see which group has the largest number of children by adding up all the children. In Group 1, there are 5 people who are childless and account for a total of (5 × 0) = 0 children. There are 5 people with 1 child (5 × 1 = 5 children) and 25 with 2 children (25 × 2 = 50 children). Notice that what we are doing for each category is multiplying the score for that category (the number of children in column I) times the number of people who have that score (column II) to get the total number of children

Table 4-2 Distributions of Number of Children and Computation of the Total Number of Children in Two Groups

	Group 1		Group 2	
(I) Number of Children	(II) Frequency	(III) Total Children	(IV) Frequency	(V) Total Children
0	5	0	22	0
1	5	5	24	24
2	25	50	120	240
3	10	30	20	60
4	2	8	9	36
5	2	10	5	25
6	1	6	0	0
TOTAL	50	109	200	385

for people in that category (column III). This is a slightly more complicated procedure for calculating the mean than the one we used before, because many cases have the same score (whereas no two countries had the same income in our previous example).

As you can see at the bottom of column III, there are a total of 109 children in Group 1. Using the same procedure for Group 2 (multiplying each frequency in column IV by the corresponding score in column I), we get a total of 385 children (bottom of column V). Group 2, then, has more than three times as many children as Group 1 (385 versus 109); but, as you may have already noticed, this is an unfair comparison, since Group 2 has four times as many parents (200 versus 50).

This presents us with the same problem that we had when trying to compare frequency distributions in Chapter 3: In order to compare the two groups, we have to allow for differences in their size. The best way to do this is to divide each total number of children by the number of cases they came from—that is, 109/50 and 385/200. This gives us 2.18 for Group 1 and 1.92 for Group 2—the number of children in each group relative to the number of parents. The figure below illustrates the mean for Group 1 using the balance-beam analogy. (Note that it is necessary to imagine standing some people on each other's heads.)

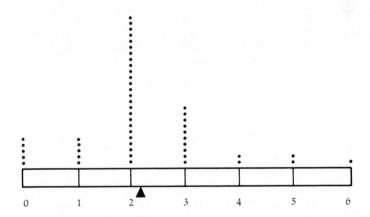

As before, notice that no member of either group has a score equal to the mean. In fact, as is often the case with discrete variables, it is impossible for anyone to have the mean number of children, since no one can have 1.92 or 2.18 children. Notice also that although Group 2 has a larger number of children overall, it has a smaller mean, telling us that it has fewer children relative to its size. Thus, the average number of children is higher in Group 1 than in Group 2.

Box 4-2 One Step Further: The Mathematics of Means with Grouped Data

With larger data sets (such as those in Table 4-2) in which many observations have the same score, there is a computing formula for the mean that is more efficient than the formula that defines the mean. In Group 1, for example, 25 people have a score of 2. Instead of adding a score of 2 in 25 times, we could just multiply the score by its frequency. This is just what we did to get the results in columns III and V. The formula that represents this operation is:

$$\frac{\sum f_i X_i}{N}$$

In this formula, f represents the frequency of a score, X represents the *score* (ranging from 0 to 7), and N represents the number of *observations* (50 in Group 1, 200 in Group 2).

When we have grouped data, such as those in Table 3-4 (p. 34), the same formula can be used. Each score (X_i), however, is the score that divides each grouped category exactly in half (called the *midpoint*). Suppose, for example, that we grouped the categories in Table 4-2 as follows:

	Group 1		Group 2	
Number of Children	Frequency	Total Children	Frequency	Total Children
0	5	0	22	0
1–2	30	45	144	216
3–4	12	42	29	101.5
5–6	3	16.5	5	27.5
TOTAL	50	103.5	200	345

Each person in the "0 children" category would be assigned a score of 0, but each person in the "1–2 children" category would be assigned a score of 1.5 children, and each in the "3–4 children" category a score of 3.5 children. In using the computing formula for the mean introduced above, each X_i is the midpoint of one of the four categories for the variable "number of children."

Notice, by the way, that the total number of children in each group differs from the total calculated from the ungrouped distributions in Table 4-2. This shows how grouping data can result in a loss of detail and a distortion of the results.

Race and Fertility: An Example A mean is a number that describes a group of numbers by adding them together and dividing by the number of observations. Its advantage is that it allows us to represent an entire distribution with a single number, and although we inevitably lose detail by doing this, it nonetheless enables us to learn a great deal about group differences.

Consider, for example, the data in Table 4-3. In 1967 and 1982 (as in other years in between), the Census Bureau interviewed a large representative sample of married U.S. women and asked them how many children they already had and how many additional children they expected to have. Adding the two numbers together results in the total number of children married women expect to have during their lives. This table shows the means for groups defined by age, race, and the year in which the data were gathered. There are 12 different groups represented in this table, one for each cell. The upper-left corner cell, for example, consists of 18 to 24-year-old white women in 1967. The numbers in the cells are means, and we can use them to compare these groups in spite of the fact that they are of quite different sizes (although it is not shown in the table, the number of blacks in each age group is only around a tenth the number of whites, for example, and the sizes of the different age groups within races also vary).

There is a great deal we can tell from this table. First, if we compare columns I and II and compare columns III and IV, we can see that regardless of age or race, women in 1982 expected to have fewer children than did comparable women in 1967. White 18 to 24-year-olds, for example, expected to have an average of 2.86 children in 1967, but the expectation had dropped to 2.09 by 1982.

Table 4-3 Mean Number of Children Expected by Currently Married Women by Age and Race, United States, 1967 and 1982

Age	Whites		Blacks	
	(I) 1967	(II) 1982	(III) 1967	(IV) 1982
18–24	2.86	2.09	2.76	2.05
25–29	3.00	2.15	3.41	2.29
30–34	3.20	2.18	4.26	2.45

SOURCE: U.S. Bureau of the Census, 1983, p. 67, Table 91.

Second, we can use this table to compare the expected fertility of whites and blacks and to see how race differences have changed. In 1967, black expectations were higher than white expectations among those 25 to 29 and 30 to 34 years old, but slightly lower among 18 to 24-year-olds (compare columns I and III—find the appropriate figures and see if I am correct). By 1982, the pattern still held, although the magnitude of the differences seems to have shrunk. Among 30 to 34-year-olds in 1967, for example, the difference between the white and black means was 1.06 (or, 4.26 − 3.20), but by 1982 the difference had shrunk to just 0.27 (or, 2.45 − 2.18). If you look at the differences for all such pairs (by subtracting the mean for whites in each year from the corresponding mean for blacks), you will find that the absolute differences have grown smaller across the board (see the right-hand pair of columns in Table 4-4).

Another way to measure changes in the racial difference in fertility is to take the ratio of black means to white means for each of the two years. As the left-hand pair of columns in Table 4-4 shows, the black/white ratios have grown closer to 1.0 (indicating equality) for each age group.

Third, notice that older women tend to have higher birth expectations than younger women (reading down each column in Table 4-3). This can mean two things. It might mean that younger women will have fewer children than women who grew up in different times (a generational difference). Second, it may also be that younger women tend to underestimate their total fertility by overestimating their ability to control births. Older women, however, who are relatively close to the end of their fertile years, are in a position to make more accurate lifetime estimates since their childbearing potential is far more limited. Only time will tell which of these interpretations is the more important, but there is evidence that suggests that both play a part—that

Table 4-4 Expected Number of Children: Ratios of Black to White Means and Differences between Means, United States, 1967 and 1982

Age	Black/White Ratio		Black-White Difference	
	1967	1982	1967	1982
18–24	0.97	0.98	−.10	−.04
25–29	1.14	1.07	.41	.14
30–34	1.33	1.12	1.06	.27

SOURCE: U.S. Bureau of the Census, 1983, p. 67, Table 91.

younger women will have fewer children than the generations of women that went before them and that they tend to underestimate their fertility.

By using a combination of techniques—cross-tabulation, means, and ratios—we can answer a variety of questions in a clear and straightforward way (as well as raise a few new ones in the process). In spite of the drawbacks of reducing data to group measures, these are very valuable tools in social research.

THE MEDIAN

The **median** is a measure of central tendency that also tries to represent the "center" of a distribution, but it differs from the mean in two ways. First, it can be used with ordinal-scale variables as well as interval and ratio scales. Second, it defines the "center" from a different point of view that creates a new interpretation.

To calculate the median for a distribution, we simply arrange the cases in order from the highest score to the lowest (or from the lowest score to the highest—it does not matter which way it is done). The median is defined as *the middle score in an ordered distribution, the one that divides it exactly in half*. To see what this means, go back to the simple set of data in Table 4-1 which showed per capita income for seven countries. The scores are already arranged in order, which makes our job easier. There are an odd number of scores (seven) which means that the middle score is going to be the fourth since there are three above it and three below it. The median per capita income in this case is the income for Syria, or, $1,254.

Notice that this is quite different from the mean for this distribution that we calculated earlier—$4,473. Why so different? The answer is that unlike the mean, the median is not affected by extremely large or small scores. It is "insensitive" to extremes in a distribution. If we raised Switzerland's income to $50,000 per capita or lowered India's income to $1, the median would still be $1,254, since that would still be the income of the middle case in the distribution. Because the median is insensitive to extreme scores, it is generally considered to be a better measure of the "typical case" than the mean is.

In this example, there were an odd number of cases, which made it relatively easy to identify the middle case. When there are an *even* number of cases, the procedure is only slightly more complicated. Suppose, for example, that we added an eighth country—the Soviet Union, with a per capita income of $4,861—to the list in Table 4-1 to get Table 4-5. Where is the center of the distribution? In this case, it will lie *between* the middle *pair* of cases or, between Mexico and Syria. There is, of course, no such case; there is only a blank space between these two cases. So, what we do is create an imaginary "case" that

Table 4-5 Per Capita Income, Eight Selected Countries, 1980

Country	Per Capita Income
Switzerland	$16,210
United States	$10,408
Soviet Union	$4,861
Mexico	$1,901
	median = $1,578
Syria	$1,254
Ghana	$846
Egypt	$485
India	$208

SOURCE: U.S. Bureau of the Census, 1983, p. 865, Table 1509.

divides the distribution exactly in half, with four countries above it and four countries below. We give an income score to this "case" that lies halfway between the middle pair, in other words, halfway between Mexico's $1,901 and Syria's $1,254. The result is a median score of $1,578 [or, ($1,901 + $1,254)/2].

The Balance Beam Revisited As with the mean, we can use a balance-beam analogy to clarify the concept of a median. Consider the 50 people in Group 1 of Table 4-2 and the number of children had by each. We arrange them from high to low and ask them to sit on a beam that is just long enough to hold them all. Everyone has the same weight. Since we have spaced them evenly across the entire length of the beam, the fulcrum will always balance the beam at its exact center. The median is the score (the number of children in this case) of the person who is sitting right above the fulcrum. In this case, it is 2 children:

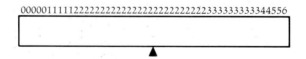

If the highest person's score is raised—say, to 10—the median will not change at all *since that person's distance from the center of the beam remains the same no matter how high his or her score gets.* Only when the score of the middle case changes will the median itself change.

> **Box 4-3 One Step Further: The Mathematics of Medians**
>
> There are two formulas that make it easy to identify which case corresponds to the median in an ordered distribution. When the number of cases (N) is odd, the middle case (M) is found by adding one to the number of observations (N) and dividing by 2:
>
> $$M = \frac{N + 1}{2}$$
>
> With 51 cases, for example, the middle case is:
>
> $$M = \frac{(51 + 1)}{2} = 26^{\text{th}} \text{ case}$$
>
> Since there is no middle case when there is an even number of cases, we create an imaginary case midway between the *pair* (M and $M + 1$) of middle cases. The lower member of this pair (M) is found by simply dividing the number of cases (N) by 2:
>
> $$M = \frac{N}{2}$$
>
> The other member of the pair is then the ($M + 1$)th case. With 20 cases, for example, the middle pair of cases consists of:
>
> $$M = \frac{20}{2} = 10^{\text{th}} \text{ case} \qquad M + 1 = 11^{\text{th}} \text{ case}$$
>
> The median is then found by adding the scores of these two cases together and dividing by 2.

The Median with Ordinal Variables Thus far we have focused on measures of central tendency for ratio-scale variables such as income and the number of children. Unlike the mean, the median can also be used with ordinal scales. Consider, for example, the distribution in Table 4-6 that compares religious attendance for U.S. adults in 1976 and 1986.

Table 4-6 Religious Service Attendance, United States, 1976 and 1986

Frequency of Attendance	1976	1986
Never	194	206
Less Than Once a Year	136	105
About Once or Twice a Year	203	183
Several Times a Year	233	175
About Once a Month	106	118
2–3 Times a Year	100	148
Nearly Every Week	90	66
Every Week	298	334
Several Times a Week	133	132
TOTAL	1,493	1,471

SOURCE: For 1972, Davis, 1982, p. 96; for 1986, Davis, 1986, p. 128. (For both years, a handful of "don't knows" are excluded because they cannot be used to calculate the median.)

How can we compare religious attendance in 1976 with attendance in 1986? We could convert each frequency distribution to percentages or proportions and compare the two years for each of the nine levels of attendance, but that would be quite cumbersome. It would be preferable to represent each distribution with a single measure of central tendency on which we could base an overall comparison.

The mean is inappropriate because we have an ordinal scale. We would have no way of assigning numerical scores, for example, to categories such as "less than once a year" or "several times a week." The median offers a way out. The categories are already arranged in order from the lowest attendance level to the highest, so all we have to do is identify the middle case. Since there are an odd number of cases for both years, we need to find the middle case. For 1976, the middle case is the 747th person (with 746 above and 746 below). If we start adding from the top of the column, we find that the first 533 cases are found in the first three categories. The fourth category contains an additional 233 cases. Since we need only 214 cases in order to get to the 747th case, we know that the median will fall in this fourth category. So, for 1976, the median level of religious attendance is "several times a year."

For 1986, the middle case is the 736th case (with 735 above and 735 below). Adding from the top we find that the first four categories include 669 cases. We need only 67 additional cases to get to the 736th case, which means that the median will fall in the fifth category, "about once a month." Using the median to represent these two distributions, we would conclude that average church attendance was higher in 1976 than in 1986.

Income, Gender, and Education: An Example Like means, medians can be used to make revealing comparisons among groups. This is especially true for variables such as income in which extremely large or small scores inflate the value of the mean while leaving the median unaffected.

To better appreciate the usefulness of medians, consider Table 4-7 which shows median income by gender and education. There are 16 medians in this table, one for each group defined by a combination of education and gender. In the upper left cell, for example, $8,400 is the median income for women with 0 to 7 years of education.

As you may have already known, the table confirms that, for both men and women, the more education you have the higher your income tends to be. It also confirms that men have higher incomes than women even when they have the same education: in each row, the male median is higher than the female. What you might not have been aware of, however, is the degree of inequality. The average for women with high school diplomas, for example, is $13,200, which is just $800 above the average income for men with less than eight years of grade school education ($12,400). Women with college degrees have a lower average income ($17,400) than men who never finished school ($17,500), and women with graduate education have an average ($21,500) that is just $200 above the median for men with high school diplomas. Not only are women far behind the average for men with comparable educations, but they are even behind men with considerably less education.

Table 4-7 Median Income for Year-Round, Full-Time Workers by Gender and Education, United States, 1982

Years of School Completed	Women	Men
Elementary: 0–7 Years	$8,400	$12,400
8 Years	$10,100	$16,400
High School: 1–3 Years	$10,700	$17,500
4 Years	$13,200	$21,300
College: 1–3 Years	$15,600	$23,600
4 Years	$17,400	$28,000
5 or More Years	$21,500	$32,300

SOURCE: U.S. Bureau of the Census, 1983.

Like means, medians represent groups, not individuals, which means that there are, of course, individuals who deviate from these patterns. There are women who make far more money than many men. What the medians show, however, is that women as a group are far behind men as a group, and that any woman who makes as much as, or even more than, a man does so in *spite* of those group differences and what they indicate about the social environment in which men and women live.

Some Cautions about Means and Medians While means and medians can be quite useful for comparing groups, an example will illustrate why they should never be relied upon too heavily. Suppose we have a distribution of income in the U.S. in one year with a mean of $15,000 and a median of $14,500. In a later year, a distribution from the same population shows that the mean and median have remained the same.

We might be tempted to conclude that income has not changed for people in the population. If the rich got richer, however, and the poor got poorer, the gains and losses might cancel each other out, leaving the mean unchanged. In addition, as long as the income of the middle case stays the same, the median will stay the same as well. Thus, even though the distribution of income has changed substantially (and, to many observers, for the worse), the measures of central tendency would indicate no change at all.

The problem, of course, is that a single number cannot possibly represent all aspects of a distribution. As we will see in later sections of this chapter, there are other aspects of distributions that we have to consider as well before reaching such conclusions.

PERCENTILES

A **percentile** refers to a score that cuts off a specified percentage of all cases in a distribution. It is used to locate an individual within a distribution or to serve as a measure of central tendency.

The median, for example, is the 50th percentile because it cuts off the lower 50 percent of all the cases. The 60th percentile is the score that cuts off the lower 60 percent of all cases, and the 90th cuts off the lower 90 percent. If the 90th percentile for College Board test scores is 650, this means that 90 percent of all those who took the test scored 650 or below, and 10 percent scored better than 650.

What is most important about interpreting percentiles is that they allow us to make ordinal comparisons *only*. If my income is in the 85th percentile and your income is in the 90th percentile, we know from this that you make more money than I do. We do *not* know, however, how *much* more you make. Depending on the distribution, we could be quite close or quite far apart. It is analogous to the results of a horse race expressed in terms of the order in which the horses finished. Knowing which came in first, second, third, and so on tells us nothing about how far apart they were at the finish or, for that matter, how quickly they ran the course.

THE MODE

The mean is appropriate only with interval and ratio scales, and the median is appropriate with ordinal scales, as well. What can we do to summarize the distribution of a nominal-scale variable?

Unfortunately, as is often the case with nominal variables, the answer is "not much." The only measure of central tendency available to us is the **mode**, which is defined as *the most common score in a distribution*. In the United States, for example, the modal race is "white," the modal gender is "female," and the modal marital status is "married." The modal political affiliation is "Democrat" and the modal religious affiliation is "Christian."

It is important to be aware that a score may constitute the mode in a distribution without a majority of the observations having that score. The modal religious affiliation in the world as a whole, for example, is "Christian," but Christians made up only an estimated 29 percent of the world population in the early 1980s. "Democrat" is the most frequent party affiliation in the United States, but 60 percent of all adults express non-Democratic preferences. What makes a score the mode is not that *most* observations have that score, but that *more* observations have that score than any other.

You should be aware that there are distributions that have no clear mode, in other words, in which the observations are distributed fairly evenly across all the categories. It is also possible to have more than one mode in a distribution, in which two or more categories share the title of "most frequent score." In the first case, the mode cannot be used as a measure of central tendency, and in the second, more than one mode must be reported.

Since the mode can be used with nominal variables, it can, of course, also be used with higher scales. In Table 4-6, for example, religious attendance is an ordinal variable, and in both years the modal level of attendance is "every week." In Table 4-2, the modal number of children is two for both groups. (In both cases, go back to the tables and confirm these statements for yourself.) Note that in Table 4-2 the mode happens to be the same as the median, but that in Table 4-6 it is very different from the medians ("several times a year" and "about once a month"). By focusing on the most frequent score, the mode conveys a very different kind of information about a distribution, one that has no necessary connection to the mean or median.

CHOOSING A MEASURE OF CENTRAL TENDENCY

Which measure of central tendency we use to describe a distribution depends on the scale of measurement involved as well as the kinds of statements we want to make. The mean can be used only with interval and ratio scales, for example; the median can be used with ordinal, interval, and ratio scales; and the mode can be used with any scale, including nominal. The closer we are to a ratio scale, then, the greater is our choice of measures.

In general, the median is the preferred measure of central tendency whenever the distribution has a small number of extreme scores. Since it is insensitive to extremes, it better represents the "typical" case in such a distribution. The mean, however, is often used because it plays a pivotal role in several important statistical applications, including the variance and standard deviation (to be discussed shortly) and correlation and regression (Chapters 7 and 8). In addition, the mean helps describe some key distributions that are used in statistical inference (Chapters 11–13). Thus, although the median is in some ways a more accurate measure of central tendency, the mean is often preferred because it gives us greater flexibility by allowing us to apply a greater range of statistical techniques.

If our goal is prediction—to guess the score for each case with the least amount of error—each measure of central tendency has its own advantages. The mode is the most frequent score, which means that if we use the modal score as our guess for each case, we will make fewer errors than we would if we used any other single score. The mode also has the advantage of representing an actual score, while the mean and median may have values that describe no real cases. The modal number of children in a family, for example, will always be a whole number of children; but the mean and median number of children can be fractional since both are computed from the data.

If we use the median score as our guess in each case, over the long run the sum of the absolute errors (the median minus the actual score, disregarding the positive and negative signs) will be smaller than with any other single score. With the mean, the sum of the errors will always be zero (the positive and negative errors will cancel each other out in the long run). From a prediction perspective, which of the measures we use depends on which kind of error we want to minimize.

MEASURING TRENDS: PANELS, COHORTS, AND LONGITUDINAL EFFECTS

Means and medians are often used to study changes in a variable over time. Two basic methods are used to measure trends. In the first, we gather information on a population at one point in time and then gather information on the same population at a later time, using the same cases in both studies. If we gather data from 50 corporations in 1986, then we must make the same measurements on the same 50 corporations when we go back at a later time. This is called a **panel study** (Lazarsfeld, Berelson, and Gaudet, 1960; Lazarsfeld and Fiske, 1938; see also Levenson, 1968).

With a second method, we gather data on a sample of cases at one time and then make the same measurements on a *different* sample drawn from the same population at a later time. The two samples are usually unique (no two cases are picked for both samples) if the populations are at all large. The result is a pair of samples each of which offers a *cross-sectional* view of the same population at two different times. Since we do not have data on the same cases at both times, we cannot measure individual changes. We cannot, for example, say that corporation A changed between the two studies. We can, however, make comparisons about the population as a whole, such as in the average number of new employees hired each year or the degree of inequality in the racial or gender composition of various occupations.

Sometimes researchers will try to use the results of a single study to draw conclusions about change over time, especially when studying people. Suppose, for example, that we find that elderly people are more conservative politically than young people. This finding could be explained in two very different ways. It might be that the process of aging tends to make people more conservative. This means that if we took the young people in our study and interviewed them at intervals throughout their lives, we would find their political views becoming increasingly conservative as they aged. This kind of change, which takes place over time, is called a **longitudinal effect** (see Coleman, 1981).

A very different explanation of the observed age difference in political views focuses on the fact that people who differ in age have not simply been alive for differing amounts of time. They have also been exposed to different social environments. Elderly Americans today differ from students of college age in more than just age. Their childhood formative years, for example, included the Depression of the 1930s and World War II. Today's college students, on the other hand, have spent their childhoods in a world living under the constant threat of nuclear war. In short, the elderly and the young belong to different **cohorts** (groups of people born at roughly the same time), and differences between them can result as much from cohort differences (known as **cohort effects**) as they can from differences in age (longitudinal effects) (Riley and Foner, 1972; Ryder, 1965). As long as we must rely on the results of a single sample, there is no way to tell which of these two kinds of effect is operating or to what degree it is a combination of the two.

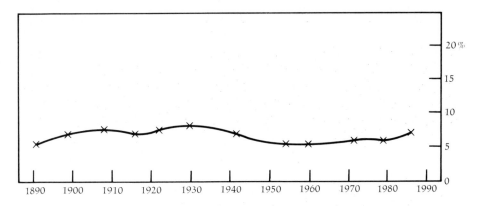

Figure 4-1 U.S. unemployment rates, selected years, 1890–1986.

SOURCE: Bureau of Labor Statistics.

There is also a potentially serious problem that arises with the way in which we present trends. Figure 4-1, for example, purports to show how unemployment rates have varied in the United States between 1890 and 1986. From this graph, it appears that conditions have been quite stable, but it is misleading because it is based on only a few measurements. If we fill in the missing years, the result is a figure we first encountered in Chapter 3, and I think you can see that the impression is radically different (Figure 4-2). By omitting the intervening years, we overlook an enormous amount of variation. Indeed, unemployment varies a great deal over time (the original points from Figure 4-1 are circled in Figure 4-2).

Whenever you find a table or graph that tries to describe a trend, make sure the points in time are not so widely spaced that considerable variation might be hidden. If you think there is too great a gap between points, then you must consider this in reaching a conclusion.

In Figure 4-3, we have a depiction of trends that quickly tells a dramatic story about demographic and social change in the United States. This figure shows historical trends in birthrates, from a high point in 1820 (the earliest year for which reasonably reliable estimates are available) to the early 1980s. You can see that birthrates dropped steadily for 120 years before rising temporarily in the post–World War II "Baby Boom" that lasted roughly 20 years before fertility resumed its long historical decline.

Trends presented in this way are particularly useful ways of showing a slice of history. If, for example, we looked at death rates for European societies over the last century, the devastation caused by World War II would be clearly and dramatically reflected in the sharply rising curve in the late 1930s and early 1940s.

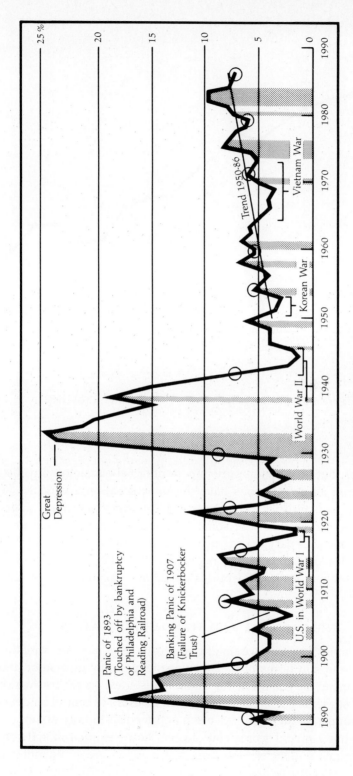

Figure 4-2 U.S. unemployment rates, 1890–1986.

SOURCE: Bureau of Labor Statistics.

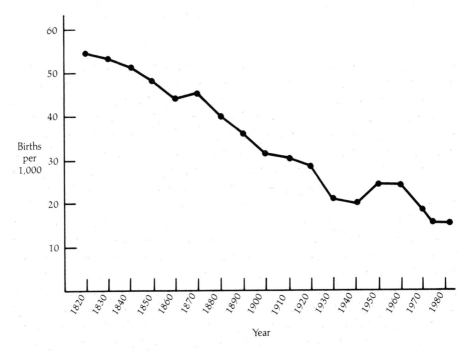

Figure 4-3   Birthrates (per 1,000), United States, 1820–1982.

Measuring Variation

An obvious, but nonetheless very important fact about almost any distribution is that not all observations have the same score on a variable. People differ in their characteristics, behavior, and the answers they give to interviewers' questions. Drugs that cure some people kill others. The cultures and structures of societies vary enormously both from one society to another and within societies over time. One might say that the simple existence of variation and difference in the world is the ultimate justification for scientific inquiry itself.

THE VARIANCE AND STANDARD DEVIATION

In the simplest sense, we might approach the idea of variation by comparing groups in terms of their homogeneity—how diverse or uniform their membership is. Table 4-8, for example, lists test scores for two school classes of 11 students each. No matter which measure of central tendency we use, both groups are identical: the mean and median in both are exactly 50, and neither has a modal score. You can see, however, that in spite of their identical central tendencies, the two distributions are quite different in

Table 4-8 Test Scores for Two School Classes of 11 Students Each, Listed in Order from Lowest to Highest Score

Class 1	Class 2
45	2
46	11
47	24
48	30
49	39
50	50
51	60
52	66
53	83
54	90
55	95

another respect. Specifically, the scores in Class 1 are tightly clustered around the mean of 50, while those in Class 2 are spread out across the entire range from 0 to 100. Class 2 is far more heterogeneous than Class 1. This is particularly clear if we arrange each class on a balance beam, as you can see below:

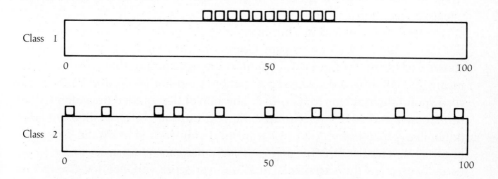

While we can say that Class 2 is more heterogeneous, we would like to be more precise, especially since in many cases the difference is not as obvious as it is here. Ideally, we would have a statistic that directly reflects the degree of heterogeneity in a distribution. In order to measure the degree to which scores in a distribution are spread out, we need some point relative to which we can measure deviation—in other words, spread out in relation to what? The conventional point is the mean, since every distribution of interval- or ratio-scale variables has one. One could always use the median, but in general practice this is not done, primarily because using the mean is more useful in other applications which we will encounter when we discuss statistical inference later on in the book.

What we will do, then, is measure the distance between each score and the mean of its distribution by subtracting the mean from each score. We call each of these differences a **deviation from the mean**. We could then add up the 11 differences for each group in Table 4-8, but there is a problem with this, which is that *the sum of the differences between the mean and each score in a distribution is always zero* (although a small residual may remain due to rounding error).

If you think of the balance-beam analogy for the mean, it will help you to understand why this is true. The mean is the point at which the beam balances on the fulcrum. The farther away each score is from the mean, the more leverage it has in pushing down the beam. In order for the beam to balance, then, the sum of the distances between the mean and scores *above* the mean has to equal the sum of the distances between the mean and the scores *below* the mean. When we subtract the mean from scores below it, we get negative differences (because the mean is larger than each score). When we subtract the mean from scores above it, we get positive differences (because the mean is smaller than each score). Since the negative and positive differences are equal, when we add them together, the result will always be zero. You can see this in Table 4-9, which shows the calculation of deviations from the mean for both groups. In each case, the sum of the positive and negative deviations is zero.

We solve this problem by simply *squaring* the deviation of each score from the mean. We have done this for Groups 1 and 2 in Table 4-10. Column I shows each score for Group 1, column II the deviation of that score from the mean, and column III the squared deviation. Columns IV, V, and VI show the corresponding numbers for Group 2. The sum of the squared deviations is 110 for Group 1 and 10,092 for Group 2. Since groups are often of different sizes (a familiar problem by now), we want to divide each sum by the size of its group (11). The result is the *average squared deviation of scores about the mean*, better known as the **variance**. Thus, the test score variance for Group 1 is $110/11 = 10.0$, and a test score variance for Group 2 is $10,092/11 = 917.5$. The symbol for a variance of a population is the Greek letter *sigma* squared (σ^2). In a sample, the symbol is simply s^2.

If we take the square root of the variance, the result is called the **standard deviation**. Its symbols are the same as those for the variance in a population

Table 4-9 Calculation of Deviations from the Mean

Group 1		Group 2	
Score	(Score − Mean)	Score	(Score − Mean)
45	−5	2	−48
46	−4	11	−39
47	−3	24	−26
48	−2	30	−20
49	−1	39	−11
50	0	50	0
51	+1	60	+10
52	+2	66	+16
53	+3	83	+33
54	+4	90	+40
55	+5	95	+45
TOTAL	0		0

Table 4-10 Calculation of Squared Deviations from the Mean

Group 1			Group 2		
(I)	(II)	(III)	(IV)	(V)	(VI)
Score	Deviation	Squared Deviation	Score	Deviation	Squared Deviation
45	−5	25	2	−48	2,304
46	−4	16	11	−39	1,521
47	−3	9	24	−26	676
48	−2	4	30	−20	400
49	−1	1	39	−11	121
50	0	0	50	0	0
51	+1	1	60	+10	100
52	+2	4	66	+16	256
53	+3	9	83	+33	1,089
54	+4	16	90	+40	1,600
55	+5	25	95	+45	2,025
TOTAL		110			10,092

or sample, without the superscript ^{2} (σ and s). In our example, the standard deviations for Groups 1 and 2 are 3.16 and 30.3, respectively.

Interpreting the Variance and Standard Deviation   You may wonder what is so special about the averaged squared deviation, not to mention its square root. Although the numbers themselves do not have much meaning in and of themselves (there is nothing inherently meaningful about a variance of 917.5, for example), they have many important uses in statistics.

As an overall measure of variation, for example, the variance has important properties that prove to be quite useful. When we square the difference between the mean and each score, we are not simply arbitrarily solving the problem of unsquared differences always adding up to zero. When added together, the squared differences (called the **sum of squares**) that make up the numerator of the variance constitute a powerful way of quantifying the total amount of variation in a set of data. This is particularly true when we try to *explain* variation—in other words, account for the fact that not all observations are the same. As we will see in Chapters 7 and 9, the sum of squares is a way of measuring variation that allows us to make very precise statements about the sources of variation—why blacks, for example, tend to have lower incomes than whites.

Variances and standard deviations are also useful for making comparisons between groups. The variance, for example, is itself a ratio scale, which means that we can take the ratio of one variance to another in order to directly compare the magnitude of variation in two groups. The variance in Group 2 divided by the variance in Group 1 is $917.5/10 = 91.8$, which tells us that there is 91.8 times more variation in scores in Group 2 than in Group 1.

Standard deviations, on the other hand, cannot be used to make this kind of precise comparison. If one standard deviation is larger than another, we can conclude only that the first indicates greater variation than the second. Thus, while we can use variances to make ratio comparisons, the standard deviation can be used only for ordinal comparisons.

Despite this limitation, the standard deviation has some important mathematical properties that are crucial for describing both the kinds of distributions we have seen in this chapter and the theoretical distributions used in statistical inference (Chapter 11). Specifically, we can use the standard deviation to determine the range within which most cases in a distribution lie, even without seeing the entire distribution. According to Chebyshev's Theorem (named for the nineteenth-century Russian mathematician), for example, regardless of what a distribution looks like, at least 75 percent of all its cases will lie within two standard deviations of the mean, either above or below. Thus, if we know that a distribution has a mean of 85 and standard deviation of 15, then we also know that at least 75 percent of the cases will lie between 55 and 115. As we will see in Chapters 11–13, the ability to make these kinds of statements about distributions lies at the heart of the statistical inference process.

Box 4-4   One Step Further: The Mathematics of Variance and the Standard Deviation

The variance is defined as the average squared deviation of scores about the mean. In mathematical notation, this definition looks like this:

$$\sigma^2 = \frac{\sum(X_i - \bar{X})^2}{N}$$

Be sure to go through the formula and see that you know what each symbol stands for as well as how the whole thing defines the variance.

The standard deviation is simply the square root of the variance:

$$\sigma = \sqrt{\sigma^2}$$

There is a simpler formula that you can use to compute the variance when you have ungrouped data such as those in Table 4-10. First you square each score (X^2) and divide by the number of scores (N). In other words, square each number in column I of Table 4-10 and divide the sum by 11. You then square the mean ($\bar{X}^2$) and subtract that from the first number you calculated. In mathematical symbols, it looks like this:

$$\sigma^2 = \frac{\sum X_i^2}{N} - (\bar{X})^2$$

When there are many observations with the same score, we multiply each score times the frequency of that score, just as we did with the mean in Box 4-2. The computing formula for the variance then becomes:

$$\sigma^2 = \frac{\sum f_i X_i^2}{N} - (\bar{X})^2$$

You can also use this formula with grouped data, except that you use the midpoint of each category as the score (X_i).

THE RANGE

The variance and standard deviation pay attention to the degree to which scores are spread out about the mean in a distribution. The **range**, on the other hand, is a cruder measure that pays attention only to the extreme high and low ends of the distribution with no concern about the scores that fall in between. The range is defined as *the difference between the highest and lowest scores*.

Suppose, for example, you graduate from college with an interest in a career in chemistry. You majored in chemistry and have a job offer from a reputable lab. You have also received a fellowship from a good graduate school and could go on to get your Ph.D. A friend points out to you that with your training, the starting salary without a Ph.D. is just a bit less than it would be if you had a Ph.D., and besides, you would be earning money instead of studying and paying tuition. Suppose the average salary for this position is $20,000 without a Ph.D. and $22,000 with a newly earned Ph.D.

Before making your decision, there is an additional fact that you might want to know about salaries. The means tell you where the two distributions (for those with and without Ph.D.s) lie on the income scale. What if you found out that the highest salary you could ever hope to achieve without a Ph.D. was $30,000, but that *with* a Ph.D. you could go as high as $60,000? This new bit of information consists of the *range* of salaries in a distribution, the difference between the highest and the lowest. While you may start off nearly as well as someone with a new Ph.D., the range tells you that the limits on how much money you can eventually earn are considerably lower without an advanced degree.

Although rarely used, the range provides a unique and sometimes valuable piece of information about a distribution, one that you should at least be aware of.

Central Tendency, Variation, and the Shape of Distributions

The importance of measures of central tendency and variation for describing variables is probably clearest when we use graphs to see the relationship between variation, central tendency, and the shape of distributions. Consider, for example, the distributions shown in Figures 4-4 to 4-7. Along the horizontal axis we have a ratio-scale variable, such as income. The height of the curve at any point corresponds to the proportion of people who have that particular income. Take a moment to study these graphs,

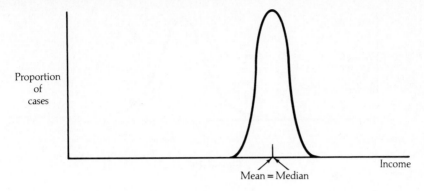

Figure 4-4

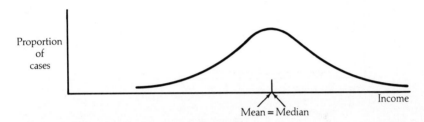

Figure 4-5

because this way of presenting ratio-scale distributions will become increasingly important in later chapters.

Notice that although the mean is the same in all four distributions, their shapes are quite different. In Figures 4-4 and 4-5, for example, the distributions are **symmetrical** (each half is a mirror image of the other), which means that the median and the mean are equal. If you imagine that the curve is resting on a balance beam, you can see that when the curve is symmetrical, it will balance if the fulcrum is placed directly beneath the middle case, which will also be the mean.

Although both distributions are symmetrical, in Figure 4-4 the cases are clustered tightly about the mean, while in Figure 4-5 they are much more spread out. In other words, the variance is greater in Figure 4-5.

If we compare Figures 4-6 and 4-7, we find that although they have the same mean and variance, their shapes are quite different. In Figure 4-6, most of the cases are clustered around and just below the mean, and a small minority of cases is spread out at the extreme high end of the distribution (also known as a *tail* of the distribution). A distribution that tends to be loaded at one end and stretches out toward the other end is called **skewed**, and Figure 4-6 is *skewed to the right*, or, **positively skewed**. Notice that in a

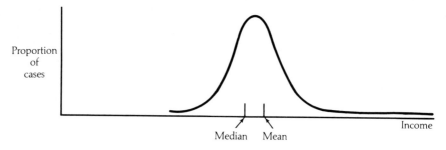

Figure 4-6

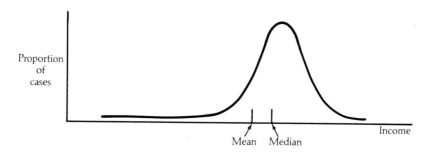

Figure 4-7

distribution that is skewed to the right, *the mean is always larger than the median*. This occurs because the mean is pumped up by the small number of extremely high scores, while the median remains insensitive to them. This is roughly the shape of the family income distribution in the United States—most families are in the middle and lower ends and only a small elite are at the upper income levels.

In contrast with Figure 4-6, the distribution in Figure 4-7 is *skewed to the left*, or, **negatively skewed**. Here the bulk of the cases lie just *above* the mean, and a minority of scores is stretched out towards the extreme *low* end of the scale. This is the kind of distribution we would probably find if we measured family income for students at an elite private college—most parents would be well-off, and only a relatively small minority would be in the middle and lower income ranges. In this kind of distribution, the mean will always be *smaller* than the median since it, once again, is pulled by extreme scores that have no effect on the median.

A good rule of thumb is that *the mean tends to follow the direction of skewness, while the median lags behind*. So, in a distribution that is skewed to the right, the mean will be farther to the right than the median and, therefore, have a numerically higher value. In a distribution that is skewed to the left, the mean will be farther to the left than the median and, therefore, of lower value.

Differences in the shape of distributions make it all the more important to be careful in drawing conclusions from means and medians alone. If, for example, we compare 1981 incomes of black and white men in the United States, the distributions might look like (a) and (b) in Figure 4-8. While some blacks make more money than some whites (where the distributions overlap), whites as a group are better off than blacks. The difference between group means ($17,000 for whites versus $11,000 for blacks) tells us that the white male distribution is located generally a considerable distance to the right of the black male distribution. But this conclusion holds *only* if the distributions are similarly shaped.

If, for example, the white distribution is skewed to the right (as in Figure 4-8), but the black distribution is skewed to the *left*, the picture might look not like Figure 4-8, but like Figure 4-9. In this case, the black mean is being pulled down by a minority of extremely low incomes while the white mean is being inflated by a minority of extremely high incomes. If we base our comparison on black and white *medians*—which are insensitive to extremely high or low scores—we can see that the difference is much smaller since the centers of the two distributions are quite close to each other.

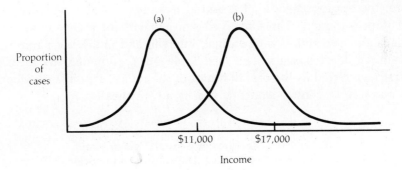

Figure 4-8

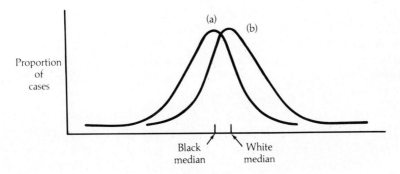

Figure 4-9

In fact, of course, black and white incomes are distributed in much the same way as in Figure 4-8, but the point of this example is to underscore the dangers of reaching conclusions based on medians and means without being aware of the shapes of the distributions they are used to summarize.

Summary

By this point, I hope you are beginning to see more clearly that variables have distributions that can be described in a variety of ways, and that once they are fully described, we can make many different kinds of comparisons between them. It is always important, however, to maintain a critical awareness of the different techniques that we can use, when each is appropriate and when it is not, and how they must be interpreted.

1. The most extreme form of data reduction is the representation of an entire distribution with a single statistic. Since even under the best circumstances data reduction results in a loss of information, several different types of statistics are used to represent different aspects of a distribution.

2. Central tendency focuses on the "center" of a distribution of scores. As before, the appropriateness of a central tendency statistic depends on the scale of measurement that is involved—means with interval and ratio scales; medians and percentiles with ordinal, interval, and ratio scales; and modes with any scale of measurement (see Table 4-11). Although the median is generally preferred as the best indicator of the typical case in a distribution, each measure of central tendency has its advantages.

Table 4-11 Measures of Central Tendency and Variation That Can Be Used Depending on the Scale of Measurement of the Variable

	Nominal	Ordinal	Interval	Ratio
Central Tendency				
Mode	Yes	Yes	Yes	Yes
Median	No	Yes	Yes	Yes
Percentile	No	Yes	Yes	Yes
Mean	No	No	Yes	Yes
Variation				
Range	No	Yes	Yes	Yes
Variance	No	No	Yes	Yes
Standard Deviation	No	No	Yes	Yes

3. Measures of central tendency are often used to show trends in distributions. Trends must be interpreted carefully in order to distinguish between cohort effects and longitudinal effects. They must also be presented carefully in order to avoid the distorting effects of missing data.

4. Variation—the fact that different observations have different scores on a variable—can be measured in several ways. The variance is the most powerful measure, enabling us to make ratio comparisons between variances. Related to the variance and also useful (but in different ways) is the standard deviation. The range is, by comparison, a relatively crude indication of variation in a distribution.

5. In order to fully understand a distribution, it is necessary to pay attention to its shape as well as its central tendency and variation. Two distributions can be identical in central tendency but quite different in variation, or identical in central tendency and variation, but quite different in shape. Symmetry and skewness are two of the most important aspects of a distribution's shape.

Key Terms

central tendency 67
cohort 83
cohort effect 83
deviation from the mean 88
longitudinal effect 83
mean 67
median 75
mode 81
panel study 83

percentile 80
range 92
skewness 93
   negative 94
   positive 93
standard deviation 88
sum of squares 90
symmetrical distribution 93
variance 88

Problems

1. Define and give an example of each of the key terms listed at the end of the chapter.
2. Use the data in Table 4-12 (p. 98) to answer parts a–h.
    a. What is the mean in this distribution?
    b. What is the median in this distribution?
    c. What is the mode in this distribution?

Table 4-12 Number of Times Unemployed for 14 People

Person	Unemployed
Benjamin	0
Joanne	1
Stuart	1
Paul	2
Emily	0
Peter	2
George	3
Julio	2
Naomi	4
Nora	1
Brent	2
Susan	1
Grace	0
Ellen	2

    d. What is the range in this distribution?
    e. What is the variance in this distribution?
    f. What is the standard deviation in this distribution?
    g. To what percentile does George's score correspond?
    h. Is this distribution skewed? Why or why not? If yes, in which direction?

3. Use Table 4-13 to answer parts a–g on p. 99.

Table 4-13 "What do you think is the ideal number of children for a family to have?" U.S. Adults, 1972 and 1986

Ideal Number	1972	1986
0	27	15
1	19	24
2	638	742
3	375	359
4	286	185
5	49	25
6	43	15
7	18	5
TOTAL	1,455	1,370

For each of the two years—1972 and 1986—calculate, compare, and interpret the meaning of your comparisons of the following:
a. mean
b. median
c. mode
d. range
e. variance
f. standard deviation
g. Given the means and medians in each year's distribution, how would you describe them in terms of symmetry and skewness?

4. In 1983, the mean number of hours of television watched by adults in the United States was 3.36. What does this mean say about the number of hours of television that *you* watched? Explain your answer and its general importance for using and understanding statistics.

5. If the median income for a group was $21,000 in 1977, and the mean income for the same group is $30,000 in 1987, can we use this information to draw conclusions about how much the income of the typical person in this group has changed? Explain your answer.

6. If the variance for one group is 4.0 and the variance for a second group is 2.0, can we say that there is twice as much variation in the first group as there is in the second? Explain your answer.
a. Could we make the same statement if the standard deviation for the first group was 4.0 and the standard deviation for the second group was 2.0? Explain your answer.

7. With which scales of measurement can each of the following be used?
a. the mean
b. the median
c. the mode
d. the range
e. the percentile
f. the variance
g. the standard deviation

8. If we double the value of the highest score in a distribution, how does it affect the value of the median? Explain your answer.

9. In a distribution with an even number of cases, which case corresponds to the median? In a distribution with an odd number of cases, which case corresponds to the median?

10. Can a distribution have more than one mode?

11. What factors affect the choice of median, mean, or mode as a measure of central tendency?

12. Why must caution be used in interpreting and displaying trends?
13. Why does the sum of the differences between the mean and each score in a distribution always equal zero (within rounding error)?
14. Under what circumstances is the mean greater than the median in a distribution? Under what circumstances is the median greater than the mean?
15. Describe each of the distributions in Figure 4-10 in terms of:
    a. the number of modes
    b. which is greater, the mean or the median
    c. skewness
    d. symmetry

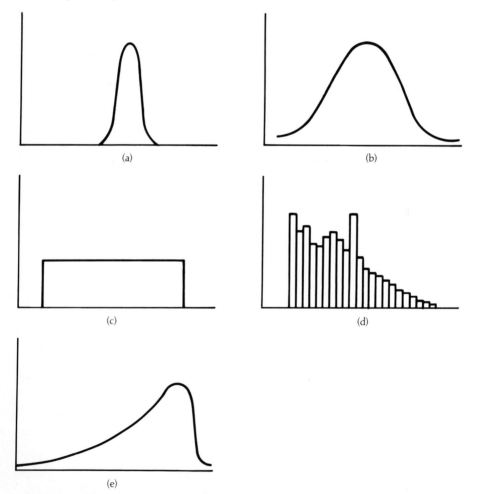

Figure 4-10

5

Elementary Probability Theory

"In this world," wrote Benjamin Franklin in 1789, "nothing is certain but death and taxes." Presumably, everything else is a matter of chance. Albert Einstein, however, believed that chance is not a natural phenomenon; rather it serves as a human substitute for precise knowledge of how everything works.

Since we are not close to certainty in most areas of knowledge, much of what we know of the world is based on the idea of chance. We must be content to estimate the relative likelihood that something will happen rather than knowing that it will or it will not. The best that we can know, for example, is that revolutions tend to occur under certain kinds of circumstances, that AIDS is most likely to be transmitted through certain kinds of behaviors, that jurors with certain social characteristics tend to acquit certain kinds of defendants more than others, that some teaching techniques tend to produce better results than others, or that subatomic particles tend to collide in certain ways at certain frequencies under certain conditions. In short, a substantial portion of our knowledge, whether derived from scientific research or everyday experience, is conditioned by varying degrees of uncertainty, a concept that is itself most precisely understood in terms of the mathematical idea of **probability**.

The Importance of Probability in Statistics

In order to understand the rest of this book, it is important that you become familiar with the basic principles of probability theory. Probability, for example, is an idea that underlies relationships between variables—conditional and unconditional probabilities define independence and dependence between variables. The concepts of "trial" and "event" underlie the sampling process, upon which statistical inference is based. Conditional probabilities are used in hypothesis testing to establish probabilities of error; compound probabilities for mutually exclusive events are used to construct confidence intervals; and expected values play a major part in the calculation of chi-square, one of the most widely used of all inferential statistics. As you study later chapters, you may not always encounter these aspects of probability in a readily identifiable form, but be assured that they play a pivotal role.

Terms such as "conditional probability" and "chi-square" are, of course, not yet part of your working statistical vocabulary, and you should not worry about understanding them at this point. It will be enough if, as you make your way through this chapter, you keep in mind that each of the concepts that make up the study of elementary probability occupies a central position in the underlying logic and procedures of statistical analysis. The more comfortable you become with the basics of probability theory, the more easily you will grasp the fundamentals of the more sophisticated statistical techniques that follow.

Most of the ideas in this chapter are comprehensible to the average college student (although not without some effort), and this chapter has been written with that kind of reader in mind. Be sure to take your time, study each idea, thoroughly familiarize yourself with new terms and symbols; and do not move on to a new idea before you feel comfortable with the ones that went before. The material in this chapter is cumulative, and you need to proceed patiently, step by step.

Some Basic Ideas

When you flip a coin, two things can happen (assuming it does not land on its edge): it will either come up heads or it will come up tails. In probability theory, each flip is called a **trial**, and each of the possible types of outcomes is called an **event**. In a single coin-flipping trial, then, there are two possible events, heads or tails.

If we draw one person at random from the U.S. population and the person is poor, then the event "drawing a poor person" has occurred. Because only one event occurs on this trial, it is called a **simple event**. If we also note that the poor person is a mother, then *two* events have occurred at the same time. This is called a **joint event**. Joint events consist of two or more events that occur on the same trial.

When we refer to the probability of a particular event, whether it is "drawing an ace from a deck of cards" or "a nuclear power plant accident," it is conventional to refer to the outcome of a trial as a **success** when that event occurs and as a **failure** when it does not. Thus, if we estimate the probability that the children of wealthy parents will become poor adults, each instance of a wealthy child becoming poor is referred to as a "success" while each instance of a wealthy child staying wealthy or dropping only as far as the middle or working classes is referred to as a "failure." As you can see, this has nothing to do with the way in which the words "success" and "failure" are commonly used, for in probability theory, being mugged during the coming year can be labeled a success, as can winning the lottery.

We can use two perspectives to understand the idea of *probability*. To illustrate the first, suppose we have ten people in a room, five men and five women. We are going to draw one person at random, and we want to know the probability that our selection will be a woman. In this case, the event "drawing a woman" is called a success and "drawing a man" is called a failure. What is the probability of a success on the first trial? There are ten people in the room, and thus ten possible outcomes. Five of these outcomes will be classified as successes (women) and five will be classified as failures (men).

The *probability* of a success is defined as *the number of potentially successful outcomes divided by the total number of possible outcomes*: in this case, 5/10 = .50. The probability of drawing a woman at random from a group of five men and five women is .50. By convention, the probability of an event is symbolized by $p(A)$, where A (or whatever letter or word you care to use) stands for the event in question. This notation—$p(A)$—is read as "the probability of event A."

In this first sense, then, probability can be thought of as the proportion of possible outcomes that are successes. In a second sense, we can define probability from what is referred to as a "long-run" perspective. Suppose we flip a coin over and over again. If the coin is fair, then a head is just as likely as a tail. If we flip the coin twice, we might nonetheless get two heads (or two tails) in a row, even though the coin is fair. If we flip it 1,000 times, however, we would expect the proportion of heads and the proportion of tails to be quite close to .50 since each event is equally likely. Thus, the probability of flipping a coin and getting a head on a single trial is represented by the proportion of heads we expect to find in the long run.

From this perspective, when we flip a coin once (or pick a person from a population) and say, "the probability is .50 of getting a head on this flip," what we are really saying is "If I flipped this coin repeatedly, in the long run I'd expect to get heads roughly half the time." After all, in the short run of a single trial, the probability of getting a head is either 1.00 or zero, since on a single trial, the outcome is either a head or it isn't. Since we do not *know* what

Table 5-1 Views on Race Discrimination in Housing by Highest Level of Educational Attainment and Age, United States, 1984

	Under 40				40 or Older				All			
	(I) Grammar School	(II) High School	(III) College	(IV) All	(V) Grammar School	(VI) High School	(VII) College	(VIII) All	(IX) Grammar School	(X) High School	(XI) College	(XII) All
Right to Discriminate												
For	33	85	111	229	130	133	107	370	163	218	218	599
Against	56	110	206	372	89	86	74	249	145	196	280	621
TOTAL	89	195	317	601	219	219	181	619	308	414	498	1220

SOURCE: Computed from 1984 General Social Survey raw data.

is going to happen on a single trial, we act as though what we expect to happen in the long run can be applied to a single trial. Thus, behind all statements about the probabilities associated with a single trial lies the idea of long-run probabilities based on large numbers of trials.

Probabilities of Simple Events

In Table 5-1 we have a cross-tabulation of three variables that describe a sample of white adults: educational attainment, age, and values about the right of homeowners when selling their houses to discriminate against blacks. The rest of this chapter will rely heavily on these data for illustrations, so begin by going over the table carefully, making sure you understand what the labels and numbers refer to.

This table contains three variables whose categories include seven simple events: having a grammar school education, high school education, or college education; being for or against the right to discriminate; and being younger than 40 years old or being 40 years old or older. To calculate the probabilities for these simple events, we divide the number of possible outcomes that are successes by the total number of possible outcomes (see Table 5-2).

Table 5-2 Simple Event Probabilities from Table 5-1

Educational Attainment

p(grammar school) = 308/1,220 = .2525
p(high school) = 414/1,220 = .3393
p(college) = 498/1,220 = .4082
TOTAL 1,220/1,220 = 1.0000

Right to Discriminate

p(for) = 599/1,220 = .4910
p(against) = 621/1,220 = .5090
TOTAL 1,220/1,220 = 1.0000

Age

p(under 40) = 601/1,220 = .4926
p(40 or older) = 619/1,220 = .5074
TOTAL 1,220/1,220 = 1.0000

If we select one person at random, for example, there are 308 ways of selecting someone with a grammar school education (bottom number in column IX). There are 1,220 people in the entire sample, and these constitute all the possible outcomes. Thus, in calculating the probability of selecting someone with a grammar school education from this sample, there are 308 ways of getting a success out of 1,220 possible outcomes, for a probability of 308/1,220 = .2525. The probabilities for the seven simple events are found in Table 5-2. Before going any further, satisfy yourself that you understand where each of these numbers came from, including the numerators and denominators in the calculations.

Box 5-1 One Step Further: Venn Diagrams and Simple Events

Venn diagrams (named for the British logician John Venn) are useful for, among other things, representing probabilities. When used with probabilities, the box represents the total number of possible outcomes, and the circle represents one event out of all those possibilities. Shading the area within the circle is a graphic way of representing the probability for that event. The area *outside* the circle represents the probability of any event *other* than A.

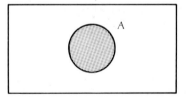

CONDITIONAL AND UNCONDITIONAL PROBABILITIES

To calculate the probability of selecting someone who favors the right to discriminate, we divide the number of those in favor (599) by the total number of possible outcomes. Whenever the denominator in a probability calculation includes all possible outcomes, the resulting probability is called **unconditional**. All of the probabilities in Table 5-2, then, are unconditional, because the denominators in each case included the entire population of possibilities from which each simple event is selected.

Suppose, however, that we want to know more than the overall probability that a U.S. adult favors segregation. Suppose that we want to compare different subgroups of the adult population. Are older and younger respondents, for example, equally likely to support the right to discriminate? To answer this kind of question, we have to calculate two probabilities of being in favor, one for those under 40 and one for those 40 and older. Among those younger than 40 (the left-hand section of Table 5-1) there are 229 who favor the right to discriminate out of a total of 601 people in that age category (be sure you see where these numbers come from). The probability of being in favor *given* that one is under 40 is, therefore, 229/601 = .3810.

Notice that the denominator represents only a subgroup of the entire sample of 1,220 adults, just as the numerator represents only a subgroup of the 599 adults in favor of the right to discriminate. Whenever we restrict ourselves to a subgroup of a population in order to calculate a probability, the result is called a **conditional** probability. In this case, we calculated the probability of being in favor *conditional* on being under 40. The conventional symbolic form for writing conditional probabilities is $p(A|B)$ which stands for "the probability of event A given event B." In this case, event A is "favoring the right to discriminate," and event B is "being under 40." The vertical line between A and B stands for "given," which is a simpler way of saying "conditional on."

To complete the comparison, we have to calculate the probability of favoring the right to discriminate *given* that one is 40 or older. There are 619 respondents 40 or older, of whom 370 favor the right to discriminate. The conditional probability here, then, is 370/619 = .5977. Older adults have a higher probability of being in favor (.5977) than do younger adults (.3810).

In order to calculate the probability of one event conditional on another, the key is to begin with the condition. In order to calculate the probability of having a college education *given* that one is under 40, we have to pay attention *only* to the 601 people who are under 40 and exclude everyone else. In other words, we will confine ourselves to the left-hand section of Table 5-1. Having done this, we simply divide the number of college-educated people who are under 40 by the total number of respondents who are under 40, or, 317/601 = .5275.

There is another way that may help you distinguish between unconditional and conditional probabilities when using data presented in a cross-tabulation. An unconditional probability *always* has the total N for the table as the denominator. A conditional probability, however, *always* has a row or column marginal total in the denominator.

PROBABILITY DISTRIBUTIONS

When we flip a coin, there are two possible events, heads or tails. When we calculate the probability for one of these events, the result is a simple probability. When we calculate the probabilities for *all* the possible events, the result is a **probability distribution**. The probability distribution for coin flipping consists of two simple probabilities: p(heads) = .50 and p(tails) = .50. Because the events in a probability distribution exhaust all the logical possible outcomes, the sum of the probabilities in the distribution must *always* equal 1.00.

Probability distributions can be either **unconditional** or **conditional**. When we select someone at random from the group in Table 5-1 and note his or her educational attainment, there are three possible events: grammar school, high school, or college. As you can see in column III of Table 5-3, the unconditional probability distribution for educational attainment lists the probabilities for each of the three possible simple events, adding to a total of 1.0000.

The first two columns, however, contain conditional probability distributions. In column I, we have the probability of each level of educational attainment *given* that someone is under 40. In column II, we have the same probabilities given that someone is 40 or older. Since within each age group one must fall in one or another of the three educational categories, the total for columns I and II must be 1.0000 (within rounding error).

Notice that the unconditional probability distribution in Table 5-3 (column III) is based on the "Total" row in the "All" section of Table 5-1. In other words, it is based on the entire sample of 1,220. The conditional distribution in column I of Table 5-3, however, is based only on the "Total" row in the "Under 40" section of Table 5-1; and the conditional distribution in column II of Table 5-3 is based only on the "Total" row in the "40 or Older" section of Table 5-1.

If you will recall Chapter 3, the distributions in Table 5-3 might look familiar. All we have done is to cross-tabulate education and age and compute the proportional distributions down each of the columns. (In terms of

Table 5-3 Probability Distribution for Educational Attainment, Unconditional (All) and Conditional on Age

Educational Attainment	Age		
	(I) Under 40	(II) 40 or Older	(III) All
Grammar School	.1481	.3538	.2525
High School	.3245	.3538	.3393
College	.5274	.2924	.4082
TOTAL	1.0000	1.0000	1.0000

SOURCE: Computed from 1984 General Social Survey raw data.

probabilities, what would be the result if you computed the proportional distributions across the rows?)

Although the technical language of probability is not usually used to describe them, the most interesting and important questions in the social sciences usually hinge on the idea of conditional probabilities. When we try to identify the social factors that make democracy most likely to flourish, for example, we are essentially looking for those *conditions* that maximize the *probability* that strong democratic institutions will exist in a society. The probability of democracy *given* a highly educated adult population, for example, is generally greater than the probability of democracy given a poorly educated population.

Whenever we compare groups, conditional probabilities play an important part. In any society, for example, the overall probability of winding up in the top 20 percent of the income distribution is, of course, 20 percent (.20). Where people wind up in the income distribution, however, depends a great deal on where they start out, which is to say, on the incomes of the families they grew up in. To see this, we compare the *un*conditional probability of a person being in the top 20 percent (.20) with the conditional probability of being in the top 20 percent *given* his or her family's income. Table 5-4 shows these conditional probabilities for the 1970s according to which tenth of the income distribution a person's family of origin is located. As you can see, the more highly placed the family of origin is, the higher is the conditional probability that sons and daughters will be in the top fifth of income as adults.

Table 5-4 Conditional Probabilities of Being in the Top 20 Percent of the U.S. Income Distribution, Given the Income of the Family of Origin

Position of Family of Origin in Income Distribution	Probability of Being in Top Fifth of Income as an Adult
Bottom 10th	.10
2nd 10th	.14
3rd 10th	.16
4th 10th	.17
5th 10th	.19
6th 10th	.20
7th 10th	.22
8th 10th	.24
9th 10th	.27
Highest 10th	.32

SOURCE: Bowles and Nelson, 1974.

Probabilities of Joint Events

The probability for the simple event of picking a college graduate from our sample in Table 5-1 is 498/1,220 = .4082, and the probability for the simple event of picking someone who favors the right to discriminate is 599/1,220 = .4910. But what is the probability of picking someone who *both* is a college graduate *and* favors the right to discriminate? Here we have the problem of calculating the probability of a pair of events occurring at the same time—what is known as a *joint event*. A joint event probability is represented by the symbol $p(A,B)$, which is read as "the probability of event A **and** event B."

A direct way of calculating this joint probability is to simply count the number of people who qualify as "successes" and divide that by the total number of people in the sample. There are 218 college graduates who favor the right to discriminate (find this in Table 5-1). Therefore, the joint probability of selecting a college graduate who is also in favor is 218/1,220 = .1787.

There is another way to calculate a joint probability, and in terms of your understanding of more sophisticated statistical techniques, it is important to understand. From this point of view, the definition of a joint probability, $p(A,B)$, is that it is equal to *the probability of event A times the probability of event B given that event A has occurred*. In symbols, it looks like this:

$$p(A,B) = p(A)p(B|A)$$

In this case, we begin with the probability of selecting a college graduate, which, as we saw above, is .4082. We then multiply this times the conditional probability of picking someone who favors the right to discriminate *given* that he or she is a college graduate. There are 498 college graduates, and of them, 218 are in favor. So, the conditional probability is 218/498 = .4378. Putting the two together, we have

$$\left(\frac{498}{1,220}\right)\left(\frac{218}{498}\right) = (.4082)(.4378) = .1787$$

where event A is "being a college graduate" and event B is "favoring the right to discriminate." Note that this is the same answer we arrived at above using the direct reading from the table. Note also that

$$p(A)p(B|A) = p(B)p(A|B) \quad \text{or} \quad \left(\frac{498}{1,220}\right)\left(\frac{218}{498}\right) = \left(\frac{599}{1,220}\right)\left(\frac{218}{599}\right)$$

Therefore,

$$p(A,B) = p(A)p(B|A) = p(B)p(A|B)$$

Before going on, be sure you understand everything we have done thus far in this section, especially the last step.

Box 5-2 One Step Further: Venn Diagrams and Joint Events

Venn diagrams can also be used to represent the probability of a joint event. There are two events, A and B, each of which has its own circle. The circles overlap, which means that a joint event between the two is possible. The shaded area where they overlap is a graphic way of representing their joint probability.

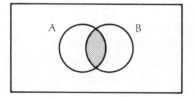

INDEPENDENCE BETWEEN EVENTS

Two events are independent of each other *if the fact that one has occurred in no way affects the probability that the other will occur.* To see what this means, suppose we want to draw *two* people from the sample in Table 5-1, and we want to know the probability of drawing someone with a high school education on *both* draws. We will call "drawing a high school graduate on the first draw" event A, and "drawing a high school graduate on the second draw" will be event B.

The situation here is comparable to the joint event in the previous section, but it is complicated by the fact that there are two ways of drawing two people from a group. We could draw one person, replace him or her in the pool and then make a second selection. This is a procedure known as **sampling with replacement**. Or, we could draw two people in a row without replacing the first selection before making the second, a procedure known as **sampling without replacement**. In the first case, the first person drawn has a chance of being selected again on the second draw; in the second case, the first person cannot be chosen twice.

How do we calculate the probability of drawing two high school graduates *with* replacement? We begin with the formula for a joint event:

$$p(A,B) = p(A)p(B|A)$$

The probability of drawing a high school graduate on the first draw is

$$p(A) = \frac{414}{1,220} = .3393$$

After the first selection, we return the chosen person to the sample and make the second selection. By replacing the first draw, *we have restored the sample to its original composition.* This means that the probability of drawing a high school graduate on the second draw is no different than it was on the first. In other words, *the outcome of the first draw in no way affected the probabilities associated with the outcome of the second draw.* This means that the probability of drawing a high school graduate on the second draw *given* that we drew a high school graduate on the first draw—$p(B|A)$—is no different than the probability of drawing a high school graduate without any conditions at all—$p(B)$. The two events are independent of each other when the conditional and unconditional probabilities are the same:

$$p(B|A) = p(B) = \frac{414}{1,220} = .3393$$

This formula is an important one in probability and statistics because it defines **independence** between events.

The probability of drawing a high school graduate and then drawing another high school graduate *with replacement* is thus

$$p(A,B) = p(A)p(B|A) = \left(\frac{414}{1,220}\right)\left(\frac{414}{1,220}\right) = .1152$$

Notice that since the conditional probability $p(B|A)$ equals the unconditional probability $p(B)$ *in the case of independence,* the formula for joint events is simplified if we substitute $p(B)$ for $p(B|A)$:

$$p(A,B) = p(A)p(B|A) = p(A)p(B)$$

This means that if two events are independent of each other, their joint probability is simply the product of their unconditional probabilities as simple events.

What if we had made our selections *without* replacement? The probability of drawing a high school graduate on the first draw remains the same: $p(A) = 414/1{,}220 = .3393$. The probability of drawing a high school graduate on the second draw, however, *given* that there is now one less high school graduate in the sample, is $p(B|A) = 413/1{,}219 = .3388$, slightly lower than the unconditional probability of .3393. Notice that both the numerator and the denominator have been reduced by 1 to reflect the selection of a high school graduate on the first draw. The joint probability for selecting two high school graduates without replacement is

$$p(A,B) = p(A)p(B|A) = (.3393)(.3388) = .1150$$

slightly lower than the probability for selecting with replacement.

With large populations to sample from, you can see that the denominators of probabilities are so large that selecting without replacement has only a tiny effect on the overall probabilities of selection. If, however, the groups are quite small, the differences between probabilities associated with replacement and nonreplacement can be substantial because the reduction of the denominator by one case will be proportionately larger. If, for example, there were only 15 people to choose from instead of 1,220 and of these 5 were high school graduates, the probability of drawing two high school graduates in a row with replacement would be $(5/15)(5/15) = .111$. If we make the selections without replacement, however, the joint probability is $(5/15)(4/14) = .095$ with both the numerator and the denominator of the second fraction diminished by 1 to allow for the selection of a high school graduate on the first draw. Notice that the joint probability is reduced by almost 15 percent in this case when selections are made without replacement, which is not a trivial difference.

Independence Is Symmetrical It is important to be aware of the fact that independence between events is a symmetrical relationship, which is to say that if B is independent of A, then A is also independent of B. In symbolic terms, if it is true that $p(B|A) = p(B)$, then it is also true that $p(A|B) = p(A)$. When we sampled with replacement, for example, knowing that we picked a high school graduate on the first draw had no effect on our ability to predict what we would get on our second draw. This also means, however, that knowing what happened on the *second* draw gives us no clues as to what happened on the *first* draw.

As you will see in Chapter 6, the symmetry of independence is important in the analysis of relationships between variables because if we want to know if variables A and B are independent of each other, we only have to test to see if A is independent of B (or, alternatively, to see if B is independent of A). If we find, for example, that the frequency of church attendance has no effect on racial prejudice, then we also know that racial prejudice has no effect on the frequency of church attendance.

Some Examples of Independence By defining independence between events in terms of probabilities, we can look at pairs of events and see if they are independent. For example, is being in favor of the right to discriminate (event A) independent of being under 40 (event B)? Put another way, if we choose someone from this sample who turns out to be under 40, does this help us guess this respondent's view of discrimination?

Without knowing the respondent's age, the unconditional probability of being in favor of the right to discriminate is $p(A) = 599/1,220 = .4910$. If we know the respondent is under 40, however, the *conditional* probability is $p(A|B) = 229/601 = .3810$ (be sure you see where these numbers come from in Table 5-1). The two probabilities are quite different. Clearly, adding the condition that the respondent is under 40 changes the probability that the respondent we select is someone who favors the right to discriminate. The two events, therefore, are not independent. They are **dependent**.

Consider a second example. If we look at newborn babies in the United States in 1980 according to their race and sex, the distribution looks like Table 5-5. As you can see, the probability that a newborn child will be a male is $1,853/3,613 = .5129$, and the probability of a newborn being nonwhite is $714/3,613 = .1976$. Are the race and sex of newborns independent events? Does, for example, knowing that a newborn is white (A) help us guess whether it is female (B)?

Table 5-5 Race by Sex of Newborns (in thousands), United States, 1980

	Sex		
Race	*Male*	*Female*	All
White	1,487	1,412	2,899
Nonwhite	366	348	714
TOTAL	1,853	1,760	3,613

SOURCE: U.S. Bureau of the Census, 1983, p. 64.

In this case, the unconditional probability for picking a female newborn is $p(B) = 1{,}760/3{,}613 = .4871$. The probability of selecting a female newborn *given* that the baby is white is $1{,}412/2{,}899 = .4871$. Thus, the probability of selecting a female newborn is the same *regardless* of race, which is to say, being female and being white are independent events.

Given that these two events are independent of each other, then the joint probability of selecting a newborn who is *both* white (A) and female (B) is

$$p(A,B) = p(A)p(B) = \left(\frac{2{,}899}{3{,}613}\right)\left(\frac{1{,}760}{3{,}613}\right) = .3909$$

MORE COMPLICATED JOINT EVENTS

A joint event is any event that involves two or more simultaneous events. Using the data in Table 5-1, for example, we could calculate the probability of selecting someone who is 40 or older (A), went to college (B), and is against the right to discriminate (C). We could calculate this joint probability directly from the table by noting that there are 74 people with this combination of characteristics (see where this comes from). The probability of selecting such a person is, then, $74/1{,}220 = .0607$.

We could obtain the same result by extending our formula for joint events to include three events:

$$p(A,B,C) = p(A)p(B|A)p(C|A,B)$$

Stop for a moment and be sure you can translate this into words and that you understand your own translation. (A sample appears after the last problem at the end of the chapter.)

The first two terms to the right of the equal sign are familiar:

$$p(A) = \frac{619}{1{,}220} = .5074$$

$$p(B|A) = \frac{181}{619} = .2924$$

The third term is the probability of being against the right to discriminate *given* that one is a college graduate who is 40 or older. There are 181 older respondents who went to college, of whom 74 are against the right to discriminate. Therefore, $p(C|A,B) = 74/181 = .4088$. The joint probability is then the product of *three* probabilities, one unconditional and two conditional:

$$p(A,B,C) = p(A)p(B|A)p(C|A,B) = (.5074)(.2924)(.4088) = .0607$$

which is exactly what we got before by calculating directly from the table.

Compound Events: Adding Probabilities

With joint events, we focus on two or more events that happen simultaneously; but with **compound events** we are interested in the probability not only that *both* occur, but that *either* occurs even in the absence of the other. Suppose, for example, that we want to know the probability of selecting someone from our group in Table 5-1 who is either under 40 *or* in favor of the right to discriminate. In probability, the word "or" is inclusive, which means that we would accept as a success someone who has *both* characteristics as well as only one of them.

We begin by finding the number of people who constitute a success. There are 601 respondents under 40, and there are 599 respondents who favor the right to discriminate (am I right?). We might be tempted to add these numbers together in order to get the total of those who have one characteristic or the other; but there is a problem with this. There are 229 people who are *both* in favor *and* under 40, and we have counted them twice. We included them among the 601 people under 40 *and* among the 599 who favor the right to discriminate. Since we have counted them twice, we have to subtract them once from the total so that they will be counted only once. The total number of potential successes is, therefore, 601 + 599 − 229 = 971. The probability of drawing someone who is either under 40 or in favor of the right to discriminate or both is 971/1,220 = .7959.

You can see this more clearly if we count the number of successes in another way. We first determine the number of those under 40: there are 601. We then determine the number of people who favor the right to discriminate; but since we have already counted those under 40 who are in favor, we need only count those *older* respondents who are in favor: 370. The total number of successes is thus 601 + 370 = 971, which is what we got before.

In formal terms, the probability of event A or event B occurring is represented by $p(A$ or $B)$. The formula reflects the kind of counting we have done above:

$$p(A \text{ or } B) = p(A) + p(B) - p(A,B)$$

COMPOUND EVENTS: ADDING PROBABILITIES / 117

For example, the probability of being under 40 years old (A) or being in favor of the right to discriminate (B) is the probability of being under 40 plus the probability of being in favor of the right to discriminate *minus* the joint probability of *both*.

MUTUALLY EXCLUSIVE EVENTS

Adding probabilities becomes simpler when we are dealing with events that are **mutually exclusive**, which is to say, events that cannot both occur at the same time. It is impossible, for example, to select someone who is both a college graduate and who never went beyond high school. Given this, suppose we want to calculate the probability of selecting a college graduate (A) or someone who never went beyond high school (B) from Table 5-1. As we saw in the previous section, the general formula for this kind of event is

$$p(A \text{ or } B) = p(A) + p(B) - p(A,B)$$

Since the two events cannot *both* occur, $p(A,B) = 0$, and the formula simplifies to the sum of two simple event probabilities:

$$p(A \text{ or } B) = p(A) + p(B) = \frac{498}{1{,}220} + \frac{414}{1{,}220} = .7475$$

Box 5-3   One Step Further: Venn Diagrams and Compound Events

In (a), we have two events, A and B, that are mutually exclusive (their circles do not overlap). Their compound probability is simply the sum of their two circles. In (b), we have two events, C and D, that are not mutually exclusive. Here their compound probability is the sum of the C circle and the D circle *minus* the shaded area where they overlap (since we include that area in C and again in D).

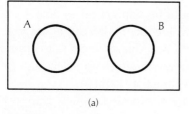

(a)

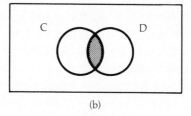

(b)

This simpler formula holds for any two mutually exclusive events and can be extended to include more than two events.

Expected Values

In calculating probabilities, we begin with frequencies. We divide the number of ways of getting a success by the number of possible outcomes, and in doing so, we calculate a proportion: the proportion of possible outcomes that are successes. We can also work in the opposite direction: we can begin with a probability and use it to estimate the frequency of successes.

Suppose, for example, that we are drawing a sample of 1,000 U.S. married adults, and we know from census data that the probability is .0139 that the people we select will be part of an interracial marriage. Therefore, the probability is (1.000 − .0139) = .9861 that the people we select will be married to someone of the same race. The first several selections might all be members of same-race couples, but according to the **law of large numbers**, the more selections we make, the more the proportions of different-race and same-race couples should approach their actual proportions in the population.

If we draw 1,000 people at random, the expected value for the proportion of people in interracial marriages is .0139. Therefore, the **expected value** for the *number* of people in interracial marriages in our sample is (.0139)(1,000) = 13.9 or, rounding off, 14. In general, *if N represents the number of trials (selections) and p represents the probability of success on any one trial, then the expected value for the number of successes is* p(N).

If we make only 100 selections, we would expect to pick only 1 person (100)(.0139) from an interracial couple. Since expected values are based on long-run probabilities, however, we probably would not get that actual outcome with only 100 selections. Notice that if we took a sample as large as 10,000 people, the expected number of interracial spouses would still be only 139, hardly enough to study in much detail.

These kinds of calculations give important guidance in deciding how large a sample to take when we want to study subgroups that are proportionately very small. If we wanted to include a sample of 200 interracial couples as part of a national sample, for example, it would be impractical to take a sample large enough to have 200 as an expected value (or, 14,388 cases). As you will see in Chapter 10, a methodological solution to this problem is to *oversample* this small subgroup of the population by identifying and selecting our 200 couples as a part of a smaller and more practical sample.

Betting also provides a clear application of expected values. Suppose we flip a coin and bet on the outcome. If it comes up heads, I pay you $5; if it comes up tails, you pay me $5. You will expect to lose $5 half the time [an expected value of $(.50)(-\$5) = -\$2.50$], but you will also expect to win $5 half the time [an expected value of $(.50)(+\$5) = +\$2.50$]. In the long run, your expected value will be the sum of these—which equals zero. In other words, you would expect to break even in the long run.

In the short run, however, you might come out behind or ahead of the game. You might win the first four tosses and be ahead by $20, or you might lose three out of four and be behind by $10. In gambling casinos, the house designs each game so that the long-run odds favor the house—which is to say, the expected value is positive for the house and negative for the players as a group. Over a long period of time and tens of thousands of betting "trials," the house wins much more than it loses. Someone who walks in and "breaks the bank," demonstrates that probabilities apply in the long run and that short-run outcomes may be far out of line with expected values. Thus individual gamblers hope that the long-run odds will apply to others and not to them, while the house thrives, secure in the knowledge that in the long run gamblers as a group will lose more than they win, and the house, therefore, will win more than it loses.

Expected values play an important part in social research, especially in using sample results to estimate the frequency of different events or characteristics in a population. Suppose, for example, that we want to estimate the number of elderly people who are living below the poverty level in order to get a better idea of the needs of this increasingly large age group. Suppose also that we select a national sample and find that 14 percent of those 65 or older have incomes below the poverty level. If we use 1982 as an example, there were 25 million elderly Americans.

If we estimate that 14 percent of all elderly people are living in poverty, then the expected value for the *number* of elderly people living in poverty is $(.14)(25 \text{ million}) = 3.5$ million. If we want to estimate the resources it will take to meet the needs of the elderly population, the number of elderly poor will be a more important figure than the percentage of poor who are elderly.

Summary

1. Much of what we know about the world and the way it works consists not of certainties, but of estimating relative likelihoods, or, probabilities.

2. In probability, we estimate the likelihood that an event will occur on a trial. Events can be simple, joint, or compound. The probability of a success on any given trial can be expressed either as the number of ways of getting a

success divided by the total number of possible outcomes or by the proportion of successes expected in the long run.

3. Probabilities and probability distributions may be either unconditional or conditional. A conditional probability is calculated *given* that another event has occurred.

4. The probability for joint events is the product of the probability of one times the probability of the second *conditional* on the first.

5. Two events are independent when the probability of one occurring in no way affects the probability that the other will occur. When two non-mutually exclusive events are independent, the probability that they will both occur is simply the product of their two probabilities.

6. Independence between events is symmetrical, which is to say, if A is independent of B, then B is also independent of A. When two events are mutually exclusive, their compound probability is simply the sum of their individual probabilities.

7. To estimate the frequency of an event given knowledge of its probability, we simply multiply the probability times the number of trials.

Key Terms

dependence 114

event 102
  compound 116
  joint 102
  simple 102

expected value 118

failure 103

independence 112

law of large numbers 118

mutually exclusive events 117

probability 103
  conditional 107
  unconditional 106

probability distribution 107
  conditional 108
  unconditional 108

sampling with replacement 111

sampling without replacement 111

success 103

trial 102

Problems

1. Define and give an example of each of the key terms listed at the end of the chapter.

Refer to Table 5-6 to answer problems 2–11.

Table 5-6 Answers to the Question, "Can you think of a situation in which you would approve of an adult male punching an adult male stranger?" U.S. Adults, 1978

Approve of Punching?	Education			All
	(C) Less Than High School	(D) High School	(E) College or More	
Approve (A)	244	563	199	1,006
Disapprove (B)	226	241	56	523
TOTAL	470	804	255	1,529

2. How many simple events are there in Table 5-6? What are they? Compute each of their probabilities.

3. How many joint events are there in Table 5-6? What are they? Compute each of their probabilities using the appropriate formula, then verify your answer using the table directly.

4. We draw two people from Table 5-6 with replacement and note their levels of education. List the possible outcomes, being sure to specify the order in which each event occurs. Compute the probability for each possible outcome.

5. Identify two compound events each of which involves two simple events. Calculate their probabilities using the appropriate formula.

6. Identify two compound events each of which involves three simple events. Calculate their probabilities using the appropriate formula.

7. Identify examples of unconditional and conditional events. Compute their probabilities.

8. Identify an example of conditional and unconditional probability distributions. Compute their probabilities.

9. Is the event "less than high school" independent of the event "approve" of punching? Support your answer with the appropriate probabilities. How about the event "college or more" and the event "disapprove" of punching? How about the event "approve" of punching and the event "disapprove" of punching?

10. If we randomly selected 100 people from this sample of 1,529 adults, how many would you expect to say that they disapprove of punching?

11. Identify two mutually exclusive events in Table 5-6.
12. What does it mean to say that independence or dependence between events is symmetrical?
13. In Table 5-1, identify a joint event that consists of three simple events. Compute its probability using the appropriate formula.

$p(A,B,C) = p(A)p(B|A)p(C|A,B)$ reads, "The probability that A, B, **and** C will occur equals the probability that A will occur times the probability that B will occur given that A has occurred, times the probability that C will occur given that both A and B have occurred."

6

Relationships between Two Variables: Nominal and Ordinal Scales

Why do blacks make less money than whites, and why do women make less money than men? Why are Canadians more likely than Americans to vote in elections? How do the perceptions that younger people have of the elderly differ from the perceptions that the elderly have of themselves? Do blacks who attend racially integrated schools perform better academically than blacks in segregated schools? How does the mode of production in a society affect religious beliefs? Under what circumstances are social movements most likely to achieve their goals?

All of these questions involve variables (such as race, nationality, voting behavior, mode of production, and religious belief) and ideas about how those variables are related to each other. In each case, we are interested in seeing the ways in which people or societies with certain characteristics tend to differ in terms of other characteristics. In short, we are interested in *relationships between variables*.

Independence and Dependence

To get a better idea of what two-variable relationships are about, consider the variables "gender" and "fear." As measured in the 1984 General Social (GENSOC) Survey, the fear variable consists of answers to the question, "Is there any area right around here—that is, within a mile—where you would be afraid to walk alone at night?" What would it mean to say that these two variables are or are not related to each other?

We could answer this in terms of the question, "Do men and women report different feelings about their safety in the world?" Or, do the social consequences of gender differences go along with a tendency to differ in levels of fear, and does this tendency take on some meaningful, identifiable pattern? If we think of a sense of personal safety as something that is socially distributed, how is its distribution related to gender? To see what this might mean, look at the data in Table 6-1.

Before we continue, you should satisfy yourself that this type of table is something you have seen before (in Chapter 3), and you should familiarize yourself with it thoroughly. Read the title; be sure you understand what the variables stand for; and note the direction of percentaging. *Before you start reading the next paragraph*, go through the table and determine what each number means, which percentages we can legitimately compare with which, and what those comparisons tell us about gender and fear. *Stop* reading until you have done this.

Now that you have done some analysis on your own, you can compare your results with mine. We percentaged the table down the columns: within each category of gender we calculated the percentage distribution for the fear variable. This means that we can compare the reported fear of men and women. We *cannot* use this table as it is to compare the gender composition of people who do and do not report fear. We cannot, for example, infer from this table that people who report fear are more likely to be female than are people

Table 6-1 Fear of Walking Alone at Night by Gender, U.S. Adults, 1984

	Gender		
Fear	(I) Women	(II) Men	(III) All
Yes	58%	19%	42%
No	42	81	58
TOTAL	100%	100%	100%
(N)*	(855)	(596)	(1451)

SOURCE: Computed from 1984 General Social Survey raw data.
*There are 22 people who gave no response from the original sample of 1,473.

who do not report fear. If you do not see why this is true, then go back to Chapter 3 and brush up on the importance of the direction of percentaging (page 39).

Table 6-1 shows that women are considerably more likely than men to report fear of walking alone in their neighborhoods (58 percent versus 19 percent) and considerably less likely *not* to report fear (42 percent versus 81 percent). There is, then, a **relationship** between gender and the *tendency* to report fear as measured in this study. Statistically, we would say that fear is *dependent* on gender, which is to say that differences in fear are associated with differences in gender. In this case, gender is called the **independent variable**, and fear is called the **dependent variable**.

The results in Table 6-1 do *not* mean that gender differences *cause* differences in fear, although one might make a strong argument that there is a causal relationship since women are often the targets of violence simply because they are women (see Brownmiller, 1975, and Russell, 1984). The data show a statistical relationship, which, by itself, cannot tell us what it means. The *interpretation* of statistical relationships such as these is not itself a statistical problem. The only way to demonstrate a causal relationship statistically is through the use of controlled experiments, a subject that is beyond the scope of this book.

Note that if we percentaged the table in the other direction (across the rows), we would be looking at how differences in gender are associated with differences in fear. In this case, we would say that gender is statistically dependent on fear. Gender would now be called the dependent variable, and fear would be the independent variable. It of course makes more sense to argue that gender causes differences in fear than it does to argue that differences in fear cause differences in gender, so we would in this case prefer to designate gender as the independent variable and fear as the dependent variable.

From this comes a general principle for percentaging tables involving relationships between variables: *percentaging is always done within the categories of the independent variable*. If we designate gender as the independent variable, then we must calculate percentages among men and among women, as we have done in Table 6-1.

A PROBABILITY APPROACH

When differences on one variable are associated with differences on a second variable, the variables have a statistical relationship. When two variables are not related, they are independent of each other. It is easiest to see what independence between variables is about if we extend the idea of independence first introduced in Chapter 5.

Consider, for example, the relationship between gender and fear. What might Table 6-1 have looked like if these two variables were unrelated to each other? If they were unrelated, then men and women, although differing on gender, would not differ in their fear of walking at night in their neighborhoods. This would mean that the percentages of men and women reporting fear would be identical, as they are in Table 6-2.

Table 6-2 is hypothetical. It shows what the cells of the table would look like if we kept the marginals (the last column) the same but created new cells that would show women to be no more likely than men to report fear. In this table, there are no differences between men and women—42 percent of both genders reported fear. Thus, the two variables—gender and fear—are independent of each other.

Notice that when there is independence, the distribution of fear within each gender category is identical to that for the sample as a whole (the "All" column). Can you see why this is true? (To see why, try to imagine it *not* being true.)

The idea of a relationship between two variables rests on the twin concepts of dependence and independence, both of which we encountered in Chapter 5 in reference to pairs of events (you may want to review conditional and unconditional events before reading on). If we put the actual cells and marginals of Table 6-1 in the form of proportions rather than percentages, we then have the probabilities shown in Table 6-3.

What does this table tell us? First, the probability that someone chosen at random from this sample will report fear is .42, and the probability of not reporting fear is .58 (in column III—be sure you see this in the table and understand why it is true). If we go into the cells, we can see the *conditional* probabilities of being afraid or unafraid *given* that one is a woman (column I)

Table 6-2 Fear of Walking Alone at Night by Gender, U.S. Adults, Hypothetical

	Gender		
Fear	(I) Women	(II) Men	(III) All
Yes	42%	42%	42%
No	58	58	58
TOTAL	100%	100%	100%

Table 6-3 Fear of Walking Alone at Night by Gender, U.S. Adults, 1984

	Gender		
Fear	(I) Women	(II) Men	(III) All
Yes	.58	.19	.42
No	.42	.81	.58
TOTAL	1.00	1.00	1.00
(N)	(855)	(596)	(1451)

SOURCE: Computed from 1984 General Social Survey raw data.

or a man (column II). When we calculate proportions within a gender category such as women, we restrict ourselves to that category of people. Since the resulting probabilities (.58 and .42) apply only to women, they are conditional on that characteristic. The "All" column thus tells us how likely *anyone* in our sample is to be afraid, while the cells tell us how the men and women compare on fear. The columns in Table 6-3 are thus probability distributions—an unconditional distribution that includes everyone and two conditional distributions, one for men and one for women.

This brings us to the definition of independence between variables in terms of probability distributions. Go back for a moment to Table 6-2. Notice again that when the variables are independent, the percentages within gender categories are the same as the marginals (column III) for both genders combined. In simple terms, gender makes no difference on fear. In terms of probabilities, this is the key to defining independence: the percentage distribution for the entire group ("All") is the same as the percentage distributions for each category of the independent variable. If we remember that the marginals represent unconditional probabilities and the cells represent conditional probabilities, the result is a more sophisticated definition of independence that is a direct extension of independence between events as discussed in Chapter 5: *Two variables are independent of each other if and only if the conditional probabilities are identical to the unconditional probabilities.* This is called **stochastic independence**.

The use of conditional and unconditional probabilities as a way of defining stochastic independence is not simply another way of looking at things. Although it may not seem particularly important to you now, in later chapters it will be crucial to your understanding of chi-square, one of the most important tools used in statistical inference. It is therefore to your advantage to spend some time getting a firm grasp on it now rather than leaving it for later.

How to Describe Relationships between Variables

Relationships between variables can be described in terms of two basic kinds of characteristics. First, we can describe the way in which differences on one variable are associated with differences on another. Blacks are more likely than whites to be unemployed, for example, and the wealthier people are the more likely they are to vote for Republicans. Second, we can describe how *much* of a difference in one variable is associated with differences in another. How much more likely are blacks to be unemployed? How much more likely are the wealthy to vote Republican? Only when we have a clear sense of both the pattern of differences and the strength of those differences do we have an effective description of a statistical relationship.

THE DIRECTION OF RELATIONSHIPS

The **direction of a relationship** between variables refers to the pattern by which the conditional distributions differ from one another. In general, direction takes one of two main forms, **linear** and **nonlinear**.

Linear Relationships A relationship is linear if, as scores on one variable increase, scores on the other variable tend to either increase (a **positive** relationship) or decrease (a **negative** relationship). The higher your education, for example, the higher your income tends to be. The older you are, the more likely you are to have high blood pressure. The higher the demand for a product, the higher its price tends to be. The more surplus goods a society produces, the more likely it is to have a stratification system. These are all examples of positive linear relationships. If my education is higher than yours (a positive difference when you subtract yours from mine), then it is likely that my income will also be higher (another positive difference). Since positive differences on one variable tend to go with positive differences on the other, the relationship is called positive.

A negative linear relationship is one in which relatively high scores on one variable are associated with relatively low scores on another variable. As societies become more complex, for example, the family as an institution tends to become less powerful. The less income you have, the more likely you are to have something stolen from your home. The more education you have, the smaller your family tends to be. People who grow up in small families tend to have higher intelligence-test scores than people who grow up in large families. If you come from a smaller family than mine (a positive difference when you subtract your family size from mine), your test score will likely be *higher* than mine (a *negative* difference when you subtract your score from mine). Since positive differences on one variable tend to go with negative differences on the other, the relationship is called negative.

Nonlinear Relationships Nonlinear relationships generally fall into one of two types. The first is called **curvilinear** because as scores on one variable increase, scores on the other variable change in one direction and then reverse direction one or more times. Figure 6-1, for example, shows the curvilinear relationship between death rates and age. Death rates are high among newborns, but quickly drop in childhood and remain quite low until after age 40. Death rates then rise with increasing steepness with advancing age.

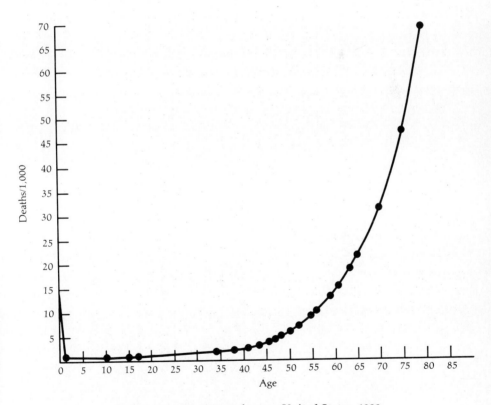

Figure 6-1 Death rates by age, United States, 1980.
SOURCE: U.S. Bureau of the Census, 1983, Table 103.

Table 6-4 ESP Experience* by Church Attendance, U.S. Adults, 1984

ESP	Church Attendance							
	Less Than 1/Year	1–2/ Year	Several/ Year	Once/ Month	2–3/ Month	Almost Weekly	Every Week	Several/ Week	All
Yes	41%	29%	28%	25%	27%	29%	35%	39%	33%
No	59	71	72	75	73	71	65	61	67
TOTAL	100%	100%	100%	100%	100%	100%	100%	100%	100%
(N)**	(291)	(171)	(201)	(109)	(122)	(69)	(339)	(133)	(1435)

SOURCE: Computed from 1984 General Social Survey raw data.
* "Have you ever felt as though you were in touch with someone when they were far away from you?"
** There are 38 people who gave no response from the original sample of 1,473.

In Table 6-4, we have a curvilinear relationship in cross-tabulation form between the frequency of church attendance and whether people report having had experiences of extrasensory perception (ESP). The top row of figures shows the percentage of people in each church attendance category who report having had an ESP experience, and you can see that as we move across the row the percentages drop from a high of 41 to a low of 25 and then climb back to 39 percent. The likelihood of ESP thus drops with increasing church attendance but then reverses itself and rises with increasing attendance. If we drew a picture of this relationship, it would look like a broad flattened U.

Although most relationships in social science research are linear, some of the most interesting are curvilinear. In his classic study of suicide, for example, the great 19th-century sociologist Emile Durkheim studied the connection between suicide and the degree of social cohesion found among members of different social groupings. One of his most interesting findings was that while suicide rates tended to be relatively high in groups with weak levels of cohesion, rates were also relatively high among those with very strong levels of cohesion. Protestants tended to have higher suicide rates than Jews because of Protestants' relatively loose-knit social ties; but groups at the other end of the spectrum—such as professional soldiers and those who live in societies that practice forms of ritual suicide—also showed relatively high suicide rates. Durkheim found the lowest rates among groups with moderate levels of social cohesion. Once again, the relationship looks more like a U than a straight line (Durkheim, 1897).

Nonlinear relationships that do not take the form of a relatively smooth curve are the most difficult to interpret and perhaps can be best described as **irregular**. This happens when there is a relationship between two variables, but there is no neat or consistent pattern to the differences. Suppose, for example, that we gather data on educational attainment and attitudes towards a political candidate and get the results shown in Table 6-5 (see Figure 3-5 for another example).

As you read across the top row of figures, you can see that it is difficult to describe this relationship. As education increases, support for the candidate goes up, and then down, up again, and then down again. Although it would be technically accurate to describe the relationship as curvilinear (in the shape of a roller coaster), the curve has so many bumps and valleys that calling it a curve tells us very little. Sometimes the "bumps and squiggles" in such a relationship are relatively small and can be ignored, but in this instance there is clearly something messy going on and we cannot make statements such as "support increases with education" in spite of the fact that the percentages on the right side of the table are generally larger than those on the left. The only

Table 6-5 Support for Candidate A by Years of Schooling, Hypothetical

	Educational Attainment							
	Elementary			High School		College		
Position	0–4	5–7	8	1–3	4	1–3	4+	All
Support	20%	25%	30%	10%	50%	75%	35%	38%
Oppose	80	75	70	90	50	25	65	62
TOTAL	100%	100%	100%	100%	100%	100%	100%	100%
(N)	(138)	(249)	(357)	(510)	(1059)	(327)	(360)	(3000)

alternative is to describe the pattern of differences between levels of education in detail. There is no shorthand substitute for this (except to call it irregular).

We could clean up the mess by collapsing the education variable into two categories. If we lumped all those with less than a high school degree into one category, for example, and everyone else into a second category, we would get the results shown in Table 6-6. In this form, the variables appear to be related positively: people with more education are more likely to support the candidate than are people with less education. This result is achieved, however, only at the expense of hiding what is in fact a very messy relationship. As Table 6-5 made clear, it would be very misleading to suggest that there is a positive relationship between education and support, which is precisely the kind of conclusion one might be tempted to draw from Table 6-6.

Table 6-6 Support for Candidate A by Educational Attainment, Hypothetical

	Educational Attainment		
Position	Less Than a High School Graduate	High School Graduate or Higher	All
Support	20%	52%	38%
Oppose	80	48	62
TOTAL	100%	100%	100%
(N)	(1254)	(1746)	(3000)

In this case, collapsing the education variable into a dichotomy is *not* a legitimate way to handle an irregular nonlinear relationship. If a relationship is already linear, it is acceptable to dichotomize one or both variables for the sake of simplicity; but it is never acceptable to dichotomize in order to commit what amounts to a misrepresentation of the direction of the relationship.

Whenever you encounter a table in which a variable has been collapsed into a dichotomy, you should ask yourself if the relationship would look different if the variable had been left alone. It is a researcher's responsibility to justify such practices, to demonstrate that more complicated findings are not being swept under the rug in favor of conclusions that may be unsupported by the full set of data.

THE STRENGTH OF RELATIONSHIPS

The **strength of a relationship** between variables refers to the degree to which conditional probability distributions differ from one another. To see what this means, consider again the relationship between fear and gender that we encountered in Table 6-1. Suppose that the relationship had in fact looked like that shown in Table 6-7. As you can see by comparing the two tables, the difference between men and women is much stronger in Table 6-7 than it is in Table 6-1. In Table 6-1, there is a difference of 39 percentage points between women's and men's tendencies to report fear (58% − 19%); but in Table 6-7, the gender difference amounts to 75 percentage points (75% − 0%). Thus, gender makes a much greater difference in Table 6-7 than it does in Table 6-1, and we describe the relationship as "stronger."

The literature of statistical techniques includes many **measures of association**, numbers that are used to reflect the strength of a relationship between

Table 6-7 Fear of Walking Alone at Night by Gender, U.S. Adults, Hypothetical

	Gender		
Fear	(I) Women	(II) Men	(III) All
Yes	75%	0%	46%
No	25	100	54
TOTAL	100%	100%	100%
(N)	(855)	(596)	(1451)

variables; and, as we have seen, there are several ways to describe the direction of relationships. As with just about everything else in statistics, which techniques we can use to describe these two aspects of relationships depend on the scale of measurement of the variables involved (a fact of statistical analysis that should be familiar to you by now). There are different measures of association for relationships involving nominal-, ordinal-, interval-, and ratio-scale variables (not to mention relationships that involve two different scales, such as a nominal-scale variable in relation to an ordinal-scale variable). In addition, nominal-scale variables present a special problem when it comes to describing them in linear terms.

As you read the next sections, keep in mind two basic principles about scales of measurement and the appropriateness of different statistical techniques. First, *any technique that can be used with a lower scale variable can also be used on all higher scales*. Thus, any measure of association that can be used with nominal-scale variables can also be used with ordinal-, interval-, and ratio-scale variables. Second, when the variables in a relationship are of different scales, *the lower scale is the one that determines which techniques are appropriate*. Thus, in describing a relationship between an ordinal variable and a ratio variable, only techniques that are appropriate for *ordinal* scales may be used.

The rest of this chapter will discuss relationships involving nominal- and ordinal-scale variables and introduce you to some of the more commonly used measures of association. A full discussion of relationships among interval- and ratio-scale variables is found in Chapter 7.

Relationships between Nominal-Scale Variables

How do we describe the direction and strength of relationships that involve nominal-scale variables? Unlike higher scales of measurement, nominal variables introduce some special problems, depending on whether they are dichotomies.

DIRECTION IN NOMINAL-SCALE RELATIONSHIPS

With nominal-scale variables, the description of direction is somewhat arbitrary, and how we go about it does not matter very much as long as we are consistent. Since nominal variables make no distinctions of "higher" or "lower," or "more" or "less," the idea of one variable taking on higher values as scores on the other increase or decrease makes little sense. With nominal scales of more than two categories, the only way to describe the direction of the relationship is to carefully describe the pattern of differences without using such summaries as "positive" and "negative."

In Table 6-8, for example, we have the relationship between political party affiliation and whether people voted in the 1980 U.S. presidential election. Political party is a nominal-scale variable with no inherent order of higher or lower, more or less to the categories. The only way to describe the direction of this relationship is to spell out the differences among the categories: "In the 1980 election, Republicans were the most likely to vote, followed by Democrats and Independents in that order."

It is worth repeating here that in a relationship between variables of different scales, the *lower* scale of the two is the one that determines how the relationship can be described. If an ordinal-scale variable is related to a nominal-scale variable with more than two categories, for example, we cannot describe it as "positive" or "negative" just as we could not use those words to describe the relationship in Table 6-8.

As you might already have guessed, describing the direction of relationships involving nominal scales can become quite cumbersome when there are several categories to compare, but this problem is simplified when the nominal variables are in the form of dichotomies. As we saw in Chapter 2, with only two categories, we are able to treat nominal scales as interval scales, which means that we can describe relationships that involve them as linear and use the terms "positive" and "negative." To do this, we arbitrarily assign an order to the two categories, usually in the form of "plus" and "minus."

To see how this is done, consider the relationship between gender and fear

Table 6-8 Vote in 1980 U.S. Presidential Election by Political Party Affiliation

	Political Party Affiliation			
Vote?	*Democrat*	*Independent*	*Republican*	All
Yes	72%	62%	81%	71%
No	28	38	19	29
TOTAL	100%	100%	100%	100%
(N)*	(504)	(462)	(349)	(1315)

SOURCE: Computed from 1984 General Social Survey raw data.
* There are 158 cases missing from the original sample of 1,473 because 106 respondents were ineligible to vote in 1980 and an additional 52 gave no answer to the question.

Table 6-9 Fear of Walking Alone at Night by Gender, U.S. Adults, 1984

Fear	Gender		All
	Women (+)	Men (−)	
Yes (+)	58%	19%	42%
No (−)	42	81	58
TOTAL	100%	100%	100%
(N)*	(855)	(596)	(1451)

SOURCE: Computed from 1984 General Social Survey raw data.
* There are 22 cases with no response from the original sample of 1,473.

that began the chapter (Table 6-9). Notice that this table is identical to Table 6-1 except that we have arbitrarily assigned plus (+) and minus (−) signs to the categories of each variable. To describe the direction of the relationship, we identify which signs tend to go with which. In this context, a positive relationship is one in which pluses tend to go with pluses more than with minuses, and minuses tend to go with minuses more than with pluses. In a negative relationship, pluses tend to go with minuses more than with pluses, and minuses tend to go with pluses more than with minuses (be sure this is clear before you go on).

In our example, women (+) are more likely to say "yes" (+) than "no" (−), and men (−) are more likely to say "no" (−) than "yes" (+). Thus, pluses tend to go with pluses and minuses tend to go with minuses, and the relationship is positive. We could describe this relationship as follows: "There is a positive relationship between being a woman and reporting fear." We could also say that "There is a positive relationship between being a man and *not* reporting fear."

Because the assignment of plus and minus signs to the categories is completely arbitrary, it is important to understand that if we change the signs for one of the variables the *verbal* description of the relationship will change even though the statistical relationship will be the same. If we assigned a plus to men, for example, and a minus to women, the relationship would become negative since minuses would now tend to go with pluses (women and "yes") and pluses would tend to go with minuses (men and "no"). In this case, we could describe the relationship *either* as "a negative relationship between being a woman and not reporting fear" *or* "a negative relationship between being a man and reporting fear."

You should be sure to understand that the four different descriptions of the relationship between gender and fear are all equivalent to one another; they all describe the same statistical relationship between these two variables.

NOMINAL-SCALE MEASURES OF ASSOCIATION

In Table 6-9 we found a positive relationship between being female and reporting fear of walking alone in one's neighborhood at night. Having described the direction of the relationship, we now want to measure how *strong* it is. Gender makes a difference, but how *much* of a difference?

To answer this question, we are going to use *measures of association*. Although many have been devised, we will concentrate on only a few in order to both introduce the idea of a measure of association and expose you to some frequently used measures that you are most likely to encounter in the social science literature.

Measures of association can be divided into two main types. With **asymmetric** measures, it is necessary to specify which variable is assumed to be independent and which is dependent. **Symmetric** measures, however, will give the same numerical results regardless of which variable is considered independent.

In general, the ideal measure of association has three qualities:

1. *A value of zero when the variables are independent* (as in Table 6-2).

2. *A value of -1.00 when there is a perfect negative relationship* (as would be the case if *all* women reported no fear and *all* men reported fear), *and a value of $+1.00$ when there is a perfect positive relationship*. Notice that the plus and minus signs describe the *direction* of the relationship and have no bearing on its strength. An association of $+1.0$ is no stronger than an association of -1.0, even though in normal usage negative numbers are "smaller" than positive numbers.

3. *A meaningful interpretation in terms of the degree of dependence between the variables.*

You might be wondering why these seemingly rather arbitrary ideals exist and why they are important. Like many scientific conventions, the standards for ideal measures of association make it easier for us to make meaningful interpretations of statistical results, especially when they involve comparisons between different measures. If, for example, the upper limit for a measure of association was unknown, it would be very difficult to know just what to make of a particular value or set of values for it. We would attach much greater importance to values of .95 and .60, for example, if the highest possible value was 1.00 than we would if it was 100 or 150. It is therefore important that the upper and lower limits for measures of association be both known and consistent from one measure of association to another.

All of the measures we will discuss have these ideal qualities, but as you will see, they differ in their interpretation. Along with the limitations imposed by scales of measurement, ease of interpretation is an additional basis for choosing one measure over another.

Yule's Q and the Logic of Paired Comparisons Yule's Q (named for its originator Yule (1912) and the 19th-century Belgian statistician Quetelet) is a symmetrical measure of association that is used to measure the strength of relationships between dichotomous variables such as those found in Table 6-9 (called a 2×2 *table* because each variable has two categories). To understand both how Q is calculated and how to interpret the result, we will use three hypothetical versions of the relationship between gender and fear. Suppose we just look at eight women and men who answered the question about fear. Instead of writing the number of cases in each cell, we instead write in the first name of each respondent.

There are three different possible outcomes. In Table 6-10a, all four men report no fear and all four women do report fear. The relationship is a perfect positive relationship since all pluses are paired with pluses and all minuses are paired with minuses. In Table 6-10b, there is a perfect negative relationship since all four men report fear and all four women do not. In Table 6-10c, the relationship is positive, although far from perfect. The women are more likely than the men to report fear.

To measure how much of a difference gender makes, Yule's Q is based on the idea of matching pairs of people who differ on gender (a man and a woman in each pair) and then seeing how they differ on fear. If we have a pair in which the woman(+) reports fear(+) and the man(−) does not(−), that pair contributes to a positive relationship between the variables *because they differ on fear in the same way that they differ on gender*. The woman is (+) on the gender variable and the man is (−); the woman is (+) on the fear variable and the man is (−). This is called a **concordant pair**. In Table 6-10a, *all* of the possible pairs of men and women who differ on fear are concordant, and the relationship is a perfect positive.

In Table 6-10b, the story is just the reverse. If we have a pair in which the woman(+) reports no fear(−) and the man(−) does report fear(+), that pair contributes to a negative relationship between the variables *because they differ on fear in the opposite way that they differ on gender*. The woman is (+) on the gender variable and the man is (−); but the woman is (−) on the fear variable and the man is (+). This is called a **discordant pair**. In Table 6-10b, *all of the*

Table 6-10a

Fear	Women (+)	Men (−)
Yes (+)	Carla Angela Margaret Carol	
No (−)		George Luis Sam Bill

Table 6-10b

Fear	Women (+)	Men (−)
Yes (+)		Luis Sam Bill George
No (−)	Margaret Carol Carla Angela	

Table 6-10c

Fear	Women (+)	Men (−)
Yes (+)	Margaret Carol	George
No (−)	Carla Angela	Luis Sam Bill

possible pairs of men and women who differ on fear are discordant, and the relationship is a perfect negative.

In Table 6-10c, the results are mixed. Margaret and Carol, for example, can each be paired with Luis, Sam, and Bill. In each case, the woman differs from the man in the same way that they differ on fear, and the pairs are concordant. Margaret is (+) on the gender variable and all three men are (−); Margaret is (+) on the fear variable and all three men are (−). All of these pairs, then, contribute to a positive relationship between the variables. The number of concordant pairs is the **positive cross-product**. It is called a cross-product because it is calculated by multiplying the number of cases in one cell times the number in the cell diagonally across from it (in this case $2 \times 3 = 6$).

Notice, however, that we can also pair Carla and Angela with George and get pairs of women and men who differ on fear in another way. Since both women can be paired with George, there are $1 \times 2 = 2$ discordant pairs. For each of these pairs, the women are (+) on the gender variable and the men are (−); but the differences on the fear variable are just the opposite: the women are (−) on the fear and the men are (+). The total number of discordant pairs (2 in this case) is the **negative cross-product**.

If the positive cross-product is greater than the negative cross-product, the relationship is positive. If the negative cross-product is greater than the positive cross-product, the relationship is negative. If the cross-products are equal, then there is no relationship and the variables are independent of each other. The greater the imbalance between the two cross-products, the stronger is the relationship.

Yule's Q is calculated by subtracting the negative cross-product from the positive cross-product and dividing the result by the *sum* of the two cross-products. For Table 6-10c, for example,

$$Q = \frac{(2 \times 3) - (1 \times 2)}{(2 \times 3) + (1 \times 2)} = \frac{4}{8} = +.50$$

In general, the formula for Q is

$$Q = \frac{(ad) - (bc)}{(ad) + (bc)}$$

where a, b, c, and d correspond to the cells as shown in Table 6-11.

Table 6-11 Fear of Walking Alone at Night by Gender, U.S. Adults, 1984

	Gender		
Fear	Women (+)	Men (−)	All
Yes (+)	a	b	a + b
No (−)	c	d	c + d
TOTAL	a + c	b + d	a + b + c + d

In terms of concordant and discordant pairs, the formula for Q becomes

$$Q = \frac{(\text{concordant pairs}) - (\text{discordant pairs})}{(\text{concordant pairs}) + (\text{discordant pairs})}$$

A convenient aspect of Q is that it can be calculated using frequencies, proportions, or percentages. If we return to our original relationship between gender and fear in Table 6-9, we get Q as follows:

$$Q = \frac{(58 \times 81) - (19 \times 42)}{(58 \times 81) + (19 \times 42)} = \frac{4698 - 798}{4698 + 798} = +.71$$

The highest value that Q can attain is 1.0, either positive or negative. This will occur whenever one or both cells in a diagonal is empty. A **mathematically perfect** relationship is one in which both cells in a diagonal are empty, as in Tables 6-10a and 6-10b. In these cases, knowing the gender of someone in the sample allows us to guess how that person answered the fear question with complete accuracy.

You will notice, however, that in Table 6-7 the value of Q will be +1.0 even though there are cases in which pluses go with minuses (the 25 percent of women who report no fear). This is called a **conditionally perfect association**. If we know that the respondent is a man, then we can guess his response to the fear variable without error; but if we know the respondent is a woman, we will not be able to make perfect guesses. Thus, conditional on being male, the relationship is perfect. Because Yule's Q can achieve a value of 1.0 even when the relationship is not mathematically perfect, it is worthwhile to examine the original table to be sure of just what the results mean for each relationship.

Interpreting Yule's Q In the relationship between gender and fear, we found a $Q = +.71$. What does the number mean? In a general sense, we can use Davis's (1971) assignment of verbal labels to different values of Q as a good guide.

Value of Q	Appropriate Phrase
$+.70$ to $+1.00$	Very strong positive association
$+.50$ to $+.69$	Substantial positive association
$+.30$ to $+.49$	Moderate positive association
$+.10$ to $+.29$	Low positive association
$+.01$ to $+.09$	Negligible positive association
$.00$	No association
$-.01$ to $-.09$	Negligible negative association
$-.10$ to $-.29$	Low negative association
$-.30$ to $-.49$	Moderate negative association
$-.50$ to $-.69$	Substantial negative association
$-.70$ to -1.00	Very strong negative association

The best verbal description of the strength of the relationship between gender and fear would then be "a very strong positive association."

In addition to verbal descriptions, there is also a precise mathematical interpretation for Q. Yule's Q is based on the idea of taking pairs of cases that differ on one variable and then seeing in what way they differ on the other. In the relationship between gender and fear, we took pairs of people who differed on gender and saw how they differed on fear. The pairs fell in one of the two diagonals depending on whether they were concordant or discordant. Suppose we take each pair of people and guess which diagonal it falls in. The logic of Q focuses on two ways of doing this. We could flip a coin and guess that half of the pairs are concordant and half are discordant. Or, we could guess that all of the pairs are one or the other. Regardless of which method we use, we will tend to make mistakes. The key to understanding Q, however, is that *the stronger the relationship is, the fewer mistakes we will make if we guess that everyone belongs in just one diagonal*. Thus, the interpretation of Q (and many other measures of association) rests on the idea of a *proportional reduction in error* of prediction (known for short as a **PRE interpretation**).

In Table 6-10a, for example, we will make no mistakes if we guess that for every pair in our sample the woman answers "yes" and the man answers "no," just as we will make no mistakes in Table 6-10b if we guess just the opposite for every pair. If we guess that half of the pairs belong in one diagonal and half in the other, however, we will be wrong half the time. Thus, by guessing that *all* pairs belong in just one diagonal, we reduce our errors by 100 percent (never wrong instead of wrong half the time). In proportional terms, we reduce our errors by 1.00, which is just the value of Q in both of these cases.

Mathematically, then, Yule's Q is the *proportional reduction in errors resulting from guessing that all pairs belong in one diagonal versus guessing that they are evenly distributed between the two diagonals.* As such, Q indicates to what degree the pairs tend to fall in one diagonal more than another. When the variables are independent, the pairs are equally likely to fall in each diagonal, the positive and negative cross-products are equal, and Q has a value of zero.

In most cases relationships fall somewhere between independence and perfection, as in Table 6-10c. Unlike Tables 6-10a and 6-10b, if we guess that every pair belongs in one diagonal (women say "yes" and men say "no"), we will make some mistakes, but *we will make fewer mistakes than if we guess that the pairs are divided evenly between the two diagonals.* As we saw earlier, in this table, Q = +.50, which means that if we guess that every woman said "yes" and every man said "no," we will make 50 percent fewer errors than if we flip a coin for each pair and base our guesses purely on chance.

The Phi Coefficient Pearson's **phi** (ϕ) coefficient (named for the great British statistician Karl Pearson) is another symmetrical measure of association used to measure the strength of relationships between nominal-scale variables. Unlike Q, ϕ can be calculated *only* with frequencies. In 2 × 2 tables (in which each variable has two categories), ϕ is calculated as follows, using the same symbols for cells and marginals found in Table 6-11:

$$\phi = \frac{(ad) - (bc)}{\sqrt{(a + b)(c + d)(a + c)(b + d)}}$$

The numerator is identical to that for Yule's Q—the difference between the positive and negative cross-products. The denominator is the square root of the product of the four marginals.

Although similar to Q, ϕ has some advantages. Like Q, ϕ will equal zero when the variables are independent of each other. Phi differs from Q, however, in that it will have a value of +1.0 or −1.0 only when the relationship is mathematically perfect—in other words, only when one diagonal is empty. In Table 6-7, for example, we had a relationship that was conditionally perfect, since 25 percent of the women did not report fear. The value for Q in this table is +1.0. To calculate ϕ, we must convert the marginals and cells of the table to frequencies, as in Table 6-12 (for practice, calculate Q

Table 6-12 Fear of Walking Alone at Night by Gender, U.S. Adults, Hypothetical

	Gender		
Fear	Women (+)	Men (−)	All
Yes (+)	641	0	641
No (−)	214	596	810
TOTAL	855	596	1451

in Table 6-12). While Q indicates a perfect relationship with a value of $+1.00$, ϕ has a value of

$$\phi = \frac{(641 \times 596) - (0 \times 214)}{\sqrt{641 \times 810 \times 855 \times 596}} = \frac{382{,}036 - 0}{514{,}372} = +.74$$

Thus, while Yule's Q is sensitive to the conditionally perfect association in Table 6-12, ϕ indicates that the relationship, while very strong, is not mathematically perfect for all categories of gender.

Interpreting Phi In addition to being insensitive to conditionally perfect relationships, ϕ has another useful characteristic. If we square the value of ϕ (ϕ^2) the resulting quantity represents the *proportion of the variation in the dependent variable that is statistically explained by its relationship to the independent variable.* What does this mean?

We know that for our sample there is variation in reports of fear—some say they are afraid and some do not. We also know that there is variation in gender—some people are women, some are men. This means that there are two kinds of *differences* (or, types of variation), differences in gender and differences in fear. The question here has to do with the degree to which differences in gender *go along with* differences in fear. In a perfect relationship, every difference in gender is matched perfectly to a particular difference in fear: every man is afraid and every woman is not; or every man is not afraid and every woman is. When the variables are independent, differences in gender are as likely to be tied to one difference in fear as they are to the other: the men and women are equally likely to report fear.

The ϕ^2 statistic tells us the exact degree to which two variables vary together. In our example, it tells us the proportion of variation in fear that can be statistically tied to differences in gender. For Table 6-9 (the actual relationship between gender and fear), $\phi = .39$ and $\phi^2 = .15$, which tells us that 15 percent of the variation in fear is statistically accounted for by the relationship between fear and gender. By subtracting the 15 percent from 100, it also tells us that 85 percent of the variation in fear has nothing to do with gender. Be careful to note as we have before that *statistically* accounting for variation is not the same as explaining the true *cause* of variation.

Phi can be calculated for tables larger than a 2 × 2, but in these cases it takes on some undesirable characteristics. First, the resulting quantity has no positive or negative sign. Second, the value of ϕ in larger tables often exceeds 1.0 which makes it all but impossible to give it a meaningful interpretation. For these reasons it is best to use ϕ and ϕ^2 only in 2 × 2 tables.

In general you will find that values for ϕ will be lower than Q for the same set of data. For example, $\phi = +.39$ in Table 6-1, lower than the value of $+.71$ we got for Q in the same table. Thus, while both measures of association have values of zero when there is independence and range between -1.0 and $+1.0$ in 2 × 2 tables, as the relationship between two variables gets stronger, the value of Q tends to rise much faster than the value of ϕ, making ϕ a more conservative and stable measure.

To see how this difference between ϕ and Q affects the range of possibilities, assume that we have a 2 × 2 table in which the marginals for both variables are split fifty-fifty. As the relationship gets stronger (in other words, as we rearrange the cells in order to produce increasingly strong relationships) the values of Q and ϕ would compare as follows:

Value of Q	Value of ϕ
.08	.04
.38	.20
.69	.40
.88	.60
.98	.80
.997	.92
.999	.96

As you can see, Yule's Q rises very sharply at first and then levels off; ϕ, however, rises more smoothly. While Q is widely used and well known, it does not discriminate between moderately and very strong relationships as well as ϕ does. For this reason, ϕ would seem to be a scientifically preferable measure of association. The statistic ϕ^2 is even more useful because, as a proportion, it is itself a ratio-scale measure that allows us to make direct ratio comparisons between different relationships. We can say, for example, that when $\phi^2 = .40$, the relationship is four times as strong as it is when $\phi^2 = .10$. We cannot make this kind of comparison with Q or ϕ. With Qs of .40 and .10, for example, we can only say that the first is larger than the second (an ordinal comparison) without being able to quantify just how much larger it is.

Goodman and Kruskal's Tau Relationships between variables can be viewed in terms of increasing our ability to guess one characteristic from knowledge of another. If income and education are related, then knowing people's education tells us something about what their incomes are likely to be. Knowing the score on the independent variable improves our ability to guess scores on the dependent variable. If there is no relationship (the variables are independent), then knowing the score on the independent variable adds nothing to our knowledge of what the score on the dependent variable is most likely to be.

Goodman and Kruskal's tau (named for the U.S. sociologist-statistician Leo G. Goodman and his student William Kruskal) is an asymmetrical measure of association designed for nominal-scale variables with any number of categories (an advantage over both Q and ϕ) (Goodman and Kruskal, 1954). It is designed to measure the strength of association by answering this question: "If we want to guess someone's score on a dependent variable (Y), how much is our accuracy improved by knowing his or her score on an independent variable (X)?" To see how tau is calculated, we will return to the relationship between political party affiliation and voting behavior first encountered in Table 6-8. As with ϕ, to compute tau we must put the cells and marginals in the form of frequencies, as in Table 6-13.

Suppose for the moment that we do not know the party affiliations of the 1,315 people in this sample. In other words, all we know is the total number

Table 6-13 Vote in 1980 U.S. Presidential Election by Political Party Affiliation

| Vote? | Political Party Affiliation | | | |
	Democrat	*Independent*	*Republican*	All
Yes	363	286	283	932
No	141	176	66	383
TOTAL	504	462	349	1315

SOURCE: Computed from 1984 General Social Survey raw data.

of people who voted and did not vote (the last column). What if we then used this information to take each of the people in the sample and randomly guess whether they voted in 1980. How many errors would we expect to make in our guesses? In other words, how inaccurate would our guesses be without any information about respondents' party affiliations?

To make our random guesses, we might assign numbers to all 1,315 respondents and then guess that all those with numbers 1 through 932 voted and all those with numbers above 932 did not. If we carried out this procedure (which we do not have to do in order to calculate tau) how many mistakes are we likely to make?

We are going to guess that 932 people voted. The probability that each of those people did *not* vote is 383/1,315 = .2913 (see where this comes from in Table 6-13). This means that we can expect that just over 29 percent of the respondents we guess to have voted will in fact not have voted and will constitute mistaken guesses on our part. We would thus expect to make (.2913)(932) = 271.5 *mistakes* in guessing which 932 people out of the 1,315 voted.

In a similar way, we are going to guess which 383 people did not vote, and for each guess, the probability is 932/1,315 = .7087 that the person in fact *did* vote. This means that we would expect to make (.7087)(383) = 271.5 *mistakes* in guessing which 383 people out of the 1,315 did *not* vote. The total number of errors we would expect to make is (271.5 + 271.5) = 543.0 errors. Therefore, in guessing which of the 1,315 people voted and which did not without knowing party affiliation, we would expect to be wrong in 543 cases. We will call this quantity E_1 (E for error).

Now consider what would happen to our accuracy if we knew what each person's political affiliation was at the time we made our guesses about their voting. To find out, we repeat the procedure of guessing whether people voted, *but now we do it within each political affiliation category in order to make use of our added information about each respondent*. Insofar as party and voting are related to each other, the distributions of voting behavior *within* each party category should differ from one party to another, and insofar as Republicans, Democrats, and Independents do differ, our ability to guess whether people voted improves. The fact that the probability of voting varies from one party to another constitutes useful information about voting, and the more information we have, the more accurate our guesses should be.

We start with Democrats. We will randomly guess which 363 Democrats voted, and on each guess the probability of error is 141/504 = .2798, giving us (363)(.2798) = 101.6 expected errors. We will randomly guess which 141 Democrats did not vote, with a probability of error of 363/504 = .7202 for each guess. This gives us (141)(.7202) = 101.5 expected errors. In guessing which of the 504 Democrats voted and which did not, we would expect to make a total of (101.6 + 101.5) = 203.1 errors.

If we repeat the process for Independents, we get

$$(286)(176/462) + (176)(286/462) = 109.0 + 109.0 = 218.0 \text{ errors}$$

and for Republicans:

$$(283)(66/349) + (66)(283/349) = 53.5 + 53.5 = 107.0 \text{ errors}$$

The total number of errors we would expect to make in guessing voting behavior *using* our knowledge of party affiliation would then be

$$E_2 = 203.1 + 218.0 + 107.0 = 528.1$$

Goodman and Kruskal's tau is calculated as the *proportional reduction in error in guessing scores on the dependent variable that results from knowing the respondents' scores on the independent variable*. The formula is

$$\text{tau} = \frac{E_1 - E_2}{E_1}$$

In this case,

$$\text{tau} = \frac{543 - 528.1}{543} = .03$$

Interpreting Tau Like Q, tau has a PRE interpretation: it represents how much of an improvement in accuracy we enjoy by knowing the scores on the independent variable. In general, tau, like ϕ, is a conservative measure of association. In a 2 × 2 table, in fact, tau and ϕ^2 are equal (you will see this if you calculate tau for Table 6-9 and compare the result with the value of ϕ^2). As with ϕ^2, we can use tau to make ratio comparisons between relationships. We can say, for example, that a tau of .06 represents a relationship that is twice as strong as a tau of .03.

Like ϕ, tau equals zero when the variables are independent and it equals 1.0 only when the relationship is mathematically perfect (calculate tau for Table 6-12 both for practice and to see that this is true). Unlike ϕ and Q, however, tau can never be negative (just as ϕ^2 can never be negative). So, tau cannot tell us anything about the *direction* of a relationship; it only indicates strength.

Unlike Q and ϕ, tau is an asymmetric measure of association. In our example, we tried to guess voting behavior using our knowledge of party affiliation. We could have done the reverse, guessing party affiliation from knowledge of voting behavior. Because tau can be computed from relationships running in both directions, two symbols are often used. If variable A is independent and variable B is dependent, the symbol is tau-*b*; if variable B is independent and variable A is dependent, the symbol is tau-*a*. Because tau is asymmetric, the numerical values for tau-*a* and tau-*b* will not often be the same.

There are other measures of association for nominal-scale variables (such as lambda). The three discussed here, however, are used widely and should serve to introduce you to one of the most important aspects of relationships between variables.

Relationships between Ordinal-Scale Variables

Since ordinal-scale variables represent a higher level of measurement than nominal scales, any measure of association that can be used with nominal scales can also be used with ordinal scales. Thus Yule's Q, ϕ, ϕ^2, and Goodman and Kruskal's tau can all be used with ordinal scales (as well as, of course, with interval and ratio scales). In the following sections we will introduce two additional measures, **Goodman and Kruskal's gamma** and **Somers' d**.

ORDINAL-SCALE MEASURES OF ASSOCIATION

Goodman and Kruskal's Gamma In relationships involving ordinal variables, one of the most frequently used measures of association is Goodman and Kruskal's gamma (Goodman and Kruskal, 1959), a symmetrical measure that can be used in tables of any size. In a 2 × 2 table, gamma and Q are identical, but in larger tables, the computation of gamma is more complex and more restricted.

Like ϕ, gamma can be computed only when the cross-tabulation is in the form of frequencies. It is also important to be aware that the results of gamma will make sense only if the categories of each variable are arranged in the proper order. Thus it would be inappropriate to put the category "lower class" between the categories "middle class" and "upper class."

Table 6-14 Arrangement of Categories for Computation of Gamma

(A) Dependent Variable	Independent Variable			
	High	Medium	Low	All
High				
Medium				
Low				
TOTAL				

(B) Dependent Variable	Independent Variable			
	Low	Medium	High	All
Low				
Medium				
High				
TOTAL				

If you are going to use the computational procedure outlined below, you should arrange the variables in a symmetrical fashion, as shown in Table 6-14. Both variables should start in the upper left corner with the same end of their ordinal scale—either both high or both low. If one variable starts at its high end and the other at its low end, and you use the computational procedures described below, the numerical result will be correct but the sign indicating the direction of the relationship will be the opposite of what it should be. This is simply a matter of adopting and following a convention.

To see how gamma works, consider the relationship between people's educational attainment and their tolerance of U.S. college teachers who admit they are communists (Table 6-15). As you can see, there is a positive relationship between education and tolerance: the higher the educational category is, the more likely people are to express tolerance.

Like Yule's Q, gamma rests on the idea of comparing pairs of cases that differ on one variable to see how they differ on a second variable. The only difference is that with tables larger than a 2 × 2, the positive and negative cross-products include more than just two cells. In the case of a 2 × 2 table, the positive cross-product goes from the upper left cell to the lower right cell, and the negative cross-product goes from the upper right cell to the lower left cell. Yule's Q is then calculated as the difference between the two cross-

Table 6-15 Tolerance of Communist College Teachers* by Educational Attainment, U.S. Adults, 1984

Tolerate Communists?	Educational Attainment			
	Some College or More	High School Grad	Less Than High School	All
Yes (+)	70%	56%	35%	52%
No (−)	30	44	65	48
TOTAL	100%	100%	100%	100%
(N)**	(545)	(464)	(376)	(1385)

SOURCE: Computed from 1984 General Social Survey raw data.
* Answers to the question: "Suppose a man who admits he is a communist is teaching in a college. Should he be fired, or not?" For this table, a "no" to this question was scored as a "yes" on tolerance, and a "yes" to this question was scored as a "no" on tolerance.
** There are 88 cases with no response from the original sample of 1,473.

products divided by their sum. The procedure for computing gamma is essentially the same, except for a larger amount of work.

We begin by putting the relationship in the form of frequencies (Table 6-16). To get the positive cross-product we start in the upper left cell and multiply it by the *sum* of all the cells to the right and below (382)(204 + 244). What we are doing in this step is taking the 382 college people who are tolerant and counting the number of pairs we can make between them and (204 + 244) people who are *both* less educated *and* less tolerant. We repeat this procedure for all cells that have a cell down and to the right. Here, there are a total of two, the second of which results in a product of (260 × 244). The positive

Table 6-16 Tolerance of Communist College Teachers by Educational Attainment, U.S. Adults, 1984

Tolerate Communists?	Educational Attainment			
	Some College or More	High School Grad	Less Than High School	All
Yes (+)	382	260	132	714
No (−)	163	204	244	671
TOTAL	545	464	376	1385

SOURCE: Computed from 1984 General Social Survey raw data.

cross-product is the sum of these two products:

$$(382)(204 + 244) + (260)(244) = 171{,}136 + 63{,}440 = 234{,}576$$

This is comparable to the *ad* cross-product in the formula for Q. To get the negative cross-product we repeat the procedure, starting in the upper *right* cell and multiplying it by the sum of the cells that are below and to the *left*:

$$(132)(204 + 163) + (260)(163) = 48{,}444 + 42{,}380 = 90{,}824$$

Once we have the two cross-products, we compute gamma in the same way that we computed Q:

$$\text{gamma} = \frac{(\text{concordant pairs}) - (\text{discordant pairs})}{(\text{concordant pairs}) + (\text{discordant pairs})}$$

$$\text{gamma} = \frac{234{,}576 - 90{,}824}{234{,}576 + 90{,}824} = \frac{143{,}752}{325{,}400} = +.44$$

Interpreting Gamma We could simply characterize the strength of the relationship between educational attainment and tolerance for communist college teachers as "moderately positive." Like Yule's Q, however, gamma also has a PRE interpretation. In this case, if for each pair of people who differ on education we always guess that the person with the higher education will be tolerant and the person with the lower education will be intolerant, we will make 44 percent fewer errors than if we randomly guess differences on tolerance.

Gamma shares with Q certain disadvantages. It can attain a value of +1.0 or −1.0 even when the relationship is not mathematically perfect, and it is a relatively liberal measure that gives higher values than more conservative measures of the same relationship. It is, nonetheless, a quite useful and widely used measure of association because it allows a PRE interpretation of relationships between variables of more than two categories even though they are only ordinal scales.

It is also important to be aware that gamma is appropriate only for describing the strength of *linear* relationships. In Table 6-4, for example, we saw a curvilinear relationship between church attendance and ESP experiences. If we used gamma to measure the strength of this relationship, the result would be .01, falsely indicating the virtual absence of any relationship at all. For this reason, it is important to always examine a cross-tabulation before selecting a measure of association.

Somers' d One criticism of gamma is that it pays attention only to pairs that contribute either to a positive or negative relationship (concordant and discordant pairs). The formula for gamma does not include other types of pairs, especially those that tend to show *no* relationship between the variables. These are pairs of cases that differ on the independent variable and yet have the *same* score on the dependent variable.

To see what this means, return to the relationship between tolerance of communist teachers and educational attainment (Table 6-16). As we saw when we computed the positive and negative cross-products for this relationship, there are 234,576 concordant pairs in which the person with a higher education is tolerant and the person with a lower education is intolerant, and there are 90,824 discordant pairs in which the person with the higher education is intolerant and the person with the lower education is tolerant. From these, we calculated a value of +.44 for gamma.

What this approach ignores is the fact that there are many pairs in which people differ on education and yet fall in the *same* tolerance category. The 382 people in the upper left cell, for example, can be paired with (260 + 132) people to the right of them who have different educations but who nonetheless agree about tolerating communists. These pairs contribute neither to a positive nor a negative relationship. Instead, they tend towards *no* relationship. By ignoring such pairs (called *tied*), gamma tends to produce higher values than we would get if we included them. Somers' d is a measure of association that takes these tied pairs into account (Somers, 1962).

Like gamma, Somers' d must be calculated with frequencies and is appropriate only for linear relationships. Unlike gamma, it is asymmetric, requiring us to specify which variable is the independent variable and which is the dependent. Somers' d is calculated in the same way as gamma except that in addition to counting the concordant and discordant pairs, we also count the number of pairs that differ on the independent variable (different educations) but are tied on the dependent variable (the same on tolerance). The number of tied pairs is then added to the denominator of the formula:

$$d = \frac{\text{(concordant pairs)} - \text{(discordant pairs)}}{\text{(concordant pairs)} + \text{(discordant pairs)} + \text{(tied pairs)}}$$

To count tied pairs, we begin in the upper left corner, just as with gamma; but this time we multiply that cell times the sum of all the cells to the right but *not* below. We do this for each cell that has another cell to its *right*. There are four such cells in Table 6-16, and the number of tied pairs is

$$(382)(260 + 132) + (260)(132) + (163)(204 + 244) + (204)(244) = 306{,}864$$

Somers' *d* for these sample data is then

$$d = \frac{(234{,}576) - (90{,}824)}{(234{,}576) + (90{,}824) + (306{,}864)} = \frac{143{,}752}{632{,}264} = +.23$$

which is considerably lower than the value of +.44 that we got for gamma.

Interpreting Somers' d Although Somers' *d* has become quite popular among social scientists, it does *not* have a PRE interpretation, which is perhaps its only major disadvantage. The best way to describe the various magnitudes of *d* is to use Davis's guide for describing Yule's *Q*.

RANK-ORDER MEASURES OF ASSOCIATION

With some sets of data we are able to rank cases from high to low on two ordinal variables, and rank-order measures of association allow us to see to what degree the two sets of ranks agree or disagree with each other. Consider, for example, the relationship between how wealthy states are and how high their infant mortality rates are. Demographers generally consider infant mortality rates to be a good indicator of the overall level of health in a given area. To what extent does a high rank on state wealth correspond to a low rank on infant mortality?

In Table 6-17, we have data for ten states, one from each of the ten major regions of the United States. Column I shows the rank of each state in terms of per capita income, with the wealthiest state getting a rank of 1 and the poorest a rank of 10. In column II we find the rank of each state in terms of infant mortality, with the highest mortality rate getting a rank of 1 and the lowest mortality rate getting a rank of 10. If infant mortality is negatively related to per capita income, we should find that those with high ratings on one should tend to have *low* ratings on the other. In the following two sections we will consider two measures of association designed to measure the strength of such relationships.

Spearman's r_s The first rank-order measure is **Spearman's r_s**. To calculate it we compare the two columns of ranks by taking the difference between each pair of ranks (column III), squaring each difference (column IV), and then

Table 6-17 Infant Mortality Rank and per Capita Income Rank for Ten U.S. States, 1980

State	(I) Per Capita Income Rank	(II) Infant Mortality Rank	(III) Difference (I) − (II)	(IV) Squared Difference
California	1	9	−8	64
Michigan	2	3	−1	1
New York	3	6	−3	9
Massachusetts	4	10	−6	36
Delaware	5	1	4	16
Florida	6	5	1	1
Texas	7	7	0	0
Iowa	8	2	6	36
Arizona	9	8	1	1
Mississippi	10	4	6	36
				$D^2 = 200$

SOURCE: U.S. Bureau of the Census, 1983, Tables 108 and 772.

adding them up (D^2 at the bottom of column IV). (Notice that the sum of the deviations in column III must always equal zero; if it does not, a mistake has been made somewhere.) Spearman's r_s is then calculated with the formula

$$r_s = 1 - \frac{6(D^2)}{N(N^2-1)}$$

where N stands for the number of cases that are being ranked. For the ranks in Table 6-17,

$$r_s = 1 - \frac{6(D^2)}{N(N^2-1)} = 1 - \frac{6(200)}{10(99)} = 1 - 1.21 = -.21$$

The result in this case indicates a relatively weak tendency for states ranked high on income to be ranked low on infant mortality. Although the relationship is in the expected direction, it is not very strong.

If the two sets of ranks had been identical, each of the differences in column III of Table 6-17 would have been zero and r_s would be

$$r_s = 1 - \frac{6(0)}{10(99)} = 1 - 0 = +1.0$$

When there is a perfect negative relationship, the second term in the formula is always equal to 2.0, making $r_s = -1.0$. (For practice, verify this statement by making column II the exact opposite of column I [rank California 10th on mortality, Michigan 9th, etc.] and recalculate r_s.) When the two sets of ranks are independent of each other, the second term in the formula will equal 1.0 and r_s will equal zero.

Occasionally cases are tied on one or both variables (two states might have identical mortality rates, for example). In this case we assign both cases the average of the two ranks involved, so that if two cases are tied for first place, we assign each a rank of 1.5 (the average of ranks 1 and 2). If the number of ties is small, r_s can then be calculated in its usual way. If the number of ties is substantial, however, a correction factor must be used (see Siegel, 1956, pp. 206–10).

Interpreting Spearman's r_s Spearman's r_s has no precise mathematical interpretation. The best way to characterize its numerical value is to use Davis's descriptions for different values of Yule's Q, as described earlier.

Kendall's Tau A second rank-order measure of association is **Kendall's tau** (not to be confused with Goodman and Kruskal's tau-*a* and tau-*b*). We can apply Kendall's tau to the kinds of data found in Table 6-17, but in a slightly different way (Kendall, 1963).

To compute Kendall's tau we first put the cases in order for one of the variables, just as we have done for per capita income in column I of Table 6-17. We then determine to what extent the cases are in the same order on the second variable (column II). To do this, we take all possible *pairs* of states and see if each pair differs on mortality in the same way that it differs on income. In other words, we see if the pair is concordant or discordant. If the pair is concordant, we give it a score of +1. If it is discordant, we give it a score of −1.

California, for example, is ranked higher than Michigan on per capita income (1 versus 2), but it is ranked lower on infant mortality (9 versus 3). This is a discordant pair and it gets a score of −1. This is also true for California and New York (look to see if this is true). With California and Massachusetts, however, California is ranked higher on *both* income *and* infant mortality. Thus the California-Massachusetts pair is concordant and gets a score of +1. We repeat this procedure for all possible pairs of states, working our way systematically through the list as in Table 6-18.

Since the sum for the pairs (P) is negative, we know that the ranks on mortality disagree with the ranks on income more than they agree. To force tau to have a range of values between −1.0 and +1.0, we then divide this sum by the maximum that it could be, which is equal to the number of pairs on our

Table 6-18 Concordant and Discordant Pairs of States

Pair of States	Type of Pair	Score
California–Michigan	discordant	−1
California–New York	discordant	−1
California–Massachusetts	concordant	1
California–Delaware	discordant	−1
California–Florida	discordant	−1
California–Texas	discordant	−1
California–Iowa	discordant	−1
California–Arizona	discordant	−1
California–Mississippi	discordant	−1
Michigan–New York	concordant	1
Michigan–Massachusetts	concordant	1
Michigan–Delaware	discordant	−1
Michigan–Florida	concordant	1
Michigan–Texas	concordant	1
Michigan–Iowa	discordant	−1
Michigan–Arizona	concordant	1
Michigan–Mississippi	concordant	1
New York–Massachusetts	concordant	1
New York–Delaware	discordant	−1
New York—Florida	discordant	−1
New York–Texas	concordant	1
New York–Iowa	discordant	−1
New York–Arizona	concordant	1
New York–Mississippi	discordant	−1
Massachusetts–Delaware	discordant	−1
Massachusetts–Florida	discordant	−1
Massachusetts–Texas	discordant	−1
Massachusetts–Iowa	discordant	−1
Massachusetts–Arizona	discordant	−1
Massachusetts–Mississippi	discordant	−1
Delaware–Florida	concordant	1
Delaware–Texas	concordant	1
Delaware–Iowa	concordant	1
Delaware–Arizona	concordant	1
Delaware–Mississippi	concordant	1
Florida–Texas	concordant	1
Florida–Iowa	discordant	−1
Florida–Arizona	concordant	1
Florida–Mississippi	discordant	−1
Texas–Iowa	discordant	−1
Texas–Arizona	concordant	1
Texas–Mississippi	discordant	−1
Iowa–Arizona	concordant	1
Iowa–Mississippi	concordant	1
Arizona–Mississippi	discordant	−1
	TOTAL $P = -5$	

list. The largest number we could get would occur either when every pair was concordant (+1 for every pair) or when every pair was discordant (−1 for every pair). A shorthand formula for calculating the number of pairs is $N(N-1)/2$, where N is the number of cases being ranked (10 in this case). You can use this formula to appreciate why rank-order associations are best left to computers when the number of cases is not small. With 10 cases, for example, there are 45 possible pairs, but with just 50 cases there are 1,225 and with 100 cases there are 4,950!

The formula for Kendall's tau is

$$\text{tau} = \frac{P}{N(N-1)/2} = \frac{-5}{10(9)/2} = -.11$$

Interpreting Tau If all pairs are in the same order on mortality as they are on income, each pair of states will be concordant and will produce a score of +1. Tau will then equal $45/45 = 1.0$. If ranks on mortality are the exact opposite of those on income, each pair of states will be discordant and will get a score of −1. Tau will equal $-45/45 = -1.0$. If there is no relationship between the two sets of rankings, there will be as many concordant pairs as discordant pairs and tau will equal $0/45 = 0$.

You may have noticed that the value of Kendall's tau is lower than that for Spearman's r_s (−.11 versus −.21). This happens because r_s relies on squared differences between ranks giving extreme differences a relatively large impact on the overall result. Tau treats all differences equally, assigning a +1 or −1 regardless of the size of the difference.

Summary

1. Relationships between variables may be thought of either in terms of differences between categories or in terms of statistical independence as defined by the comparison of conditional and unconditional probabilities.

2. Relationships between variables can be described in terms of their direction and their strength.

3. The direction of relationships is generally either linear or nonlinear. Linear relationships can further be divided into those that are positive and those that are negative. Nonlinear relationships are usually either curvilinear or irregular.

4. In describing relationships between nominal dichotomies, it is necessary to pay careful attention to how plus and minus signs have been assigned

Table 6-19 Summary of Measures of Association

Measure	Scales	PRE?	Symmetry	Restrictions
Yule's Q	All	Yes	Symmetric	2 × 2 tables only
phi	All	No	Symmetric	2 × 2 tables only, Frequencies only
phi^2	All	Yes	Symmetric	2 × 2 tables only, Frequencies only
Goodman & Kruskal's tau	All	Yes	Asymmetric	Any size table, Frequencies only
Goodman & Kruskal's gamma	Ordinal, Interval, Ratio	Yes	Symmetric	Frequencies only, Linear relationships only, Properly ordered categories
Somers' d	Ordinal, Interval, Ratio	No	Asymmetric	Frequencies only, Linear relationships only, Properly ordered categories
Spearman's r_s	Ordinal, Interval, Ratio	No	Symmetric	Ranks with few ties
Kendall's tau	Ordinal, Interval, Ratio	No	Symmetric	Ranks with few ties

to categories of each variable. When there are more than two categories, relationships cannot be described as "positive" or "negative."

5. Ideally, measures of association should have a value of zero when there is no relationship and values of +1.0 and −1.0 when the relationship is perfect positive or perfect negative. In addition, the measure should have a meaningful interpretation.

6. There are a variety of measures of association that can be used with nominal and ordinal scales (see Table 6-19 for a summary). In general, they differ in whether or not they have a PRE (proportional reduction in error) interpretation and whether they are symmetric or asymmetric. Any measure that can be used with a lower-scale variable can also be used with variables of a higher scale.

Key Terms

cross-product 140
  negative 140
  positive 140
dependent variable 125
direction of a relationship 128
  linear 128
    negative 128
    positive 128
  nonlinear 129
    curvilinear 129
    irregular 131
independent variable 125
measure of association 133
  asymmetric 137
    Goodman and Kruskal's tau 146
    Somers' d 153

symmetric 137
  Goodman and Kruskal's gamma 149
  Kendall's tau 156
  phi 143
  Spearman's r_s 154
  Yule's Q 138
pairs 138
  concordant 138
  discordant 138
PRE interpretation 142
relationship 125
stochastic independence 127
strength of a relationship 133
  conditionally perfect 141
  mathematically perfect 141

Problems

1. Define and give an example of each of the key terms found at the end of the chapter.

2. When an independent variable is related to a dependent variable, does this mean that one causes the other?

3. Do we do percentaging within the categories of the independent variable or the dependent variable? Why?
4. Explain the probability approach to statistical relationships between variables.
5. Ideally, what characteristics should all measures of association have?

Use Table 6-20 to do problems 6-8.

Table 6-20 Support for Gun Control Laws by Gender, United States, 1984

Control	Gender		All
	Men (+)	Women (−)	
Favor (+)	371	663	1034
Oppose (−)	218	178	396
TOTAL*	589	841	1430

SOURCE: Computed from 1984 General Social Survey raw data.
* There are 43 cases of nonresponse from the original sample of 1,473.

6. Is position on the gun control issue independent of gender? Use both approaches to the idea of relationships between variables.
7. How many different ways are there to describe the relationship between position on gun control and gender? Describe each of them.
8. Measure the strength of the relationship between position on gun control and gender using
   a. Yule's Q
   b. phi
   c. phi^2
   d. Goodman and Kruskal's tau
   For each measure,
   (1) describe any limitations on its use
   (2) describe it in terms of symmetry
   (3) provide an interpretation of the result
   (4) compare its strengths and weaknesses with the other measures
9. Under what circumstances is it justified to collapse a variable into a dichotomy? Under what circumstances is it not justified?

Table 6-21 Tolerance of Homosexual College Teachers*
by Education, United States, 1984

	Educational Attainment			
Tolerate Homosexuals?	*Some College or More*	*High School Grad*	*Less Than High School*	All
Yes (+)	437	268	159	864
No (−)	121	198	225	544
(N)**	558	466	384	1408

SOURCE: Computed from 1984 General Social Survey raw data.
* Answers to the question: "Suppose a man who admits he is a homosexual is teaching in a college. Should he be fired, or not?" For this table, a "no" to this question was scored as a "yes" on tolerance, and a "yes" to this question was scored as a "no" on tolerance.
** There are 65 cases of nonresponse from the original sample of 1,473.

Use Table 6-21 to do problems 10–14.

10. Is tolerance of homosexuals independent of education? Use both approaches to the idea of relationships between variables.

11. Describe the relationship between education and tolerance in terms of linearity and direction.

12. Measure the strength of the relationship between tolerance and education using
    a. Goodman and Kruskal's gamma
    b. Somers' d
    For each measure,
    (1) describe any limitations on its use
    (2) describe it in terms of symmetry
    (3) provide an interpretation of the result
    (4) compare its strengths and weaknesses with the other measure

13. Could you use Goodman and Kruskal's tau to measure the strength of the relationship between education and tolerance? Why or why not?

14. What would you have to do to use phi to measure the strength of the relationship between tolerance and education? Would it be justified in this case?

15. Five people are competing for awards. Their performances have been ranked by two judges. Use both Spearman's r_s and Kendall's tau to measure the strength of the relationship between the two sets of rankings (on p. 163). Be sure to interpret your results.

Contestant	Judge 1	Judge 2
Sue	4	5
Marc	1	4
Bill	2	1
Peggy	5	2
John	3	3

16. In a relationship between a nominal-scale variable with three categories and an ordinal-scale variable, which measures of association are appropriate?

17. In a relationship between an ordinal-scale variable and a ratio-scale variable, which measures of association are appropriate?

18. In a relationship between two ordinal-scale variables, which measures of association are appropriate?

19. Which measures of association give values of +1.0 or −1.0 *only* when the relationship is mathematically perfect?

20. Which measures of association have a PRE interpretation?

↘ 7 ↙

Relationships with Interval and Ratio Scales: Regression and Correlation

In Chapter 6 we talked about the strength and direction of relationships involving nominal and ordinal variables. Although many of the measures of association introduced there had PRE interpretations, all that we could say about the direction of relationships was that they were positive, negative, or curvilinear. In contrast, **regression analysis** is a set of techniques that allows us to describe relationships involving interval- and ratio-scale variables with greater precision and meaning than was available in the analysis of nominal and ordinal relationships.

If we divided levels of education and income into categories and cross-tabulated the two variables, for example, we would find that being in a higher educational category tends to be associated with being in a higher income category. We could then compute a measure of association such as gamma to indicate the strength of the relationship. With regression analysis, we can be much more precise about the relationship between these two ratio-scale variables. We can, for example, determine how many dollars of income are, on the average, associated with each additional year of schooling. This not only gives us a more precise idea of the effect of education on income, but also allows us to make important kinds of comparisons between groups. Does education have a higher income "payoff" for some groups than for others?

Does getting a college degree result in equal income gains for men, women, whites, and blacks? Or does the social mechanism through which education affects income differ from one group to another?

Volumes have been written about regression and correlation. The area is much too large to be covered thoroughly in this book. Nonetheless, you will find a greater quantity of mathematical notation in this chapter than in any other in the book, so it is particularly important that you take your time and be sure you understand each step before proceeding to the next.

The Basics of Linear Regression

We start with two variables both of which are at least an interval scale. Consider, for example, the hypothetical relationship between the number of streetlights on a residential block (X) and the number of crimes that occur on that block in a year (Y) for a sample of five blocks in a neighborhood we will call the "East End" (Table 7-1). The mean of X is 2.0, as is the variance. The mean of Y is 6.0, and the variance is 2.0.

Figure 7-1 shows a scatterplot of the relationship between streetlights and crime in the East End. On the horizontal axis we have the number of streetlights (X), and on the vertical axis we have the number of crimes (Y). Each point on the plot represents one of the five blocks in our sample.

As you can see just by looking at the numbers and their scatterplot, the relationship between X and Y is perfect and negative: for each increase of one streetlight there is a drop of exactly one crime per block. Notice that the points all lie on a straight line. How do we go about describing this relationship in more detail?

Table 7-1 Crimes per Block by Number of Streetlights for Five Hypothetical Blocks in the East End Neighborhood

Block	Number of Streetlights	Number of Crimes
A	0	8
B	1	7
C	2	6
D	3	5
E	4	4

The Regression Line

Linear regression analysis tries to describe a set of data such as those in Table 7-1 and Figure 7-1 with a straight line. Rather than simply describing the relationship between streetlights and crime as "negative," for example, regression analysis goes a step further by identifying a straight line that comes the closest to including all of the data points. This, of course, makes sense only for relationships that are at least roughly linear to begin with.

We "fit" a line to the set of points in such a way that the average distance between the points and the line is minimal. To measure the average distance between the points and the regression line we take the vertical distance between each point and the line and square it. (If we did not square each of these differences between the points and the line, the sum would be zero just as it was when we subtracted each point in a distribution from its mean as part of computing a variance.) The regression line, then, is the unique line that minimizes the squared distances between itself and the points in a scatterplot. For this reason, it is called a **least-squares regression line**. There is only one such line for any given set of data.

In algebra, a straight line is represented by a formula, called a *function*. The general form of a straight-line function is $\hat{Y} = a + bX$, where $\hat{Y}$ is the predicted value of the dependent variable (on the vertical axis) and X is the independent variable (on the horizontal axis). The "hat" symbol over the Y signifies that the value of Y is the predicted value that corresponds to each value of X. In this case, we are not using "predicted" in the sense of guessing

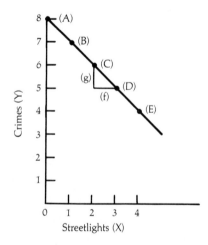

Figure 7-1 Crimes by number of streetlights in the East End.

about the future. Rather, we are referring to what values of Y will most likely go along with a particular value of X in this set of data.

The a in the regression equation is the point where the line crosses the vertical (Y) axis and represents the value of Y when X equals zero. It is called the **regression constant** (also known as the **Y-intercept**). The b is the number of units that Y changes for every unit change in X. It is called the **slope** or the **regression coefficient**. (In math you may have learned to write a linear function as $Y = mX + b$, where X and Y are the independent and dependent variables, respectively, m is the slope, and b is the regression constant. As you can see, in the social sciences a different set of symbols is used to represent the same quantities.)

In our example of streetlights and crime, the slope is the change in the number of crimes associated with each additional streetlight. Look at the small triangle in Figure 7-1. The bottom of the triangle (f) represents a distance of one streetlight, and the left side of the triangle (g) represents the drop in crimes associated with the addition of one streetlight. As you can see, the straight line that best fits these data shows a drop of one crime per streetlight. The slope is thus -1.0.

A negative slope goes from the upper left towards the lower right of the graph (as in Figure 7-1). A positive slope would go from the lower left towards the upper right. A slope of zero is a straight horizontal line parallel to the X-axis. A slope that is perfectly vertical (perpendicular to the X-axis) has a slope equal to infinity, but this occurs only when there is only one value of X (X is a constant, not a variable). In this rare case one would not even try to describe the relationship between X and Y since Y would be the only variable in the relationship, and there is no such thing as a relationship between a variable and a constant.

CALCULATING THE SLOPE

To calculate the slope and regression constant, we begin with the data in Table 7-1. The slope of a least-squares regression line is defined by the formula

$$b_{y.x} = \frac{\sum(X - \bar{X})(Y - \bar{Y})}{\sum(X - \bar{X})^2}$$

The subscript $y.x$ for the symbol b is used to indicate which variable is considered independent and which is considered dependent (the slope, then, is an asymmetric statistic). In this case, y is the dependent variable, and we refer to the regression analysis as "the regression of Y on X." Notice that the symbol for the dependent variable always occurs first in the subscript. If we designated X as the dependent variable, for example, the symbol for the slope for the regression of X on Y would be $b_{x.y}$.

The numerator in the formula is the **covariance**, which measures the extent to which X and Y vary together. By "vary together," I mean that as X changes there is a pattern to the way Y changes. Look at the two parts of the covariance. The first consists of the difference between each X score and its mean. The second takes each corresponding Y score and finds the difference between it and the mean of Y. If the relationship is positive, then when X is above its mean, Y also will tend to be above its mean; when X is below its mean, Y will tend to be below its mean. In both cases, the product $(X - \bar{X})(Y - \bar{Y})$ will be positive most of the time (as the product of either two positive or two negative numbers), and the slope will be positive.

If the relationship is negative, then when X is below its mean, Y will tend to be above its mean, and when X is above its mean, Y will tend to be below its mean (as in Figure 7-1). In both cases the product will tend to be negative (as the product of one negative and one positive number), making the slope negative as well. When there is no relationship, then no matter what the value of X is, Y will have an equal tendency to be above or below its mean. The product $(X - \bar{X})(Y - \bar{Y})$ will be positive as much as it is negative, and the sum of those products (the covariance) will be zero. This will, of course, make the slope equal zero as well since the covariance is the numerator in the formula for the slope.

If you study the denominator for the slope closely, you may notice that the term is the same as the numerator in the formula for calculating the variance of X: the squared difference between each X score and the mean of X. In fact, this is just what we have. The formula given above for $b_{y.x}$ is a somewhat simplified version of the following:

$$b_{y.x} = \frac{\frac{\sum(X - \bar{X})(Y - \bar{Y})}{N}}{\frac{\sum(X - \bar{X})^2}{N}}$$

Since the Ns cancel out, the formula for the slope is usually given in the more simplified form. In general, however, remember that the slope is the covariance between X and Y divided by the variance of X.

The computation of the slope for the relationship between streetlights and crime is shown in Table 7-2. As you can see, the formula confirms what we could, in this case, tell by just examining the data: $b_{y.x} = -1.0$. Before going

Table 7-2 Calculating the Regression Slope for the East End

Block	X	$(X - \bar{X})$	$(X - \bar{X})^2$	Y	$(Y - \bar{Y})$	$(X - \bar{X})(Y - \bar{Y})$
A	0	−2	4	8	2	−4
B	1	−1	1	7	1	−1
C	2	0	0	6	0	0
D	3	1	1	5	−1	−1
E	4	2	4	4	−2	−4
TOTAL	10	0	10	30	0	−10
Mean	2.0			6.0		
Variance	2.0			2.0		

$$b_{y.x} = \frac{\Sigma(X - \bar{X})(Y - \bar{Y})}{\Sigma(X - \bar{X})^2} = \frac{-10}{10} = -1.0$$

on, however, take a moment to go through Table 7-2 and be sure you see where each quantity comes from and how it fits in the formula for the slope. (For a more convenient computational formula for the slope, see Box 7-1.)

CALCULATING THE REGRESSION CONSTANT

Having calculated the slope, our formula for the linear regression line is now $\hat{Y} = a - 1X$. All that is missing is the value for a, the regression constant. In this we are aided by the fact that the least-squares regression line will always pass through the point whose coordinates are the mean of X and the mean of Y, or $(\bar{X}, \bar{Y})$. Since we already know the means for both variables (from Table 7-2), we simply substitute these values into the regression equation:

$$\bar{Y} = a - 1\bar{X}$$
$$6 = a - 2$$
$$8 = a$$

The regression constant (the Y-intercept) for this set of data is 8, and the complete regression equation for the regression of Y on X is now

$$\hat{Y} = 8 - X$$

> **Box 7-1 One Step Further: Computing the Slope**
>
> With a small number of cases, the formula that defines the slope can be used to calculate it. With larger numbers of cases, however, it quickly becomes very cumbersome and time-consuming to subtract the mean from each X and Y score. To make matters simpler, statisticians have developed a computational formula for the slope:
>
> $$b_{y.x} = \frac{N(\sum XY) - (\sum X)(\sum Y)}{N(\sum X^2) - (\sum X)^2}$$
>
> where N is the number of cases. We would use this new formula to compute the slope for Table 7-1 as follows:
>
> | Block | X | X^2 | Y | XY |
> |-------|---|-------|---|-----|
> | A | 0 | 0 | 8 | 0 |
> | B | 1 | 1 | 7 | 7 |
> | C | 2 | 4 | 6 | 12 |
> | D | 3 | 9 | 5 | 15 |
> | E | 4 | 16 | 4 | 16 |
> | TOTAL | 10 | 30 | 30 | 50 |
>
> $$b_{y.x} = \frac{5(50) - (10)(30)}{5(30) - (10)^2} = \frac{250 - 300}{150 - 100} = -1.0$$

This formula and the regression line that it represents describe the *direction* of the relationship between the number of crimes and the number of streetlights (negative for East End). It also tells us the average impact of adding a streetlight—a reduction of one crime per year. As you can see, by giving us a prediction rule, this description allows us to go considerably beyond the relatively crude "positive" and "negative" descriptions of relationships involving nominal and ordinal scales. What, however, about the strength of such relationships? For this we turn to the correlation coefficient.

Correlation: How Strong Is the Relationship?

When we fit a regression line to a set of data points, we are identifying the unique straight line that, on the average, fits the points with the least amount of error. If we look at this way of describing a relationship in terms of prediction, we are trying to find the most accurate way of predicting scores on one variable (Y) from knowledge of another (X). The least-squares regression line represents the best way of making those predictions using a linear function, but we also need a measure of association that tells us just how good our predictions are. Or, to put it in more familiar terms, we need some indication of how strong the relationship between the two variables is.

To describe relationships involving interval- or ratio-scale variables, we use a symmetrical measure of association known as a **correlation coefficient** (or correlation, for short). Its more technical name is the **Pearson product-moment correlation coefficient**, and it is symbolized by the small letter r. Like all good measures of association, Pearson's r ranges from -1.0 to $+1.0$ and has a value of zero when the variables are independent. The higher the value of r, the closer is the fit between the least-squares regression line and the set of points in the scatterplot (see Figure 7-2).

It will help your understanding of r if we begin with the formula that defines it:

$$r = \frac{\sum(X - \bar{X})(Y - \bar{Y})}{\sqrt{[\sum(X - \bar{X})^2][\sum(Y - \bar{Y})^2]}}$$

You may have recognized the numerator of the formula for r: it is the same as that in the formula that defines the slope. It is the covariance, and just as it indicates the direction of the relationship in the formula for the slope, so too does it indicate the direction of the relation for r. This is, of course, as it should be: since both r and b refer to the same relationship, they should have the same sign.

The denominator for r is different from that for the slope, however, in that it has an added term for the variance of Y. As with the slope, the first term is the same as the numerator in the formula for calculating the variance of X: the squared difference between each X score and the mean of X. The second term is the numerator for calculating the variance of Y. As with the slope, the

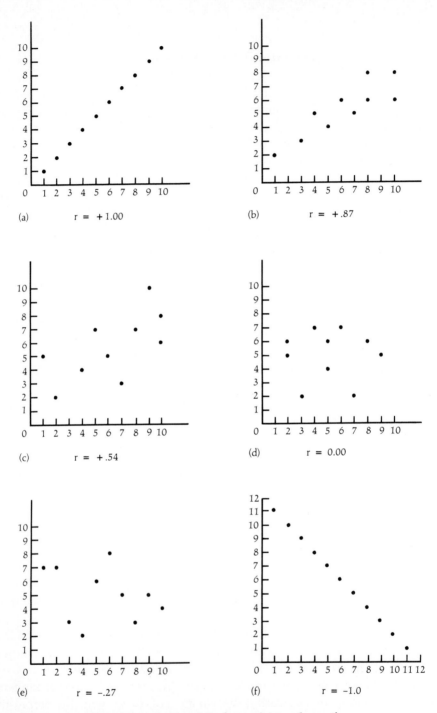

Figure 7-2 Scatterplots for various values of r.

formula given above for r is a somewhat simplified version of the following:

$$r = \frac{\frac{\sum(X - \bar{X})(Y - \bar{Y})}{N}}{\sqrt{\left[\frac{\sum(X - \bar{X})^2}{N}\right]\left[\frac{\sum(Y - \bar{Y})^2}{N}\right]}}$$

$$= \frac{\text{(covariance of } X \text{ and } Y)}{\sqrt{(s_x^2)(s_y^2)}}$$

The correlation coefficient, then, is in fact the covariance between X and Y divided by the square root of the product of their variances, which, when we take the square root, becomes the product of the two standard deviations. Since the Ns in the above formula cancel out, the formula used to define Pearson's r omits them for the sake of simplicity. In sum, then, Pearson's r consists of

$$r = \frac{\text{(covariance of } X \text{ and } Y)}{\text{(standard deviation of } X)\text{(standard deviation of } Y)}$$

Applying the first formula for r as in Table 7-3, we get $r = -1.0$. Before going on, go through the calculations in Table 7-3 to be sure you understand how to go about it. (See also Box 7-2 for a simplified computational formula for r.)

Table 7-3 Computing Pearson's r for Streetlights and Crime, East End Neighborhood

Block	X	$(X - \bar{X})$	$(X - \bar{X})^2$	Y	$(Y - \bar{Y})$	$(Y - \bar{Y})^2$	$(X - \bar{X})(Y - \bar{Y})$
A	0	−2	4	8	2	4	−4
B	1	−1	1	7	1	1	−1
C	2	0	0	6	0	0	0
D	3	1	1	5	−1	1	1
E	4	2	4	4	−2	4	−4
TOTAL	10	0	10	30	0	10	−10

$$r = \frac{\sum(X - \bar{X})(Y - \bar{Y})}{\sqrt{[\sum(X - \bar{X})^2][\sum(Y - \bar{Y})^2]}} = \frac{(-10)}{\sqrt{(10)(10)}} = -1.0$$

> **Box 7-2 One Step Further: Computing the Correlation Coefficient**
>
> As was the case with slopes, with a small number of cases the formula that defines the correlation coefficient can be used to calculate it. With larger numbers of cases, however, it quickly becomes very cumbersome and time-consuming. To make matters simpler, statisticians have developed a computational formula for the correlation coefficient:
>
> $$r = \frac{N(\sum XY) - (\sum X)(\sum Y)}{\sqrt{[N(\sum X^2) - (\sum X)^2][N(\sum Y^2) - (\sum Y)^2]}}$$
>
> where N is the number of cases. We would use this new formula to compute the correlation for Table 7-1 as follows:
>
> | Block | X | X² | Y | Y² | XY |
> |-------|---|----|---|----|-----|
> | A | 0 | 0 | 8 | 64 | 0 |
> | B | 1 | 1 | 7 | 49 | 7 |
> | C | 2 | 4 | 6 | 36 | 12 |
> | D | 3 | 9 | 5 | 25 | 15 |
> | E | 4 | 16 | 4 | 16 | 16 |
> | TOTAL | 10| 30 | 30| 190| 50 |
>
> $$r = \frac{5(50) - (10)(30)}{\sqrt{[5(30) - (10)^2][5(190) - (30)^2]}} = \frac{250 - 300}{\sqrt{(50)(50)}} = \frac{-50}{50} = -1.0$$

As we have seen, the relationship between streetlights and crimes in the East End is perfect, but in the real world we rarely encounter relationships that perfectly fit a straight-line function. In most cases, many (if not most) points do not sit directly on the least-squares regression line. Suppose, for example, that we compare the data in Table 7-1 with those for another hypothetical neighborhood called the "West End" with results shown in Table 7-4.

The relationship between crime and streetlights is quite different in the West End. Although the relationship is still negative, it is no longer perfect,

Table 7-4 Crimes per Block by Number of Streetlights for Five Hypothetical Blocks in the West End Neighborhood

Block	Number of Streetlights	Number of Crimes
F	0	6
G	1	5
H	2	6
I	3	4
J	4	4

and both the mean and variance of Y are smaller. This is especially apparent in scatterplot form (Figure 7-3).

To fit a regression line to the West End data, we first calculate the slope (Table 7-5), for which we get a value of $b_{y \cdot x} = -0.5$. Thus, each additional light reduces crime by only one-half crime per year on the average, only half of what we found in the East End. Substituting the means for X and Y into the

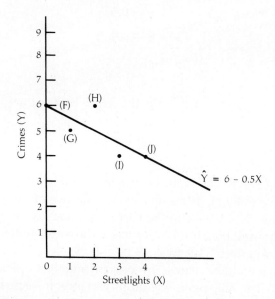

Figure 7-3 Crimes by number of streetlights in the West End.

Table 7-5 Calculating the Regression Slope for the West End

Block	X	$(X - \bar{X})$	$(X - \bar{X})^2$	Y	$(Y - \bar{Y})$	$(X - \bar{X})(Y - \bar{Y})$
F	0	−2	4	6	1	−2
G	1	−1	1	5	0	0
H	2	0	0	6	1	0
I	3	1	1	4	−1	−1
J	4	2	4	4	−1	−2
TOTAL	10	0	10	25	0	−5
Mean	2.0			5.0		
Variance	2.0			0.8		

$$b_{y.x} = \frac{\Sigma(X - \bar{X})(Y - \bar{Y})}{\Sigma(X - \bar{X})^2} = \frac{-5}{10} = -0.5$$

equation for a straight line, we get the regression constant for the West End:

$$\bar{Y} = a - 0.5\bar{X}$$

$$5 = a - 0.5(2)$$

$$6 = a$$

The least-squares regression line for the West End is then

$$\hat{Y} = 6.0 - 0.5X$$

and is drawn on Figure 7-3. Notice that although this is the line that fits the points the most closely of all possible lines, only two of the points (F and J) actually lie on the line. Unlike in the East End, there is error in the fit of the line to the points, and the error is reflected in the correlation coefficient: r has dropped from a perfect −1.0 to −.79 (see Table 7-6).

Notice also that both the regression constant and the slope are smaller than in the East End. The regression line drops less steeply because although the range of X scores is the same (0–4), the variance of Y is smaller. This means that instead of Y dropping four points (from a high of 8 to a low of 4) over the entire range of X, it now drops a total of 2 points (from a high of 6 to a low of

Table 7-6 Computing Pearson's r for the West End

Block	X	$(X - \bar{X})$	$(X - \bar{X})^2$	Y	$(Y - \bar{Y})$	$(Y - \bar{Y})^2$	$(X - \bar{X})(Y - \bar{Y})$
F	0	−2	4	6	1	1	−2
G	1	−1	1	5	0	0	0
H	2	0	0	6	1	1	0
I	3	1	1	4	−1	1	−1
J	4	2	4	4	−1	1	−2
TOTAL	10	0	10	25	0	4	−5

$$r = \frac{\sum(X - \bar{X})(Y - \bar{Y})}{\sqrt{[\sum(X - \bar{X})^2][\sum(Y - \bar{Y})^2]}} = \frac{(-5)}{\sqrt{(10)(4)}} = -.79$$

4). Thus, the slope is less steep to reflect the smaller resulting drop in Y for every unit drop in X. Since the slope tilts less sharply downward to the right, the left end of the line will cross the Y-axis at a lower point.

EXPLAINED VARIANCE

We can interpret Pearson's r in the same way that we interpret phi—we can characterize it in terms such as "strong," "moderate," and "weak," but we cannot go much further than that. Also like phi, however, r takes on a much more precise meaning when we square it: r^2 *represents the proportion of the variation in Y that is statistically accounted for by its relationship with X.* This quantity is known both as r^2 and as the **coefficient of determination**.

To get a better idea of what this means, first consider the total variance for Y in the West End, the number of crimes. The variance in crime reflects the fact that blocks differ in the number of crimes that occur on them. The purpose of relating Y to another variable in this case is to explain this variation: why is there more crime on some blocks than others? In the relationship between X and Y, we can think of the total variation in Y as consisting of two portions, that which is accounted for by X (**explained variance**) and that which is not (**unexplained variance**).

As you will remember from Chapter 4, the variance for Y is the average squared difference between each score and the mean. In Figure 7-4, we have drawn a horizontal line across the graph, which represents the mean for Y (5.0). Each of the five points on the graph is connected to this new line by a vertical arrow, and the length of each of these arrows is the difference between each point and the mean of Y.

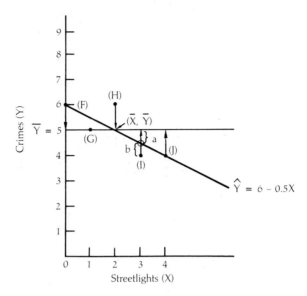

Figure 7-4 Crimes by number of streetlights, showing explained and unexplained variance.

If there were no relationship between the number of streetlights and the number of crimes, our best guess for the number of crimes on each block would simply be the mean of Y, or 5.0. But, there is a negative relationship between Y and X, which is to say, when X is greater than its mean, Y tends to be less than its mean, and when X is smaller than its mean, Y tends to be greater than its mean. (The larger the number of streetlights, the smaller the number of crimes. The smaller the number of streetlights, the larger the number of crimes.) To some degree, then, we do better if we use the least-squares regression line to make predictions about the value of Y than we do if we always use the mean of Y.

Consider, for example, block *I* in Figure 7-4. Its contribution to the total variance of Y is represented by the distance between it and the horizontal line for the mean of Y. If we guessed that $Y = 5.0$ (the mean) when $X = 3$, our guess would be much too high. *Using the mean of Y as our prediction accounts for none of the variation in Y.* What if we use the least-squares regression line? For $X = 3$, we would predict a value of $\hat{Y} = 6.0 - 0.5\,X$, or, $\hat{Y} = 6.0 - 1.5 = 4.5$ (circled on the regression line directly above point *I*). Our prediction is still too high, but it does account for *some* of the difference between *I* and the mean of Y. Specifically, the regression line takes into account that portion of the difference between I and the mean that lies between the circled point on the line and the mean of Y (segment a). This piece of the distance between I and the mean of Y represents that portion of the variation in Y that is *explained* (or, accounted for) by the regression line. It consists of the difference between the predicted value of Y and the mean of Y (or, $\hat{Y} - \bar{Y}$). If we repeat this procedure for all the data points, the total amount of variance in Y that is explained by its relationship with X is

$$\text{Explained variance} = \frac{\sum(\hat{Y} - \bar{Y})^2}{N}$$

(Stop for a moment and think about the connection between this formula and Figure 7-4.)

When we square Pearson's r, the result represents the proportion of the total variance in Y that consists of explained variance (it is equivalent to ϕ^2 in a 2×2 table). If we write this as a formula, it looks like this (since both variances are divided by N, they cancel out and do not appear):

$$r^2 = \frac{\sum(\hat{Y} - \bar{Y})^2}{\sum(Y - \bar{Y})^2}$$

Notice that if there were no relationship between X and Y, the regression line would coincide with the horizontal line for the mean of Y, and since each predicted value of Y would equal the mean of Y, the numerator would be zero, as would r^2. Notice also that if the relationship is perfect and each point lies exactly on the regression line, the distance between the points and the mean of Y would be the same as the distance between the *line* and the mean of Y (as in Figure 7-1). In this case, both numerator and denominator of the formula will be the same, and the value for r^2 will be 1.0. In our example in Figure 7-4, $r^2 = (-.79)^2 = .62$, which means that 62 percent of the variation in Y can be statistically accounted for by the least-squares regression line for its relationship with X.

It is important to keep in mind that the use of the term "explained" does not imply that a causal relationship exists between X and Y. To say that X accounts for 62 percent of the variance in Y does not mean that X has any causal effect on Y. It only means that as X changes Y tends to change in a predictable way. At best, this kind of statistical result may lend support to a theoretical argument about cause and effect, but it cannot prove it true. As we will see in Chapter 9, for example, there is a positive relationship between the occupational prestige of fathers and the occupational prestige of their adult sons. While it is plausible that this might indicate a causal relationship (sons following "in their fathers' footsteps," for example, or high-prestige fathers having more "pull" with potential employers of their sons), the evidence shows that the relationship is not directly causal. Rather, the sons of fathers with high-prestige jobs tend to get better educations, and it is education, not father's prestige, that has a direct effect on the prestige of the son's job.

In a similar way, it is important to be clear about the distinction between explanation and prediction. With explanation, we want to discover the causal mechanism through which differences on one variable produce differences on another. With prediction, we want to discover how differences on X allow us to predict differences on Y regardless of whether X causes Y. Married people, for example, are more likely than single people to have high blood pressure. This knowledge gives us a prediction tool that, nonetheless, turns out to have nothing to do with cause and effect. Married people are more likely to have high blood pressure only because they tend to be older than single people, and it is age, not marital status, that has a causal effect on blood pressure. In Chapter 8, we will go into such problems of interpretation in much greater detail.

UNEXPLAINED VARIANCE AND THE STANDARD ERROR OF ESTIMATE

Return now to the total variance for Y and Figure 7-4. For block I, its contribution to the total variance is the distance between its score and the mean of Y. The portion of that distance that lies between the regression line and the mean of Y consists of explained variance—variance that is accounted for by the line.

You will notice, however, an additional piece of the total distance that is not accounted for by the regression line, a piece that runs between the point and the line (segment b which is called a **residual**). The distance between each point and the corresponding point on the regression line, then, consists of unexplained variance. It is that portion of the difference between the point and the mean of Y that is not accounted for by the regression line. In the case of point I, the regression line correctly predicts that when X is above its mean (greater than 2.0), Y is below its mean (less than 5); but the line does not predict a value of Y that is far *enough* below the mean of Y. This error in prediction represents a piece of unexplained variance. When we repeat this process for all the data points, we find that the unexplained variance is

$$\text{Unexplained variance} = \frac{\sum (Y - \hat{Y})^2}{N}$$

While the explained variance indicates the degree to which the points hug the regression line, the unexplained variance is a measure of the degree to which the points *deviate* from the regression line. Since the regression line is often thought of as a tool for estimating Y from knowledge of X, the square root of the unexplained variance is often referred to as the **standard error of estimate** (symbolized as $s_{est.y}$). The higher the standard error of estimate is, the *worse* is the regression line as a tool for estimating values of Y. When $r = 1.0$, the standard error of estimate equals zero because there is no error in predicting Y from knowledge of X. When $r = 0$, however, the standard error of estimate equals the standard deviation for Y.

Since r^2 measures the proportion of the total variance in Y that is explained by its relationship with X, then the proportion of the total variance that consists of unexplained variance is equal to $(1 - r^2)$.

$$(1 - r^2) = \frac{\sum(Y - \hat{Y})^2}{\sum(Y - \bar{Y})^2}$$

If we take the square root of the unexplained variation, the result ($\sqrt{1 - r^2}$) is called the **coefficient of alienation**. It measures the degree to which X and Y are *un*related in the same way that r measures the degree to which they are related. Put from a slightly different perspective, while r measures the degree to which Y is related to X, the coefficient of alienation measures the degree to which Y is related to all variables *other than* X. Using the coefficient of alienation, we can calculate the standard error of estimate quite simply:

$$s_{est.y} = s_y \sqrt{1 - r^2}$$

where s_y is the standard deviation for Y. For the West End, then, we get the following results:

Least-squares regression line: $\hat{Y} = 6.0 - 0.5X$

Correlation coefficient $= r = -.79$

Coefficient of determination = explained variance = $r^2 = .62$

Unexplained variance = $(1 - r^2) = (1 - .62) = .38$

Coefficient of alienation = $\sqrt{(1 - r^2)} = \sqrt{.38} = .62$

Standard error of estimate = $s_y \sqrt{1 - r^2} = 0.8(.62) = 0.496$

By comparison, the results for the East End are

Least-squares regression line: $\hat{Y} = 8.0 - 1.0X$

Correlation coefficient $= r = -1.0$

Coefficient of determination = explained variance = $r^2 = 1.00$

Unexplained variance = $(1 - r^2) = (1 - 1) = 0.00$

Coefficient of alienation = $\sqrt{(1 - r^2)} = \sqrt{0.00} = 0.00$

Standard error of estimate = $s_y\sqrt{1 - r^2} = 2(0.00) = 0.00$

If you compare these two sets of figures carefully, you will notice some sharp differences between them, differences that can result in considerable mistakes if they are not interpreted properly.

Interpreting Slopes and Correlations

It is important to keep in mind that slopes and correlations describe very different aspects of linear relationships, and although under certain rare circumstances they can have the same value (see Box 7-3), they are usually not equal. The correlation tells us how accurately we can predict Y from knowing X. The slope tells us how much of a change in Y is associated with each unit change in X. The correlation thus reflects the usefulness of the linear regression line as a prediction rule. The slope, however, is a direct measure of the average *impact* of X on Y.

To see what a difference it can make if we confuse the two, consider our two versions of the relationship between streetlights and crime for two different neighborhoods:

East End	South End
$b = -1.0$	$b = -1.5$
$r = -1.0$	$r = -.30$

If we compare the slopes, we find that in the East End the number of streetlights has a smaller impact on crime than it does in the South End (-1.0 versus -1.5). This means that, on the average, each increase of one streetlight is associated with a smaller absolute decline in crime in the East End than in the South End. The correlation coefficient for the East End, however, is larger than the coefficient for the South End, which means that we can predict crime from knowledge of the number of streetlights with greater accuracy in the East End than in the South End.

It is very important that you understand that these are two very different kinds of statements. If we used the correlations to compare the relative impact of streetlights on crime in the two neighborhoods, we would arrive at the wrong conclusion, just as we would if we used the two slopes to compare the strength of the relationship between streetlights and crime in the two

Box 7-3 One Step Further: Variance and the Relative Size of Slopes and Correlations

We have seen that when we compare the full formulas for the slope and correlation, the numerators of both contain the covariance. The denominator of the slope consists of the variance of X, and the denominator of the correlation coefficient consists of the product of the standard deviations for X and Y.

$$b_{yx} = \frac{\frac{\sum(X-\bar{X})(Y-\bar{Y})}{N}}{\frac{\sum(X-\bar{X})^2}{N}} = \frac{\text{covariance of } X \text{ and } Y}{s_x^2}$$

$$r = \frac{\frac{\sum(X-\bar{X})(Y-\bar{Y})}{N}}{\sqrt{\left[\frac{\sum(X-\bar{X})^2}{N}\right]\left[\frac{\sum(Y-\bar{Y})^2}{N}\right]}} = \frac{\text{covariance}}{\sqrt{(s_x^2)(s_y^2)}} = \frac{\text{covariance}}{(s_x)(s_y)}$$

It takes only a little rearranging to see that if we multiply r by the standard deviation of Y, (s_y), then the standard deviation of Y in the denominator is cancelled out. In addition, if we divide r by the standard deviation for X, then the denominator becomes s_x^2, which is the variance for X. The result, then, looks exactly like the formula for the slope.

$$r\left(\frac{s_y}{s_x}\right) = \left(\frac{\text{covariance}}{(s_x)(s_y)}\right)\left(\frac{s_y}{s_x}\right) = \frac{\text{covariance}}{s_x^2} = b$$

The implication of all this is that the slope and correlation are closely related to each other, as follows:

$$b_{y.x} = r\left(\frac{s_y}{s_x}\right)$$

This tells us that when X and Y have the same standard deviations, the slope and the correlation will have the same value. When there is more variation in X than in Y, the correlation will be larger than the slope. And when there is more variation in Y than in X, the value for the correlation will be smaller than the slope.

neighborhoods. Whenever you encounter a statement that refers to the *impact* of one variable on another in a regression analysis, you should immediately start looking for information about slopes. Any use of correlations to support statements about the impact of one variable on another is absolutely inappropriate.

By the same token, whenever you encounter a discussion of the relative strength of relationships in a regression analysis, the evidence must be in the form of correlation coefficients or something based on them (such as coefficients of determination, alienation, or the standard error of estimate). Slopes may not be used for such comparisons. Slopes are used to make guesses about the value of Y for each value of X. Correlations indicate how accurate the guesses are.

STOCHASTIC AND MEAN INDEPENDENCE

You will recall from previous chapters that when two variables, X and Y, are stochastically independent of each other, the distribution of Y is the same within all categories of X. With the exception of Yule's Q, when measures of association had a value of zero, this meant that the two variables were stochastically independent.

In regression analysis, however, there is a second type of independence. If we fit a least-squares regression line to the scatterplot in Figure 7-5, both the slope and the correlation will be zero. The mean for Y is 5, and, for each value of X, the values of Y are just as likely to be above the mean of Y as they are to be below (look at the figure and verify this for yourself). The result is that the mean of Y is the same for all values of X. This indicates that knowing the score on X does not improve our ability to predict Y, since the score we will predict for Y will always be equal to the overall mean of Y. In this case, when the *mean of Y is the same for all values of X*, we say the variables are **mean-independent**.

What is the difference? With stochastic independence, the conditional probability distributions for Y are the same for all values of X; with mean independence, however, the conditional *means* which summarize conditional probability distributions are the same for all values of X *even though the distributions they summarize may be quite different*. Each column of points in Figure 7-5 has the same mean, but they differ enormously in their variance. Cases with a score of 2 on X, for example, have scores on Y that range from 1 to 9; but those with a score of 9 on X have scores ranging only from 4.5 to 5.5 on Y. By relying on the means to summarize the conditional distributions of Y, we lose information. Therefore, knowing that the *mean* of Y is the same for all values of X does not tell us whether the *distribution* of Y is the same for all values of X.

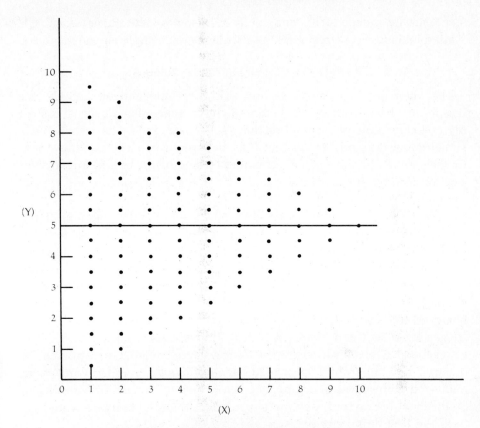

Figure 7-5 Example of mean independence and stochastic dependence.

You can see in Figure 7-5 that although knowing X makes no difference in our predictions of the *mean* of Y, the distributions of Y for each value of X are quite different. Although X and Y are mean-independent, they are not stochastically independent. If we know the score on X, we *do* improve our ability to predict specific values of Y: if $X = 10$, we would never predict $Y = 9$, but we might if we knew that $X = 1$ or 2.

If X and Y are mean-independent, this does not imply that they are stochastically independent; but if they are stochastically independent, then they are mean-independent. (Do you know why?)

You can now see that regression analysis uses the independent variable to predict mean values for the dependent variable. If the slope and correlation coefficient are zero, we know only about the ability of X to improve our predictions of the mean of Y. This does not mean that X and Y are

A NOTE ON CURVILINEAR REGRESSION

The difference between mean and stochastic independence is only one reason to check scatterplots before doing a regression analysis. Linear regression rests on the assumption that a set of data can best be described as a straight line; but not all relationships between variables take this form. Consequently, there is a set of techniques for finding the best-fitting *curve* for a set of data, but it is used much less often in the social sciences and we will not go into it here.

It is important, however, to avoid the mistake of blindly assuming that a relationship is linear and going ahead and trying to fit a straight-line function to it. The relationship in Figure 7-6, for example, represents a perfect curvilinear relationship in that all of the points lie on a curve that can be described with an algebraic function, although not a linear one of the form $Y = a + bX$. If we nonetheless try to fit a least-squares line to this curve, however, the results will indicate a slope and correlation coefficient of zero, suggesting that X and Y are unrelated. This is clearly not the case.

In general, a relationship has to be sharply curved in order to justify the use of curvilinear rather than linear regression techniques. In other words, with mildly curvilinear relationships, the least-squares straight line results in predictions that are only slightly less accurate than the ones we would get from the best-fitting curve.

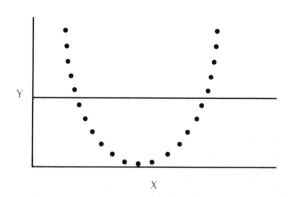

Figure 7-6 A curvilinear relationship with the least-squares *linear* regression line superimposed.

The lesson here is that it is, as with all statistical techniques, important to carefully examine a set of data before trying to apply particular tools to it, for only in this way can we be sure that the tools we apply are appropriate.

Summary

1. Regression and correlation analysis are a set of techniques for precisely describing relationships involving interval- or ratio-scale variables.

2. In regression analysis, a straight-line function is fitted to a set of data points in such a way that the squared vertical distances between the points and the line are minimized. The result is a least-squares regression line, of which there is only one for each set of data. The line takes the general form of $\hat{Y} = a + bX$, where $\hat{Y}$ is the predicted value for the dependent variable, a is the regression constant, b is the slope, and X is the independent variable.

3. Pearson's r is a symmetrical measure of association that indicates the degree to which the data points tend to hug the least-squares regression line. When r is squared (r^2) the result is the coefficient of determination, which indicates the proportion of the variance in Y that is accounted for by its linear relationship with X.

4. The unexplained variance is the proportion of the total variation in Y that is unaccounted for by the least-squares regression line. The square root of the proportion of variance that is unexplained is called the coefficient of alienation. When the coefficient of alienation is multiplied by the standard deviation of Y, the result is called the standard error of estimate. The larger the standard error of estimate is, the weaker is the relationship between X and Y.

5. Slopes and correlation coefficients may not be used interchangeably. The slope indicates how much change in Y is associated with each change in X. The correlation indicates the degree to which the least-squares regression line produces accurate predictions of Y from knowledge of X.

6. When the slope and correlation in a regression analysis equal zero, the variables are mean-independent, which is to say, the mean of Y is the same for every value of X. This does not mean that the variables are stochastically independent, however. The distribution of Y for different values of X can be quite different even though the means are the same.

7. Linear regression techniques should not be used to describe relationships that are markedly curvilinear. If linear techniques are used in such cases, they will tend to understate the strength of the relationship and misrepresent its direction.

Key Terms

coefficient of alienation 181
coefficient of determination 177
correlation coefficient 171
covariance 168
explained variance 177
least-squares regression line 166
mean independence 184
Pearson product-moment correlation coefficient 171

regression analysis 164
regression coefficient 167
regression constant 167
residual 180
slope 167
standard error of estimate 181
unexplained variance 177
Y-intercept 167

Problems

1. Define and give an example of each key term at the end of the chapter.
2. If we do a regression analysis for the number of years of schooling children have (Y) in relation to the number of years of schooling their mothers had (X_1), we get the following results for the U.S. in 1984:

$$\hat{Y} = 8.43 + .42X_1 \qquad r = +.48$$

If we do the same analysis except substitute the fathers' educations (X_2) for the mothers', we get

$$\hat{Y} = 9.45 + .35X_2 \qquad r = +.48$$

The standard deviation for Y is 1.1 years.
a. State the meaning of each equation as a whole.
b. State the meaning of each part of each equation.
c. Which is the better predictor of years of schooling, the father's education or the mother's? How do you know?
d. Which has the greater impact on years of schooling, the father's education or the mother's? How do you know?
e. What is the coefficient of determination for the first equation? For the second? What does it mean?

f. What is the coefficient of alienation for the first equation? What is the coefficient of alienation for the second? What does it mean?
g. If the standard deviation for Y is 1.1 years, what is the standard error of estimate for the first equation? For the second?
h. How much of the variation in years of schooling is explained by father's education? By mother's?
i. If a woman's mother had 12 years of schooling, how many years of schooling would you predict for her?
j. If a woman's father had 12 years of schooling, how many years of schooling would you predict for her?

3. Given the data in Table 7-7 for the relationship between the years of schooling and the number of children a person has, calculate *and state the meaning of*
    a. the slope
    b. the regression constant
    c. the correlation
    d. the coefficient of determination
    e. the coefficient of alienation
    f. the proportion of variance that is unexplained
    g. the standard error of estimate

    and answer the following questions:
    h. What does the least-squares regression equation look like?
    i. For people with 15 years of schooling, how many children would you predict they would have on the average?
    j. Suppose I in fact have six children. How would you account for the difference between this fact and your prediction?

Table 7-7 Number of Children (Y) by Years of Schooling (X)

Years of Schooling (X)	Number of Children (Y)
10	4
12	3
2	5
4	6
20	1
16	3
10	1

4. If the mean of X is 10 and the mean of Y is 15, and if the slope for the regression of Y on X is −4, what is the value of the regression constant? What does the resulting regression equation look like?

5. Is it possible to have a regression analysis in which there is a relationship between the variables and yet *none* of the actual data points lies on the regression line? Why or why not?

6. If X and Y are independent of each other, what are the values of the following?
   a. $b_{y.x}$
   b. $s_{est.y}$
   c. r
   d. the coefficient of alienation
   e. the coefficient of determination
   f. the regression constant

7. If X and Y are independent of each other, what value would you predict for Y for any given value of X?

8. If you wanted to determine the impact that one variable has on another in a regression analysis, what statistic would you use?

9. Explain the difference between stochastic and mean independence. Which is indicated when $r = 0.00$?

10. If you wanted to determine the strength of the relationship between two variables in a regression analysis, what statistic would you use?

11. What kinds of errors can result from using linear regression techniques to describe curvilinear relationships? Be specific.

⋋8⋌

Relationships of More Than Two Variables: Controlling

Chapters 6 and 7 introduced many ways to describe relationships between variables. In some instances, research questions can be answered without going beyond a two-variable relationship. In medicine, for example, showing that death rates increase dramatically among people who smoke cigarettes was enough for the U.S. Food and Drug Administration to issue warnings to the general public about the dangers of smoking. In law, the relationship between the background characteristics of potential jurors and how they tend to vote in different kinds of cases is sometimes used by attorneys to affect the outcome of jury selection and, therefore, their cases. Many discrimination cases have been fought and won on the basis of a statistical pattern of inequality in jobs and pay between racial or gender categories. And there are always cases in which finding *no* relationship between variables can be highly significant. If, for example, researchers found no relationship between IQ scores and race, it would, in itself, be an important finding.

In many other cases, however, the discovery of a relationship between two variables is only the beginning of the process of analyzing data in order to answer scientific questions. In general, there are three reasons for going beyond two-variable relationships to more complicated analyses that involve additional variables.

First, we need to be sure that the statistical association between two variables actually represents a meaningful relationship and not just the coincidental interplay of other factors. If X and Y are statistically related, does this mean that X causes Y? If people who go to college are more liberal politically than those who do not, does this mean that college changes political values? Or are both college attendance and political values caused by other variables such as the characteristics of the families we come from?

Second, if we are interested in identifying the causes of the dependent variable (as we usually are), we want to understand the causal mechanism, thus making it necessary to introduce additional variables into the explanation. Does X cause Y directly? Or are there variables that are affected by X which, in turn, affect Y? Education is positively related to income; but is it education per se that causes differences in income, or is it other variables such as occupation?

Third, even if we have no reason to test or explore a two-variable relationship in greater detail, we often want to know if that relationship holds under different conditions or among different groups of observations. Does an observed gender difference, for example, hold for all social classes? Does a particular teaching technique have the same result among all kinds of students and at different grade levels?

To pursue any of these three general research goals, we must *control* for additional variables.

The Logic of Controls

There are two ways of adding variables to the analysis of what begins as a two-variable relationship. First, as we will see in the following chapter, we can use **multivariate analysis**, a technique that shows how well several independent variables are able to predict a dependent variable all at once. How, for example, can we use such characteristics as occupation, race, gender, education, and family background to predict differences in income? Although related to the logic of multivariate analysis in many ways, the techniques we will introduce in this chapter are unique and important enough to stand alone. In fact, they are a cornerstone of scientific research (see Rosenberg, 1968). They rest on the central ideal of achieving a more precise understanding of a two-variable relationship by controlling for one or more additional variables.

To get a clearer understanding of controlling, consider the relationship between occupation and income: white-collar workers make more money than blue-collar workers. To control for gender, we look at the relationship between occupation and income *within* each gender category—among males and among females. We hold gender constant by seeing how occupation affects income among people who are of the same gender. We do this for the same reason that a chemist tries to control the physical conditions in which an experiment is conducted: the chemist wants to be sure that only certain identifiable variables are affecting (and thus causing) the results. Since gender certainly has something to do with the jobs people have, the income they receive, and the relationship between the two, controlling for this variable is important for a fuller and more precise understanding of the original relationship.

In an important sense, controlling for additional variables is the next step in the process we began in Chapter 3. We began with simple distributions of single variables—such as income. We then complicated things a bit by introducing a second variable—such as occupation. This meant that we saw how the distribution of one variable such as income differed among the categories of a second variable such as occupation. This is how we arrived at relatively simple relationships between two variables.

In this chapter, we take such relationships and see how *they* differ among the categories of still more variables. For example, instead of looking at how occupations differ from one gender to the other, and instead of looking at how income differs from one gender to the other, we now look at how the *relationship* between occupation and income differs from one gender to the other. The number of variables makes things increasingly complicated, but the logic is the same.

THE ORDER OF RELATIONSHIPS

Statistical relationships can be described in terms of the number of variables that are used as controls. This is called the **order** of a relationship. A relationship between an independent variable (X) and a dependent variable (Y), for example, has no controls and is therefore called a **zero-order** relationship. In the relationship between occupation and income, when we introduce gender as a control variable (Z), the relationship becomes a **first-order** relationship. (So as to avoid confusion, we will always use X to represent independent variables, Y to represent dependent variables, and Z to represent control variables in this chapter.) If we control for gender (Z_1), race (Z_2), and education (Z_3), we then have a **third-order** relationship.

In this chapter, we will look at a variety of first-order relationships in order to illustrate the different kinds of results that justify introducing control variables. In each case, it is important to pay attention to the logic that underlies the decision to control for a particular variable and how this affects the interpretation of the results. In some cases, for example, identical *statistical* results can be interpreted in quite different ways depending on the variables involved and the way we think about their relationship. As we have seen before, statistics by themselves can tell us relatively little about how the world works. In the end, we have to think about how the world works and use statistics to test our ideas. As you will see shortly, the interpretation of the results must always take place in the context of those ideas.

Spuriousness: Is the Relationship Real?

When we gather data on two variables and find a statistical association, the first question we have to ask is, "Is this more than a statistical relationship?" Does this relationship tell us something true and valid about a cause-and-effect relationship? Or does it reflect the influence of other factors beyond the two variables involved in the relationship?

If we want to see if one variable causes another, the first thing we have to check for is the time-ordering of the variables: *for one variable to cause another, it must precede that variable in time.* It is ridiculous to argue, for example, that a variable such as educational attainment can cause a variable such as race ("going to college makes you more likely to be born white") since educational attainment is a characteristic that is determined long *after* we are born.

Although an appropriate time-ordering of variables is necessary for a cause-and-effect relationship to exist, this does not of course mean that any variable that precedes another in time is therefore a cause of it. It is here that the use of controls becomes an important tool for testing ideas about causal relationships.

Consider, for example, the finding that birth rates are higher among those who live at high altitudes in certain South American countries than they are among those who live at low altitudes (Table 8-1) If we want to lower birth rates, should we then encourage people to move to lower altitudes? There were more than a few people who seriously considered such an implication until someone pointed out that this might not be a fascinating insight into the dynamics of fertility. Instead, the relationship between altitude (X) and fertility (Y) depends on a third variable, race (Z). Those who live at high altitudes tend to be Indians, and Indians, for a variety of social reasons having

Table 8-1 Fertility by Altitude, Hypothetical

Fertility	Altitude		
	High (+)	Low (−)	All
High (+)	500	400	900
Low (−)	300	800	1100
TOTAL	800	1200	2000
$Q = +.54$			

nothing to do with altitude (such as the value placed on large families), tend to have unusually high fertility rates. If this is true, then it would be **spurious** to interpret the statistical relationship between altitude and fertility as causal.

We can represent this argument in a type of diagram (see below) often used in the social sciences. Each arrow represents a causal relationship (either one that is known or one that is hypothesized).

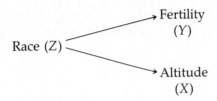

Table 8-1 shows that the positive statistical relationship between altitude and fertility is substantial ($Q = +.54$). Given that the idea of altitude causing higher fertility rates might strike us as a bit suspect, we might then ask, "Are people who live at high altitudes different from those who live at low altitudes, and might those differences account for differences in fertility?"

Since it has often been found that Indian populations have higher fertility than non-Indian populations, we would then check to see if Indians tend to live at higher altitudes than non-Indians. In Table 8-2 we can see that this is, indeed, the case. There is a strong positive relationship between being an Indian and living at high altitudes ($Q = +.88$).

Do Indians have higher fertility than non-Indians? Table 8-3 shows that they do. There is a strong positive relationship between race and fertility ($Q = +.80$).

We have now found relationships among all three variables, a condition that must be met for controlling to make any sense. If, for example, we found

Table 8-2 Altitude by Race, Hypothetical

	Race		
Altitude	Indian (+)	Non-Indian (−)	All
High (+)	600	200	800
Low (−)	200	1000	1200
TOTAL	800	1200	2000
$Q = +.88$			

Table 8-3 Fertility by Race, Hypothetical

	Race		
Fertility	Indian (+)	Non-Indian (−)	All
High (+)	600	300	900
Low (−)	200	900	1100
TOTAL	800	1200	2000
$Q = +.80$			

that Indians and non-Indians are *equally* likely to live at high altitudes, we could not argue that altitude is associated with higher fertility because Indians are *more* likely to live at higher altitudes.

TESTING FOR SPURIOUSNESS

The argument for spuriousness is that altitude is related to race, and since race is related to fertility, then altitude is also related to fertility *statistically*. If, however, it is race and not altitude that causes differences in fertility, then if we look at people who are alike on race, altitude should no longer make any difference. In short, if we *control* for race, then the relationship between altitude and fertility should disappear. By "disappear," I mean that if we look at the relationship between altitude and fertility *among* Indians and *among* non-Indians we should find that the two variables are independent of each other (Table 8-4).

Table 8-4 consists of two **subtables**, one for Indians and one for non-Indians. As you study it, notice that if we add the comparable cells of the two subtables, the result is Table 8-1 (try it). The upper-left cell in the Indian subtable plus the upper-left cell in the non-Indian subtable equals the number of cases in the upper-left cell of Table 8-1. Thus, 8-4 simply represents the data in Table 8-1 sorted out between two racial categories. Most important, notice that in both racial categories there is no relationship between altitude and fertility ($Q = 0.00$ in both cases). Once we hold race constant, the relationship disappears. This is what we would expect to find in the case of spuriousness, and these statistical findings support our reasoning.

It is important to keep the distinction between statistical findings and theoretical interpretations clear, for the data cannot tell us how to interpret them. You will soon see, for example, that there are situations in which we control for a third variable, find that the relationships disappear within the categories of the control variable, and conclude something *other* than

Table 8-4 Fertility by Altitude, Controlling for Race

	Indians (+)			Non-Indians (−)		
	Altitude			Altitude		
Fertility	High (+)	Low (−)	All	High (+)	Low (−)	All
High (+)	450	150	600	50	250	300
Low (−)	150	50	200	150	750	900
TOTAL	600	200	800	200	1000	1200
		Q = 0.00			Q = 0.00	

spuriousness. The interpretation we give to a set of statistical results ultimately depends on the variables involved and the theoretical reasoning through which we connect them.

You might be wondering at this point how we know that the relationship between fertility and race is any more "real" than the relationship between fertility and altitude. Perhaps altitude really does have a causal effect on fertility (by affecting the physiology of conception), and race is statistically related to fertility only because it is related to altitude. We can test this alternative interpretation by rearranging the data in Table 8-4 so that we examine the relationship between race and fertility, controlling for altitude.

Table 8-5 shows the relationship between race and fertility within each category of altitude. (You should go to Table 8-4 and satisfy yourself that Table 8-5 is simply a rearrangement of the cells in Table 8-4. Also add together

Table 8-5 Fertility by Race, Controlling for Altitude

	High Altitude (+)			Low Altitude (−)		
	Race			Race		
Fertility	Indian (+)	Non-Indian (−)	All	Indian (+)	Non-Indian (−)	All
High (+)	450	50	500	150	250	400
Low (−)	150	150	300	50	750	800
TOTAL	600	200	800	200	1000	1200
		Q = +.80			Q = +.80	

the appropriate cells to see that Table 8-5 is a more detailed breakdown of Table 8-3.) If altitude is the cause of differences in fertility, and if the relationship between race and fertility is not causal, when we control for altitude the relationship between race and fertility should disappear among those who are alike on altitude. As you can see, however, this does not happen: the relationship between race and fertility is as strong as ever ($Q = +.80$). Controlling for altitude makes almost no difference. Therefore, altitude cannot be used to demonstrate that a causal interpretation for the relationship between race and fertility is spurious.

The Importance of Testing for Spuriousness There are many examples of spurious interpretations in science, some of them silly, some of them not. If we kept track of the number of fire fighters present at a fire, for example, as well as the amount of damage done, we might be tempted to conclude that the more fire fighters there are, the more damage there is, so perhaps we should keep the fire fighters away. Or on noticing that as ice cream sales increase in seaside communities, so do the number of drownings, we might be tempted to limit ice cream sales in order to prevent drownings (what would you control for here to explain this relationship?).

In other cases, the possibility of spurious interpretations can be quite serious. Many studies, for example, have found a positive relationship between educational attainment and liberal values and beliefs about racial and gender equality. Reinforced by the cultural belief that education results in enlightenment, this might lead us to believe that a partial answer to the problems of racism and sexism is a greater investment in higher education. Before leaping to such a solution, we would do well to first consider the possibility that a causal interpretation of the statistical effects of education is spurious.

As we mentioned earlier, one possible explanation for these findings is that schooling has nothing to do with it. Rather, college graduates are more liberal because they are more likely to be selected from relatively liberal families and are thus predisposed to liberal views long before they go to college.

Another possible explanation is that while educated people are more likely to give liberal responses in an interview situation, they are, in fact, just as likely to be racist or sexist as everyone else. In other words, educated people are more likely to give socially acceptable responses to an interviewer even though the responses may not represent their true beliefs and values. To test this idea, we would have to find a way to measure tendencies to reveal unpopular values and beliefs to interviewers (Crowne and Marlowe, 1980).

If the statistical relationship between education and racial and gender values and beliefs is not causal, then we would expect to find that among

Box 8-1 One Step Further: Understanding Spuriousness with Concordant and Discordant Pairs

You might be wondering just how it is possible that two tables in which fertility is *in*dependent of altitude can be added together to make one table in which fertility is *de*pendent on altitude. The answers lies in the relationship between race and fertility on the one hand and race and altitude on the other. Go back for a moment to Table 8-1 (represented below with letters in the cells). If we think of this relationship in terms of concordant and discordant pairs, the fact that the relationship is positive means that there are more concordant pairs than discordant pairs. In other words, the positive cross-product (ad) is much larger than the negative cross-product (bc).

Fertility by Altitude, Hypothetical

	Altitude		
Fertility	*High* (+)	*Low* (−)	All
High (+)	a	b	$a + b$
Low (−)	c	d	$c + d$
TOTAL	$a + c$	$b + d$	$a + b + c + d$

According to the spuriousness argument, the only reason fertility and altitude are statistically related is that Indians tend to live at high altitudes *and* tend to have high fertility. In terms of the cells and cross-products, this means that cell a is relatively loaded with high-fertility Indians, and cell d is relatively loaded with low-fertility non-Indians. The result is a large positive cross-product and a large positive value for Q.

When we control for race, we produce two subtables, one for Indians and one for non-Indians. To get the Indian subtable, we remove all *non*-Indians from Table 8-1. This means that while all the cells get smaller, cell d drops far more than the others because it had so many low-fertility non-Indians in it. Cell d gets so much smaller, in fact, that the positive cross-product (ad) is reduced to the point where it is no larger than the negative cross-product. The result: $Q = 0.00$ among Indians.

In the same way, to get the subtable for non-Indians, we remove all of the Indians from Table 8-1. This means, however, that while all the cells get smaller, cell a drops far more than the others since it is so disproportionately loaded with high-fertility Indians. The result is, once again, a disproportionately large reduction in the size of the positive cross-product (ad) to the point where it is no larger than the negative cross-product, and $Q = 0.00$.

people who tell the truth to interviewers, education has no effect. This would show that the zero-order relationship appeared not because education causes differences in such beliefs and attitudes, but because education causes people to differ in how concerned they are about the impression they create in interview situations. If, however, the effects of education persist among those who tell the truth, this would support the argument that education's statistical relationship with racial and gender beliefs and attitudes is causal.

PARTIALS AND EXTRANEOUS VARIABLES

We found spuriousness in the example of altitude and fertility by controlling for race. In this case, race is called an **extraneous** variable. We examined the relationship between altitude and fertility among Indians and among non-Indians and found that within both racial categories there was no relationship between altitude and fertility. Each of these controlled relationships—within categories of the control variable—is called a **partial relationship**. The values of Yule's Q in Tables 8-4 and 8-5 are called **partial associations**.

Understanding Causal Relationships: Intervening Variables

Having satisfied ourselves that a causal interpretation of a statistical relationship is not spurious, the next step is to understand the two-variable relationship in greater detail. The use of **intervening** variables allows us to understand the mechanism through which an independent variable affects a dependent variable.

As an example, recall from Chapter 1 the Supreme Court case in which a Georgia man, sentenced to death for killing a white policeman during a robbery, appealed his conviction. His defense argued that the Georgia courts discriminate against those who kill whites, and cited as evidence the positive statistical relationship between killing whites (X) on the one hand and being sentenced to death (Y) on the other. In a study of 2,484 murders in Georgia, Baldus (1987) found that blacks who murder whites are 22 times more likely to be executed than are blacks who murder blacks; and whites who murder whites are almost 3 times more likely to be executed than are whites who murder blacks.

The state of Georgia replied, however, that although there is a statistical relationship between the race of the victim and the severity of the punishment, this does not mean that the courts are using the race of the victim to determine punishment. Rather, the state argued, another variable intervenes between race and punishment, and it is this variable and not race that is directly considered by the courts. This third variable is the circumstances in which a murder occurs (Z).

Blacks, the state argued, are more likely than whites to be killed in family and other informal disputes that do not otherwise involve criminal activity. Whites, however, are more likely to be murdered as part of the commission of a crime such as armed robbery. Such crimes, the state contended, have a higher tendency to provoke the outrage of the community and, therefore, harsher punishment. Thus the state contended that it is not the race of the victim that directly causes harsher punishment. Rather, race has a causal effect on the circumstances of the crime, and it is the circumstances in which murder occurs that directly influence the punishment.

The state argued that the relationships among the three variables look like the following diagram.

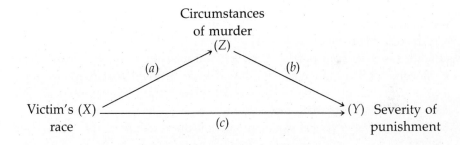

We know that the victim's race (X) and punishment (Y) are related (arrow c). In its argument against discrimination, the state maintained that differences in the circumstances (Z) of murders may play a role. For this to be true, the control variable must be related to both punishment (arrow b) and race (arrow a). If either of these relationships does not exist (or is very weak), then the state's argument falls apart. In other words, if whites and blacks do *not* tend to be murdered in different circumstances *or* if the circumstances of murders do not affect the severity of punishment as the state contends, then the state's argument cannot hold.

If murder sentences differ in severity because of differences in the circumstances in which they occur, then, once we confine our attention to murders that occur under similar circumstances, the relationship between victim's race and punishment should disappear. If Z fully explains the relationship between X and Y, *then the partial association, controlling for Z, should be zero (or very close to it)*. This would strongly support the state's case. If, on the other hand, the race of the victim still makes a difference in the severity of punishment *among murders that are committed in similar circumstances,* then the defense argument is the one that is supported. In this case, the partial association would be as strong as (or nearly as strong as) the zero-order association.

Be careful to note that even if the partial relationship dropped to zero, we would not conclude that the causal interpretation for the relationship between race and punishment is spurious. *Statistically*, we have the same kind of result here as we did when we examined the relationship between altitude and fertility, controlling for race. The interpretation of results, however, depends on the variables involved and how we think about their relationship to one another. In the fertility example, we considered two alternatives: either altitude had some unknown effect on fertility, or some other factors such as race were related to both fertility and altitude in a *plausible* way which thereby demonstrated that the zero-order relationship between altitude and fertility was not causal.

In the case of race and punishment, however, we have more than a statistical accident. The theoretical argument that links these three variables is a plausible one that is consistent with what we already know about how the world works. The only way to tell the difference between these two analytical situations is to use our heads. Although you may agree that this explanation is plausible and reasonable, you should keep in mind that there are often several different plausible explanations for the same phenomenon. As you read research reports that offer explanations, it is always important to try to think of alternative logical explanations that the researchers may have overlooked.

Notice also that the altitude-fertility and race-punishment problems differ in the time-ordering of the variables. In the case of race and punishment, the victim's race is a social characteristic determined at birth, followed by the circumstances of his or her murder and the punishment of the murderer, in that order. In this case, the control variable (Z) is called an intervening variable because, theoretically, it is causally located in between the two variables in the zero-order relationship. This is never the case with spurious relationships.

When we control for Z, two things are likely to happen to the partial association between X and Y. If the partial goes to zero, then we have statistically explained the relationship between X and Y, and we can erase arrow c in our diagram. Does this mean, however, that race has nothing to do with punishment? Not at all. Insofar as the victim's race has some direct effect on the circumstances of his or her murder, then it has **indirect** effects on punishment *through* the intervening variable.

What do we conclude if the partial association between race and punishment goes down when we control for Z, but not all the way to zero? This means that some of the association is due to the intervening effects of Z, and some is not. At this point there are two possibilities. First, there may be some **direct** effect of race on punishment, supporting the defense's contention that discrimination plays a part. Second, there may be other intervening variables that account for the unexplained portion of the relationship between X and Y. If we want to explain the original relationship, we must search for additional intervening variables that have a theoretically plausible place in the phenomenon we are studying. Selecting control variables is as much a theoretical activity as a statistical one. We cannot just go out and control for every variable we can measure. If we controlled for every variable known, the partial association between *any* pair of variables would disappear, but we would not understand what was going on. Statistics can never be an adequate substitute for thinking.

How did the Supreme Court case turn out? In a controversial 1987 decision, the Court decided in favor of the state in spite of the fact that the statistical evidence supported the defendant's argument. The decision was controversial because the court held that the statistical evidence, although clearly showing a pattern of discrimination, could not be used as evidence in the specific case of an individual defendant because it could not by itself show that the overall pattern of discrimination against those convicted of murdering whites actually affected this (or any other) particular defendant's sentence. It is a decision that raises important questions not only about the effect of statistical evidence on social policy, but also the connection between the behavior and fortunes of individuals and the social environments in which they live—both of which are, unfortunately, beyond the scope of this book.

INTERVENING VARIABLES: AN EXAMPLE WITH DATA

To see what statistical results can look like in a test for an intervening variable, consider the relationship between race (X) and college attendance (Y) among 18 to 24-year-old high school graduates who are still living as dependent members of their families (Table 8-6). As you can see, within this population there is a positive relationship ($Q = +.26$) between being white and being enrolled in college. (Note that the association would be considerably stronger if the population was not so narrowly defined by age and educational background.)

Suppose we now ask, "Why are young white high school graduates more likely than comparable blacks to be in college?" One possibility is that white families are generally better off financially than black families, so that, regardless of motivation or ability, blacks will be less likely to go on to college. To explore this idea, we first examine the relationship between race (X) and

Table 8-6 College Enrollment by Race among High School Graduates, 18 to 24 Years Old, Who Are Dependent Family Members, United States, 1982

	Race		
	White	Black	
College	(+)	(−)	All
Yes (+)	475	62	537
No (−)	586	130	716
TOTAL*	1061	192	1253
$Q = +.26$			

SOURCE: Adapted from U.S. Bureau of the Census, 1983, Table 264.
* Numbers are in 10,000s.

family income (Z) and the relationship between income (Z) and college attendance (Y).

In Table 8-7 we can see that there is a strong positive relationship between family income and being white ($Q = +.71$); and in Table 8-8, we can see that there is also a positive relationship ($Q = +.37$) between income and going to college. Thus, among young high school graduates, being white is associated

Table 8-7 Income by Race among High School Graduates, 18 to 24 Years Old, Who Are Dependent Family Members, United States, 1982

	Race		
	White	Black	
Income	(+)	(−)	All
High (+)	634	39	673
Low (−)	427	153	580
TOTAL*	1061	192	1253
$Q = +.71$			

SOURCE: Adapted from U.S. Bureau of the Census, 1983, Table 264.
* Numbers are in 10,000s.

UNDERSTANDING CAUSAL RELATIONSHIPS: INTERVENING VARIABLES / 205

Table 8-8 College Enrollment by Family Income among High School Graduates, 18 to 24 Years Old, Who Are Dependent Family Members, United States, 1982

	Income		
College	High (+)	Low (−)	All
Yes (+)	346	191	537
No (−)	327	389	716
TOTAL*	673	580	1253
Q = +.37			

SOURCE: Adapted from U.S. Bureau of the Census, 1983, Table 264.
* Numbers are in 10,000s.

with relatively high family income, and high income is associated with going to college. This completes the following hypothetical chain.

Race ⟶ Family income ⟶ College enrollment
(X)   (Z)   (Y)

To see if family income acts as an intervening variable between race and college enrollment, we look at the partial association between race and college within the categories of the income variable. In other words, we look to see if race still makes a difference once we look at people who are similar on income. As you can see in Table 8-9, the partials are considerably lower than the zero-order association, telling us that within family income categories, race makes much less of a difference in college enrollment than it appeared to do before.

Since controlling for family income fails to reduce the partial relationship between race and colllege enrollment to zero, the most accurate representation of the relationships among these three variables is shown below.

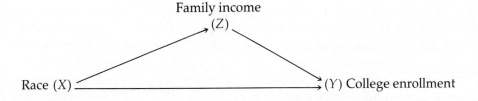

Table 8-9 College Enrollment by Race, Controlling for Family Income, among High School Graduates, 18 to 24 Years Old, Who Are Dependent Family Members, United States, 1982

	Income					
	High (+)			Low (−)		
	Race			Race		
College	White (+)	Black (−)	All	White (+)	Black (−)	All
Yes (+)	328	18	346	147	44	191
No (−)	305	22	327	281	108	389
TOTAL*	633	40	673	428	152	580
	$Q = +.14$			$Q = +.12$		

* Numbers are in 10,000s.

This model indicates that some of the effect of race is direct and that some of it is indirectly exerted through family income. If, of course, we introduced additional control variables, we might find that none of race's effects on college enrollment is direct.

Notice again that these results do not lead us to conclude that a causal interpretation of the relationship between race and college enrollment is spurious, even though the statistical results are the same as they would be in the case of spuriousness. The difference lies in our understanding of the variables and their relationships with one another. Race does affect college enrollment in this case, but indirectly, through differences in family income. The fact that the partials are not zero strongly suggests that race either has direct effects on college enrollment or that there are other intervening variables that we have not controlled for.

Component Variables

Thus far we have discussed two types of control variables, extraneous variables that demonstrate spuriousness and intervening variables that help explain causal mechanisms. In some cases, however, we are interested not only in understanding the relationship between two variables, but the variables themselves. Some variables are complex in that they in fact stand for several different characteristics at once.

Consider, for example, the negative relationship between the number of years of schooling women have (X) and the number of children they bear (Y), a relationship that is found in most societies in which formal schooling exists as a social institution. Given such a relationship, we might then ask, "What is it about schooling that leads to lower fertility? What happens in high school or in college, and how does this affect the number of babies women have?" What we are looking for here is a fuller and more precise understanding of just what it is about schooling that produces this effect. Education reflects many different variables, such as objective knowledge, the ability to reason, values, and aspirations. None of these variables is an intervening variable since each is a *part* of the schooling process. If we imagine schooling as a sort of black box, we know that people who have spent more time in the box differ consistently from those who have spent less time. We want to know what it is about being in the box that makes people different. Until we do, we will not know what "schooling" stands for, and we will not understand how it affects fertility or anything else.

To explore the meaning of schooling, we might control for variables that are a part of the schooling experience. This type of control variable is called a **component** variable because we regard it as a piece of another variable, in this case, schooling. If we assume that having a baby is in part a rational decision, we might test people's abilities to reason and control for that variable. If the partial relationship between fertility and schooling grew weaker, we might then reasonably conclude that to be educated *means* to be more rational, which leads to lower fertility (assuming, of course, that in that particular society having smaller families is in fact the more rational thing to do).

Social class is another variable that is a composite of several variables, including education, occupation, income, and a variety of values and attitudes that tend to go along with them. Many studies show that social class is positively related to high school students' aspirations to achieve academically and go on to higher education even when the level of parental aspiration for these children is controlled (see Boocock, 1972, p. 63). In this case, the independent variable (social class) is a general characteristic that includes many components. In controlling for these component variables (such as family income and parents' occupations and educations) we are trying to find out *what* it is *about* class differences that produces different levels of aspiration. By understanding what the independent variable in a relationship stands for in more precise terms, we then understand the relationship itself more precisely.

The statistical result we expect when we test for a component variable is the same as that when we control for extraneous and intervening variables. If parental occupation (Z), for example, is a component variable of social class (X) that affects children's aspirations (Y), then when we control for parental occupation the partial association between class and children's aspirations should grow weaker. The argument is represented in the following diagram.

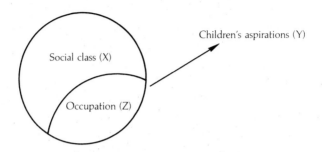

Antecedent Variables

A fourth type of control variable is the **antecedent** variable. In an earlier section, we started with a zero-order relationship between race and college enrollment and tested the idea that family income acts as an intervening variable. Suppose, however, that we had started instead with the positive relationship between income and college enrollment ($Q = +.37$ in Table 8-8). Given the fact that higher family income makes it more likely that children will go on to college, we might then want to work our way *backward* in the causal chain to identify variables that produce differences in family income.

In the diagram for this problem, the question mark represents unknown

$$? \longrightarrow \text{Family income} \longrightarrow \text{College enrollment}$$
$$(Z) \qquad\qquad\qquad (X) \qquad\qquad\qquad (Y)$$

variables that account for differences in income which in turn cause differences in college enrollment. As you might have guessed, an obvious possibility is race. If race has a causal effect on income, then we have

$$\text{Race} \longrightarrow \text{Family income} \longrightarrow \text{College enrollment}$$
$$(Z) \qquad\qquad\qquad (X) \qquad\qquad\qquad (Y)$$

which should look familiar to you. If this argument holds, then if we control for *income*, the partial association between race and college should go to zero (as it almost does in Table 8-9).

In addition, however, if we control for *race*, the relationship between income and college enrollment should *not* go down. If controlling for race weakened the relationship between family income and college enrollment, this would indicate that race was an extraneous variable and that the

Table 8-10 College Enrollment by Family Income, Controlling for Race, among High School Graduates, 18 to 24 Years Old, Who Are Dependent Family Members, United States, 1982

	Race						
	Whites (+)				Blacks (−)		
	Income				Income		
College	High (+)	Low (−)	All		High (+)	Low (−)	All
Yes (+)	328	147	475		18	44	62
No (−)	305	281	586		22	108	130
TOTAL*	633	428	1061		40	152	192
		Q = +.35				Q = +.34	

* Numbers are in 10,000s.

relationship between family income and college enrollment was spurious. If it is family income and not race that directly affects college enrollment, then when we look at people who are similar on race, it should not diminish the effect of income differences on college enrollment. As you can see in Table 8-10, this is precisely what happens. When we control for race, the relationship between income and college enrollment is every bit as strong as it was before (the slight drop from .37 to .35 and .34 is not meaningful).

In this case, race is called an *antecedent* variable to family income because it occurs before income ("ante") in the causal chain. We started with the relationship between income and college enrollment and then added race as a control variable. Earlier, when we started with the relationship between race and college and then controlled for income, income was an intervening variable (Z), race was the independent variable (X), and college was the dependent variable (Y). The use of the terms "antecedent" and "intervening" depends on which pair of variables we start with and what kinds of variables we then add as controls.

Suppressor Variables

With extraneous, intervening, component, and antecedent variables, we started with a zero-order relationship between X and Y and controlled for Z in

Table 8-11   Racial Prejudice by Social Class

| | Social Class | | | |
Prejudice	*Upper*	*Middle*	*Lower*	All
Yes (+)	a	b	c	a + b + c
No (−)	d	e	f	d + e + f
TOTAL	100%	100%	100%	100%

order to test for spuriousness, to explain a relationship by revealing causal mechanisms, to explore the meaning of independent variables, or to trace backwards through a causal chain. In each case, we looked for partial associations to grow weaker after introducing the control variable.

Sometimes, however, we begin by finding *no* relationship between two variables that we expect for theoretical reasons to be related, or we find a relationship that is weaker than we expected. Suppose, for example, that we gather data on social class (X) and racial prejudice against blacks (Y) (see Rosenberg, 1968). From previous research, we might expect to find a negative relationship between class and expressions of prejudice; but suppose we found no relationship. Does this mean that lower- and working-class people are as tolerant as middle- and upper-class people? Or is there in fact a negative relationship that is being hidden by the uncontrolled influence of additional factors?

In searching for clues, consider the cross-tabulation between the variables, using letters instead of percentages in the cells (Table 8-11). With a negative relationship, the percentages in the first row would get higher as we went from upper- to lower-class people, and the percentages in the second row would get smaller as we went from upper- to lower-class people (be sure you see this before going on). If we find *no* relationship, however, this means that $a = b = c$ and $d = e = f$ (be sure you see this, too). This means that either too many upper-class people are prejudiced (making percentage a too large) or that too many lower-class people are not prejudiced (making percentage f too large). In either case, the positive cross-product is made so large that it cancels out the negative cross-product and results in an association of zero.

At this point, the solution to the puzzle may have already occurred to you. What would happen if our sample were taken in an area with a large percentage of blacks? Blacks tend to belong to the lower class much more than whites, and they are certainly less likely than whites to be prejudiced against blacks. If there are a lot of blacks in the sample, and if they fall disproportionately in the lower class, then the lower class is going to have a larger percentage of people unprejudiced against blacks than it would otherwise have. If we remove blacks from the sample by controlling for race (Z), then the percentage in the f cell is going to fall, and the more lower-class blacks there are in the sample, the smaller it will get. As f gets smaller, of course, c is going to increase (since they must add up to 100 percent in the marginals). Then the negative cross-product will be larger than the positive cross-product, and the relationship will emerge as a negative one among whites.

When we control for race, the relationship between social class and racial prejudice against blacks "emerges." In this case, we call the control variable, race, a **suppressor** variable because the failure to control for it had the effect of suppressing the zero-order association, making it appear to be weaker than it was. Race is able to mask the relationship between social class and racial prejudice because of the way it is related to those two variables. Being black is negatively related to social class, but (and this is crucial) *being lower class is positively related to prejudice against blacks among whites, but not related to such prejudice among blacks*. Including blacks in the original sample—combining them with lower-class whites—covers up the tendency of lower-class whites to be more likely than other whites to be prejudiced. The only way to allow the relationship between class and prejudice to emerge is to remove the blacks from the sample by controlling for race.

SUPPRESSOR VARIABLES: AN EXAMPLE WITH DATA

To see how statistics can reveal suppressor variables, consider the relationship between occupation (X) and income (Y) in Table 8-12. As you can see, there is a positive relationship between X and Y, just as we would expect; but the relationship is not very strong.

Table 8-12 Income by Occupation, United States, 1984

	Occupation		
Income	White-Collar (+)	Blue-Collar (−)	All
$15,000 or More (+)	281	132	413
Less Than $15,000 (−)	262	258	520
TOTAL	543	390	933
$Q = +.35$			

SOURCE: Computed from 1984 General Social Survey raw data.

Table 8-13 Occupation by Gender, United States, 1984

	Gender		
Occupation	Women (+)	Men (−)	All
White-Collar (+)	335	208	543
Blue-Collar (−)	143	247	390
TOTAL	478	455	933
$Q = +.47$			

SOURCE: Computed from 1984 General Social Survey raw data.

If we think of the strength of the association ($Q = +.35$) in terms of cross-products, the relationship would be stronger if the positive cross-product was larger or if the negative cross-product was smaller. In looking for a suppressor variable, we need to think of categories of people whose combination of characteristics tends to dilute the association between occupation and income. Put differently, can we think of people who tend to be positive on occupation (white-collar) and yet negative on income?

Women are such a group, for they are heavily concentrated in poorly paid white-collar jobs as secretaries and office clerks. As you can see in Table 8-13, there is a substantial positive relationship between being a woman and having a white-collar job ($Q = +.47$), and yet there is an even stronger *negative* association between being female and having a high income (Table 8-14). Thus women tend to be *positive* on occupation (in terms of being white-collar workers) and yet *negative* on income, which means that women

Table 8-14 Income by Gender, United States, 1984

	Gender		
Income	Women (+)	Men (−)	All
$15,000 or More (+)	139	274	413
Less Than $15,000 (−)	339	181	520
TOTAL	478	455	933
$Q = -.57$			

SOURCE: Computed from 1984 General Social Survey raw data.

Table 8-15 Income by Occupation, Controlling for Gender, United States, 1984

	Gender					
	Women (+)			Men (−)		
Income	White-Collar (+)	Blue-Collar (−)	All	White-Collar (+)	Blue-Collar (−)	All
$15,000 or More (+)	120	19	139	161	113	274
Less Than $15,000 (−)	215	124	339	47	134	181
TOTAL	335	143	478	208	247	455
		$Q = +.57$			$Q = +.60$	

SOURCE: Computed from 1984 General Social Survey raw data.

tend to contribute *dis*cordant pairs to the association between income and occupation.

If our reasoning is correct, if we control for gender the partial relationship between occupation and income should be *stronger* than before. As you can see in Table 8-15, this is exactly what happens. Both among women and among men, the association between occupation and income is considerably stronger than the original value of +.35. Gender acts as a suppressor variable that, when controlled, reveals a stronger relationship between X and Y.

Notice that if we consider the relationships between X, Y, and Z as a set, their *direction* is quite different in this case than it was for all previous types of control variables. In the other cases, if we multiply the signs of the three relationships together, the result is always positive. Either all three relationships are positive, or two are negative and one is positive. With suppressor variables, however, the product of the three signs is *negative*, which is to say, two of the three relationships are positive, and one is negative, or all three are negative. In our example above, occupation and income are positively related, as are gender and occupation; but gender and income are negatively related.

Distorter Variables

In the relationship between class and race, if our sample had been mostly black, the relationship between class and race might have been *positive*, since the lower class would be so loaded with blacks that as a group it would appear less prejudiced than the middle and upper classes. In this case, controlling for race would make the relationship *change direction*, from positive to negative. Here we would call the control variable (Z) a **distorter** variable because the failure to control for it makes the relationship between X and Y appear to be in the wrong direction.

In his classic sociological study of suicide, Emile Durkheim (1897) found a positive relationship between marital status and suicide: suicide rates for married (+) people were higher than those for single (−) people. Rather than simply accepting the findings, Durkheim tested the idea that a third variable—age—was affecting the relationship between marital status and suicide rates. Specifically, married people tend to be older than single people, and age is positively associated with suicide. Thus, although married people in general have higher suicide rates than single people, *among people who are alike on age*, single people have higher suicide rates than married people. Without controlling for age, the relationship between being married and committing suicide is positive; but when age is controlled, the relationship between being married and committing suicide is *negative*.

In this case, age is a distorter variable that causes a relationship to change direction. Notice that the product of the signs of the three relationships is positive; being married is positively related to suicide and to age, and age is positively related to suicide.

The presence of suppressor variables makes relationships appear weaker than they are. The presence of distorter variables makes relationships appear to be in the wrong direction. Controlling for suppressor and distorter variables enables the true magnitude and direction of the relationships between X and Y to appear.

Specification and Interaction Effects

In each of the preceding sections we discussed the use of control variables to test ideas about a relationship between variables. In each case, we looked for partial associations to change uniformly in one direction or another—to grow weaker when we control for extraneous or intervening variables, for example, or to grow stronger when we control for suppressor variables.

In some instances, however, we control for a third variable simply in order to see if the relationship between X and Y differs from one category of Z to another. This procedure is known as **specification**. Blue-collar workers, for example, have been found to be more in favor of racial segregation than white-collar workers. We might be tempted to simply leave it at that and conclude that there is a moderate occupational difference in racial values among U.S. adults.

The problem with "leaving it at that," however, stems from the fact that this simple two-variable relationship represents an enormous variety of people who have been lumped together. We run the risk of wrongly concluding that a relationship found in the U.S. adult population exists in all *subgroups* of that population as well. The relationship between occupation and racial values for the entire population is in fact the sum of an enormous number of potential subtables. We could look, for example, at the relationship between occupation and racial values among blacks, whites, men, women, college graduates, illiterates, married people, widowed people, single people, divorced people, the middle class, the upper class, the working class, the lower class, urban dwellers, rural dwellers, Catholics, Jews, Protestants, Republicans, Democrats, and Independents.

If we specify for the variable "region," for example, and compare the south and the rest of the country, we would find that the relationship between occupation and support for segregation is much stronger in the south than it is outside the south (Table 8-16). In other words, occupation makes a larger difference in the south; whereas outside the south people tend to be opposed to segregation regardless of their occupation. In *neither* region is the relationship the same as it is for the country as a whole. Whenever the strength or direction of a relationship differs from one category of a control variable to another, we call this an **interaction effect** and say that the relationship between X and Y *interacts with* the control variable. In this case,

Table 8-16 Support for Racial Segregation by Occupation and by Region, United States, 1968

	Region					
	South			Nonsouth		
	Occupation			Occupation		
Position on Segregation	White-Collar (+)	Blue-Collar (−)	All	White-Collar (+)	Blue-Collar (−)	All
For (+)	18	47	65	30	36	66
Against (−)	107	60	167	267	189	456
TOTAL	125	107	232	297	225	522
		$Q = -.65$			$Q = -.26$	

SOURCE: Institute for Social Research, 1968.

for example, we would say that the relationship between occupation and racial values interacts with region, which is to say, the relationship differs from one region to another.

If we want to talk about a relationship between two variables, we have to be aware of just what group our data refer to. If the relationship is based on a national sample, we know there are many different kinds of people involved and that the relationship might look quite different from one group to another. If we want to understand fully the relationship between occupation and support for segregation, at some point we would have to explain why the association is stronger in the south.

A second example shows how specification can reveal important insights about how the world works. There is, for example, a moderately strong relationship between education and income ($Q = +.48$) when we dichotomize education as "less than college" and "some college or more" and income as "less than \$15,000" and "\$15,000 or more." That education has a positive effect on income is a finding that should surprise no one; but what happens when we examine this relationship among subgroups of the population? Does it hold equally well, for example, among whites and blacks? To see, we specify the relationship for race in Table 8-17.

As you can see, the relationship between education and income interacts with race: it is considerably stronger among blacks ($Q = +.79$) than it is among whites ($Q = +.42$). Why? To understand the meaning of these results, recall that the stronger a positive relationship is, the greater is the relative size of the positive cross-product. In other words, if the positive relationship is weaker among whites, this is because there are too many discordant pairs among whites. Either there are too many less-educated whites who nonetheless have good incomes, or there are too many highly educated whites with relatively low incomes (or both).

If you look carefully at each cell in the subtable for whites and compare it with the corresponding cell in the subtable for blacks, you will be able to see the answer to our puzzle. Whites with less education are nonetheless more than twice as likely as comparable blacks to have good incomes (35 percent versus 16 percent). This has the effect of greatly increasing the number of discordant pairs (low on education, high on income) among whites and, therefore, the negative cross-product. The result is a much smaller difference between the positive and negative cross-products and a lower value of Q for whites.

In substantive terms, the specification for race tells us that for blacks, it is very hard to get a good income without a high education; but whites are more likely to have a good income regardless of education. Thus, the lower association for whites shows that among whites there are many more

Table 8-17 Income by Education, Controlling for Race, United States, 1984

	Race						
	Blacks				Whites		
	Education				Education		
Income	Less Than College (−)	Some College or More (+)	All		Less Than College (−)	Some College or More (+)	All
Less Than $15,000 (−)	84%	38%	66%		65%	43%	54%
$15,000 or More (+)	16	62	34		35	57	46
TOTAL	100%	100%	100%		100%	100%	100%
(N)*	(79)	(52)	(131)		(415)	(386)	(801)
		Q = +.79				Q = +.42	

SOURCE: Computed from 1984 General Social Survey raw data.
* There are 541 cases missing from the original sample of 1,473 because of no income.

"exceptions to the rule" that income depends on education, while blacks are held more rigidly to the general rule that if you want a good income, you have to get a good education.

Notice that with specification we are not really interested in comparing the partial associations with the zero-order associations. Instead, we want to compare one partial with another partial to see how the relationship changes from one category of the control variable to another. This enables us to answer very different kinds of questions than those we encountered in earlier sections. Specification is thus one more valuable research tool in the search for more detailed understandings of two-variable relationships.

Controls and Small Samples: Running Out of Cases

When analyzing data that involve a relatively small number of cases, we run into problems when we introduce successive control variables. In particular, each control variable breaks the sample into increasingly small subgroups, and we are in danger of not having enough cases to represent such small subgroups accurately.

Suppose, for example, that we have data from 200 respondents. We first examine the relationship between social class and support for Republican presidential candidates. If social class has four categories (lower, working, middle, upper) and support has two (support, oppose), we will have a 4×2 table with 8 cells. If the cases were distributed evenly among the cells, this would result in an average of $200/8 = 25$ cases per cell.

What happens, however, if we control for race? Then the number of cells doubles to 16 (one 4×2 subtable for whites and one for nonwhites), and the average number of cases per cell drops to 12.5. If we then further control for region of the country (with, let us say, 7 categories), then the number of cells jumps to 112 and the average number of cases per cell drops to less than 2. With only 2 control variables we do not have a particularly complex analysis, and yet in this case we are already going to encounter a lot of empty cells simply because our sample is too small to accurately represent these relatively small subgroups (middle-class nonwhites living in the southwest or upper-class whites living in the southeast, for example). Many cells will be empty even though we know there are people who possess those combinations of characteristics.

There are three ways to deal with this problem. The first is to gather large sets of data—to draw larger samples or select larger populations for study. The second is to lower the number of cells in the table by collapsing variables into smaller numbers of categories; but as we have seen earlier, this must be done very carefully so as not to distort the direction of the relationship. The third is to refrain from controlling for additional variables. The best solution is the first, which should be taken into account when selecting a population or deciding how large a sample to select. This kind of planning in the early stages of research allows investigators to avoid either collapsing their data or limiting their analysis later on.

As we have seen many times before, the research process is full of decisions that strongly affect the kinds of results that are possible. Data cannot tell us what to do with themselves. Their analysis depends on common sense, rational thinking, theoretical reasoning, intuition, luck, and the prior knowledge the researcher brings to the problem. It is important to keep in mind that any research effort, including the analysis of the resulting data, is based on a series of decisions—which variables to control, how to interpret the results—and that every decision means that alternatives have been rejected. Only when authors report how such judgments were made are readers in a position to make up their own minds about the meaning and validity of statistical findings.

Summary

1. Control variables are introduced into two-variable relationships for a variety of reasons—to see if causal interpretations are appropriate, to explore causal mechanisms and the meaning of independent variables, to explain unexpected results, and to see if relationships hold under different conditions (see Table 8-18).

2. The order of a relationship refers to the number of control variables that are involved. Thus, zero-order relationships become higher-order relationships when control variables are introduced. In order for controlling to be used, all three variables—X, Y, and Z—must be related to one another.

3. When a causal interpretation is unjustified, it is called spurious. To reveal spuriousness, we control for extraneous variables and look for partial associations that are zero or at least lower than the zero-order association.

4. Intervening variables are those that are caused by X and in turn cause Y. When we control for intervening variables, the partial association between X and Y becomes weaker.

5. A component variable is a variable that is included in X (as the variable "income" is part of the variable "social class") and causes Y. When we control for a component variable, the partial association between X and Y becomes weaker.

6. It is often the case—as with extraneous, intervening, and component variables—that identical statistical results have different interpretations

Table 8-18 Summary of Situations in Which Control Variables Are Used and Interpretation of Results

When a causal interpretation of the relationship between X and Y is spurious

Model: $Z \begin{smallmatrix} \nearrow Y \\ \searrow X \end{smallmatrix}$

Control variable (Z): Extraneous
Direction of relationships: Product of the three relationships
(X & Y, X & Z, Y & Z) is positive (three positives or two negatives and one positive).
Partial association between X & Y when Z is controlled: Weaker

---

When a control variable helps explain how X affects Y

Model: $X \longrightarrow Z \longrightarrow Y$

Control variable (Z): Intervening
Direction of relationships: Product of the three relationships
(X & Y, X & Z, Y & Z) is positive (three positives or two negatives and one positive).
Partial association between X & Y when z is controlled: Weaker

---

When a variable that is included as part of X has an effect on Y

Model:

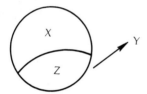

Control variable (Z): Component
Direction of relationships: Product of the three relationships
(X & Y, X & Z, Y & Z) is positive (three positives or two negatives and one positive).
Partial association between X & Y when Z is controlled: Weaker

---

When a variable causes X

Model: $Z \longrightarrow X \longrightarrow Y$

Control variable (Z): Antecedent
Direction of relationships: Product of the three relationships
(X & Y, X & Z, Y & Z) is positive (three positives or two negatives and one positive).
Partial association between X & Y when Z is controlled: No change
Partial association between Z & Y when X is controlled: Weaker

Table 8-18 *(continued)*

When a variable is masking the relationship between X & Y
  Control variable (Z): Supressor
  Direction of relationships: Product of the three relationships (X & Y, X & Z, Y & Z) is negative (three negatives or two positives and one negative).
  Partial association between X & Y when Z is controlled: Stronger

---

When a variable is changing the direction of the relationship between X & Y
  Control variable (Z): Distorter
  Direction of relationships: Product of the three relationships (X & Y, X & Z, Y & Z) is positive (three positives or two negatives and one positive).
  Partial association between X & Y when Z is controlled: Opposite sign

---

When the relationship between X & Y varies among categories of Z
  Control variable (Z): Specification or interaction
  Direction of relationships: No particular combination of directions
  Partial association between X & Y when Z is controlled: No uniform pattern of change. It differs from one category of Z to another.

SOURCE: Adapted from Rosenberg, 1968.

depending on the variables that are involved. Interpretation cannot be based on statistics alone.

7. Antecedent variables have a causal effect on X. When testing for antecedent variables, controlling for Z has no effect on the relationship between X and Y, but controlling for X weakens the relationship between Z and Y.

8. Suppressor variables, when uncontrolled, weaken the relationship between X and Y. Suppressor effects are detected when, by controlling for Z, the relationship between X and Y grows stronger.

9. Distorter variables actually make the relationship between X and Y appear to be in the opposite direction to what it should be. When Z is controlled in this situation, the relationship between X and Y changes direction.

10. Specification refers to the process of seeing how the relationship between X and Y varies from one category of the control variable to another. When this happens, we call Z an interaction variable and say that the relationship between X and Y interacts with Z.

11. Especially in analyses that use cross-tabulations, it is important to be sure to include enough cases so that small subgroups can be adequately represented.

Key Terms

control variable 192
   antecedent variable 208
   component variable 207
   distorter variable 214
   extraneous variable 200
   intervening variable 200
   suppressor variable 211
direct effect 203
indirect effect 202
interaction effect 215
multivariate analysis 192

order of relationship 193
   first-order 193
   third-order 193
   zero-order 193

partial association 200

partial relationship 200

specification 214

spuriousness 194

subtable 196

Problems

1. Define and give an example of each of the key terms listed at the end of the chapter.

2. What are the three major reasons for adding control variables to two-variable relationships?

3. If we have a relationship between X and Y, what does it mean to control for Z? See if you can explain the logic of it to someone else.

4. Explain, with examples, the meaning of the statement, "The same statistical results can have very different interpretations."

5. Using the data in Table 8-19, showing the relationship between education (X) and income (Y), controlling for occupational prestige (Z), do the following:
   a. Rearrange the cells in order to find the zero-order relationship between education and income. Compute a measure of association.
   b. Rearrange the cells in order to find the zero-order relationship between education and occupational prestige. Compute a measure of association (the same as in part a).
   c. Rearrange the cells in order to find the zero-order relationship between income and occupational prestige. Compute a measure of association (the same as in part a).

Table 8-19 Income by Education, Controlling for Occupational Prestige, United States, 1984

	Occupational Prestige					
	Low(−)			High(+)		
	Education			Education		
Income	Less Than College (−)	Some College or More (+)	All	Less Than College (−)	Some College or More (+)	All
Low (−)	141	30	171	194	155	349
High (+)	29	11	40	130	242	372
TOTAL	170	41	211	324	397	721

SOURCE: Computed from 1984 General Social Survey raw data.

d. In this case, what kind of control variable is occupational prestige? What does the model look like?

e. Compute the partial association between income (Y) and education (X) controlling for occupational prestige (Z). Interpret your results.

f. If Z was an intervening variable, how would the results have compared with those in part e? Why?

g. If Z was an antecedent variable, how would the results have compared with those in part e? Why?

h. If Z was an extraneous variable, how would the results have compared with those in part e? Why?

i. If Z was a suppressor variable, how would the results have compared with those in part e? Why?

j. If Z was a distorter variable, how would the results have compared with those in part e? Why?

k. If Z was a component variable, how would the results have compared with those in part e? Why?

l. If Z was an antecedent variable, what would happen to the partial association between Z and Y if we controlled for X?

m. If Z was an interaction variable, how might the results have compared with those in part e?

6. Show that Table 8-9 comes from Table 8-6.

7. Rearrange the cells in Table 8-16 to find:
   a. The zero-order relationship between region and position on segregation.
   b. The zero-order relationship between occupation and position on segregation.
   c. The zero-order relationship between occupation and region.
8. Why do small samples make it difficult to use control variables? Be specific and give an example.
9. "People who are in hospitals are more likely to die than are people who are not in hospitals. There is thus a positive relationship between hospitalization (X) and death (Y)." Does this mean that people should stay away from hospitals?
   a. How would you use a control variable to explore the relationship between X and Y?
   b. What kind of variable (intervening, suppressor, etc.) would you control for?
   c. What idea would you be testing about the relationship between X and Y?
   d. What kinds of statistical results might you get, and how would you interpret them?
10. In general, time spent studying (X) is positively related to grades (Y). However, the relationship between studying time and grades depends on when you study (Z). The same amount of studying will produce better results if you space it out over several days before the exam than it will if you do it all at the end and study right up to the moment of the exam.
    a. What kind of control variable is Z?
    b. What should happen to the partial relationship between X and Y when you control for Z?
11. Suppose we find a negative relationship between social class (X) and the tendency to vote for Democratic candidates (Y). We want to discover what it is about social class that accounts for such differences in voting behavior.
    a. Suppose we control for income (Z). What kind of control variable is this?
    b. What kinds of statistical results might you get when you control for Z, and how would you interpret them?
12. We find a positive relationship between juvenile delinquency (Y) and feelings of alienation (X). We suspect that extreme poverty during childhood (Z) is a source of alienation during adolescence.

a. What kind of control variable is Z?
b. What kinds of statistical results might you get when you control for Z, and how would you interpret them?

13. We find a relationship between X and Y ($Q = -.20$). When we control for Z, the partial association between X and Y is $Q = -.68$. What kind of control variable is Z?

14. We find a relationship between X and Y ($Q = -.54$). When we control for Z, the partial association between X and Y is $Q = +.22$. What kind of control variable is Z?

9

Multivariate Analysis

When we include more than one independent variable in trying to predict or explain variation in a dependent variable, we are using **multivariate analysis**. Technically, the material in Chapter 8 deals with multivariate analysis, since controlling involves the addition of one or more variables to an original zero-order relationship. While Chapters 8 and 9 have a great deal in common, the material here is enough of an extension beyond the previous chapter to call for a separate treatment.

This chapter discusses some major techniques that allow us to deal with complex causal models. For the most part, we will confine ourselves to techniques applicable to interval- and ratio-scale variables only. The mathematics in this chapter is the most advanced thus far, but even here I have tried to include only that which is necessary to understand the logic and results of multivariate analytic techniques. This is the kind of chapter you should first read through quickly and then go back through for a careful study.

We use multivariate analysis to accomplish several kinds of tasks. First, it allows us to use several independent variables to explain variation in a dependent variable. Second, it allows us to use many independent variables to predict the value of a dependent variable. Third, it allows us to determine which independent variables have the greatest impact on a dependent variable (for example, which is the most important factor in getting a job—skill, intelligence, race, sex, age, or educational degrees?). Fourth, it allows us to specify the relationships among the *independent* variables in order to discover the mechanisms by which they relate causally to the dependent variable. In short, multivariate analysis represents a sophisticated statistical approach to the problems of scientific explanation and prediction.

Multiple Regression and Correlation

In Chapter 7 we discussed simple linear regression (one independent and one dependent variable). **Multiple regression** is an extension of two-variable regression. The underlying logic is the same with some embellishments. You might want to return to the discussion in Chapter 7 and refresh yourself before proceeding with multiple regression.

It is often the case that we want to see how well a set of independent variables can jointly predict or explain a dependent variable. For example, we might want to see how well we can explain income by using education, occupational status, race, family background, and intelligence-test scores, *all working together*. This means that knowing people's educational attainment helps us predict their income; if, in addition, we know their occupation, this improves our accuracy of prediction because we know more about characteristics that are related to income.

However, if education, all by itself, explains 25 percent of the variation in income (that is, $r^2 = .25$) and occupation explains 20 percent, this does *not* mean that knowing educational attainment *and* occupation will explain 45 percent (that is, 25% + 20%) of the variation in income. Why? Remember that the independent variables provide us with information that helps us predict the dependent variable. If I know your education, I *automatically* know something about your occupation as well, and vice versa. If you have a Ph.D., you are most likely not a garbage collector, and if you have only a grade school education, you are certainly not a nuclear physicist. In other words, the two independent variables are related *to each other*. This "overlap" means that the information given by knowing both is less than the *sum* of the information conveyed by each alone.

The general formula for a multiple regression is

$$\hat{Y} = a + b_1 X_1 + b_2 X_2 + b_3 X_3 + b_4 X_4 + \cdots$$

where each X represents an independent variable.

The slopes associated with each independent variable are somewhat different from those in linear regression. In multiple regression, each slope represents the impact of the independent variable *while all other variables in the equation are held constant*. It is as if we looked at the two-variable relationship between, say, X_1 and Y among people who are identical on all the other independent variables. This tells us the impact that each independent variable has *over and above the effects of the remaining independent variables*. In our above example, the partial slope for occupation tells us what information about income we get from knowing occupation *in addition* to what education has already told us. These slopes are in the units of the independent and dependent variables (for example, change in dollar income per year of additional schooling) and are called **partial slopes** (or **partial regression coefficients**).

As with slopes, we can also compute **partial correlation coefficients**. These tell us how accurately we can predict Y from knowing the value of an independent variable (say, X_1) while holding the other variables in the equation constant. We can also compute a **multiple correlation coefficient** (R), which indicates how accurately we can predict Y by using all the independent variables at once. R^2 has the same interpretation as r^2, except that it applies to the explanation of Y by more than one independent variable.

Both partial slopes and correlations are used in testing models such as the simple ones we mentioned in Chapter 8. We could look at the relationship between income and education and ask, "Does education directly affect how much money one makes, or does it indirectly affect income through its effects on occupation?" In other words, once we *control* for occupation (examine the relationship between income and education among those who have similar jobs), does education still have a relationship with income? To answer this question, we perform a multiple regression analysis with income as the dependent variable and with education and occupational status as the independent variables:

$$\hat{Y} = a + b_1 X_1 + b_2 X_2$$

where X_1 is years of school completed and X_2 is an occupational status score.

If the effects of education on income are indirect, then once we hold occupation constant, education's impact should be zero (or close to zero). This means that b_1 should be zero in the above equation. On the other hand, if education has effects on income *independent* of those of occupation, b_1 will not be zero. We can perform the same kind of analysis using partial correlation coefficients, although, as I have mentioned, the interpretation is in terms of prediction accuracy, not impact.

Multiple regression can be used also for prediction and estimation. For example, the U.S. Bureau of the Census tries to estimate the population of states and counties during years that fall between decennial censuses. One of the methods used is a multiple regression equation which estimates population (the dependent variable Y) with four independent variables (X_1 to X_4). The equation developed for New York State counties and metropolitan areas looks like this:

$$\hat{Y} = .0203 + .1847X_1 + .2283X_2 + .1528X_3 + .4406X_4$$

where X_1 = elementary school enrollment for grades 1 through 8 plus elementary special and elementary ungraded
X_2 = the 2-year average of resident births
X_3 = the 2-year average of resident deaths
X_4 = the number of registered automobiles

Note that in this case the substantive or causal connection between the dependent and independent variables is irrelevant to the task, and there is no interest in determining which independent variables have the greatest effects on population; the sole purpose is to predict population size accurately.

Where do such equations come from? When the Census Bureau conducts a census, there is a population count for each county in the state of New York (in 1980, for example), as well as records on births, deaths, school enrollments, and automobile registrations during the census year. Using the county as a unit of analysis, researchers perform a multiple regression analysis, yielding equations like the above. To use it in succeeding years, they must assume that the pattern of relationships among those variables will remain the same throughout the decade (U.S. Bureau of the Census, 1986).

STANDARDIZED SLOPES

A problem arises in comparing the slopes of independent variables in a multiple regression analysis. In one of the examples above, we predicted income with education and occupational status. Suppose X_1 is education and X_2 is the occupational status score, and the partial slope for education is greater than for status. Our first impulse might be to conclude that education has a greater independent impact on income than occupational status has, but what do the results really tell us?

The partial slope for education (b_1) tells us how many dollars are associated with an increase of one year of schooling; the comparable slope for occupation tells us how many dollars are associated with each increase of one point on the occupational status scale. However, it is not fair to compare the impact of one year of schooling with that of one point on a rating scale; since the units are different, the slopes cannot be compared. How many status points equal one year of schooling? It is like trying to translate from apples to oranges.

We have encountered this problem in a variety of forms already. In frequency distributions we found that we could not make comparisons if the totals were not the same; the same was true of means and variances. In each case we converted the data to some common denominator. By making the units *standard*, we made meaningful comparisons possible.

In the case of partial slopes, we make the following kind of conversion. If one year of schooling (independent of occupation) means $500 in additional income, the **unstandardized** slope is $b_1 = \$500/\text{year}$. If we divide $500 by the size of the standard deviation for income (s_Y) and one year by the standard deviation of years of schooling (s_{X_1}), we have the change in income expressed in standard deviations associated with each increase of one standard deviation of years of schooling.

$$\frac{\$500/s_Y}{1 \text{ year}/s_{X_1}} = \beta_1 = b_1 \left(\frac{s_{X_1}}{s_Y}\right)$$

We have converted both numerator and denominator of the partial slope from the original units (dollars and years of schooling) to numbers of standard deviations, by dividing each variable in the equation by *its own* standard deviation. The symbol for the **standardized partial slope** is the Greek letter *beta* (β).

For example, if the standard deviation of income is $6,000, then the numerator of the partial slope is $500/6,000 = 1/12$. If the standard deviation of years of schooling is four years, then the denominator becomes 1/4. The *standardized* partial slope is then

$$\frac{\frac{1}{12}}{\frac{1}{4}} = \frac{1}{3} = .33$$

This tells us that, with occupational status held constant, for every increase of one standard deviation in educational attainment there is an increase of .33 standard deviation of income. If we do the same thing for occupational status, and get a partial slope of, say, .44, then we know that occupational status produces a greater change in income than educational attainment does. When we translate the partial slopes into this form, we can compare them without worrying about the fact that education scores may run from 0 through 16 and occupational status may run from 1 through 100.

These special partial slopes go by several names: **standardized slopes**, **standardized regression coefficients**, or **beta weights**. All three terms refer to the same thing.

AN EXAMPLE OF MULTIPLE REGRESSION ANALYSIS IN ACTION

The combination of interesting and important problems with sophisticated, properly used statistical techniques is a social scientist's dream, and not encountered as often as one would like. Otis Dudley Duncan (1969) has produced a small masterpiece, "Inheritance of Poverty or Inheritance of Race?" in which he uses multiple regression analysis to study the problem of income diffferences between American blacks and whites. In 1962 the differences between black and white mean incomes amounted to $3,790. He is concerned with the commonly held position, among those interested in racial problems, that poverty breeds poverty, that blacks are poor because their parents were poor and disadvantaged and passed on their poverty to their sons via poor educations, occupations, and mental abilities. Duncan suggests that the inheritance of disadvantaged backgrounds does not tell the whole story, that the inheritance of race is very important as well. In other words, if blacks had the same backgrounds as whites, they would still make less money because they are black.

Duncan uses data from the 1962 study of Occupational Changes in a Generation. The respondents used by Duncan were restricted to native-born men 25 to 64 years old with nonfarm backgrounds. The study is based on a national probability sample.

Duncan performs a multiple regression analysis using the respondent's income as the dependent variable and five background characteristics as independent variables (shown below with the symbols we will use later in the actual regression equations):

E_f = respondent's fathers's education

O_f = respondent's father's occupation

S = number of siblings in respondent's family of origin

E_r = respondent's education

O_r = respondent's occupational prestige

I = respondent's income for 1961

Education is measured on a scale with eight categories ranging from "none" to five or more years of college. Occupational prestige is measured with the Duncan socioeconomic index of occupational status which indicates the percentage of people who would rate an occupation's prestige as "good" or "excellent." Income is measured in thousands of dollars.

Before you go on, familiarize yourself with the variables involved and their symbols. Read the descriptions of the variables and be sure you understand what they mean. We are going to use these symbols in a pair of multiple regression equations which are intimidating enough without your being thrown by the meaning of the variable symbols. Take your time.

As we saw in Chapter 8, it makes little sense to introduce control variables unless they are all related to the independent variable. In addition, we are also interested in the associations among the independent variables. To make this clearer in multiple regression analyses, researchers often begin by looking at the correlations for every pair of variables in an analysis. This is presented in what is known as a **correlation matrix**. Table 9-1a shows the matrix for the whites in Duncan's sample, and Table 9-1b shows the corresponding matrix for blacks.

It is easiest to read a matrix by going systematically across the rows or down the columns. The first row, for example, shows the correlations between father's education (E_f) and each of the other variables. The correlation

Table 9-1a Correlation Matrix for Six Variables in Duncan's Analysis, Whites Only

Variable	E_f	O_f	S	E_r	O_r	I
Father's Education (E_f)	....	.40	−.06	.36	.22	.14
Father's Occupation (O_f)		....	−.11	.21	.12	.04
Number of Siblings (S)			....	−.17	−.09	−.09
Respondent's Education (E_r)				...	.41	.30
Respondent's Occupation (O_r)					...	.27

SOURCE: Duncan, 1969, Table 4-2.

Table 9-1b Correlation Matrix for Six Variables in Duncan's Analysis, Blacks Only

Variable	E_f	O_f	S	E_r	O_r	I
Father's Education (E_f)	....	.51	−.27	.39	.31	.17
Father's Occupation (O_f)		....	−.24	.42	.37	.24
Number of Siblings (S)			....	−.33	−.25	−.14
Respondent's Education (E_r)				...	.61	.31
Respondent's Occupation (O_r)					...	.39

SOURCE: Duncan, 1969, Table 4-2.

between father's education and father's occupational status (O_f) is .40, the correlation between father's education and the number of siblings (S) in the respondent's family is $-.06$, and so on. Notice that among blacks, the corresponding correlations are often quite different (take a few moments to go through the two matrices and compare each pair of correlations).

Duncan performs the analysis separately for blacks and whites (the white category includes the small number of nonwhites who are not blacks). The form of the two equations is as follows:

$$\text{Whites: } \hat{I} = a + b_1 E_f + b_2 O_f + b_3 S + b_4 E_r + b_5 O_r$$

$$\text{Blacks: } \hat{I} = a + b_1 E_f + b_2 O_f + b_3 S + b_4 E_r + b_5 O_r$$

Before we go on, stop to look at these equations. Both equations include the same variables, but for two different categories of people (whites and blacks). Predicted income ($\hat{I}$) is the dependent variable. The a is the regression constant: when all the independent variables have a value of zero, income equals the value of a.

The bs (b_1 through b_5) are the unstandardized partial regression coefficients (slopes) for each independent variable. These tell us what the impact of each independent variable is on income *independent of the effects of the other independent variables*. If we had wanted to see which variables had the greatest independent effects on income, we would have used standardized regression coefficients, and the equations would look like this:

$$\text{Whites: } \hat{I} = a + \beta_1 E_f + \beta_2 O_f + \beta_3 S + \beta_4 E_r + \beta_5 O_r$$

$$\text{Blacks: } \hat{I} = a + \beta_1 E_f + \beta_2 O_f + \beta_3 S + \beta_4 E_r + \beta_5 O_r$$

All these equations represent the same set of relationships; the respondent's income is seen as the product of five factors: his father's education and occupation, the size of his family of origin (number of siblings), and the respondent's own education and occupational status. The equations allow us to deal with several kinds of problems, but before you go on, be sure you feel comfortable with them. Be sure you can start at the income end of the equation and work your way through, explaining each term's meaning as well as that of the equation as a whole.

When Duncan computed the actual values of the unstandardized partial regression coefficients, he got the following results:

$$\text{Whites: } \hat{I} = 1.90 - .008 E_f + .022 O_f - .016 S + .299 E_r + .071 O_r$$

$$\text{Blacks: } \hat{I} = 2.08 + .043 E_f - .005 O_f - .025 S + .249 E_r + .021 O_r$$

If you study these equations (please do), something may seem strange. The impact of the number of siblings is negative for both whites and blacks ($-.016$ and $-.025$), reflecting the negative impact of large family size on both educational attainment and occupation (see Matras, 1984). However, the partial slope for the respondent's father's education is *negative* for whites ($-.008$), which says that the higher the father's education is, the lower the son's income is; this does not make any sense. The catch to this will be explained in material we will cover later, but crudely put, this partial regression coefficient is *not statistically significant from zero*. This means that the slope of $-.008$ represents random error that comes from the fact that we are drawing a sample and not conducting a census. It could just as well have been $+.008$ in another sample. Do not worry about this just now. You might want to return to this after studying Chapter 13.

In most research reports that use regression analysis, the partial regression coefficients are presented in a form such as Table 9-2. Note the footnote in the table, which tells us we can*not* be confident that the starred partial slopes are *not* zero in the population. Thus, Duncan's first finding is that a number of factors thought to have direct effects on income apparently do not: among blacks, father's education and occupation as well as the size of the family do not have significant *direct* effects on the son's income; among whites, the same is true for the father's education and the size of the family. Remember

Table 9-2 Partial Regression Coefficients for the Regression of Respondent's Income on a Series of Stratification Variables, for Native Men 25 to 64 Years Old with Nonfarm Backgrounds and in the Experienced Civilian Labor Force, by Race, March 1962

	Partial Regression Coefficients	
Independent Variable	Blacks	Whites
Father's Education (E_f)	.043*	−.008*
Father's Occupation (O_f)	−.005*	.022
Number of Siblings (S)	−.025*	−.016*
Respondent's Education (E_r)	.249	.299
Respondent's Occupation (O_r)	.021	.071
Regression Constant	2.08	1.90

* Coefficient less than its estimated standard error in absolute value.

that these coefficients refer to *direct* effects (not through an intervening variable), which is not to say that these variables have nothing to do with income. We will see more about this later on.

Using Duncan's Equations Duncan approaches the problem of "inheritance of poverty or inheritance of race" by using the entire regression equations in an ingenious and revealing way. Remember that the perspective of regression analysis is one of prediction: if we plug in the values for the five independent variables for a respondent, the resulting value of income is the best available prediction of what his income actually is. If we plug in the *means* for each of the five independent variables (for whites, say), we get the mean income for that group. Thus, the mean income for whites is a direct function of the means for these five variables multiplied by the appropriate regression coefficients. The same is true for the black equation.

The inheritance-of-poverty (versus inheritance-of-race) argument suggests that the gap between white and black incomes is a function of background differences. In other words, black men make less money because their fathers had inferior educations and occupations, they came from larger families, and they themselves have inferior educations and occupations. To test this idea, Duncan formulates the problem in terms of the question: "If black men had the same background characteristics as whites, would they then have the same incomes?" If, in fact, the income gap is caused by inferior backgrounds (as the inheritance-of-poverty argument suggests), then equalizing backgrounds should equalize incomes.

Duncan tests this idea by plugging the *white* means for the five independent variables into the equation for *blacks*. Be careful to note his intention: he is keeping the white means as they are, but subjecting whites to the same set of partial regression coefficients as apply to blacks. Remember what the regression coefficients tell us: how much impact does a variable have on income? How much of a payoff is associated with each additional year of school, each additional point of occupational status, and so on? Thus, the whites still have the same level of attainment (both of fathers and of respondents), but now they must, in Duncan's words, play by the rules that apply to blacks.

Keep in mind that a regression equation has two basic kinds of numbers: scores on variables and a set of regression coefficients. In this case, the scores tell us how privileged one's background is; the set of regression coefficients tell us how effectively that background is converted into income. The inheritance-of-poverty argument suggests that the income gap between whites and blacks is a function of the former; the inheritance-of-race argument suggests that the latter set of factors is involved; it is not enough for a black man to have a socioeconomic background equal to that of whites, because being black is a liability in itself. In blunter terms, blacks with high qualifications are discriminated against because of their race.

When Duncan inserted the whites means into the black regression equation, the resulting predicted mean income for whites was lowered by $2,360, closing the gap between blacks and whites by that amount. This leaves $1,430 still unaccounted for. Thus, if blacks and whites were identical on these five important background characteristics, there would still be a gap of $1,430 in mean incomes. In Duncan's words:

> ... there remains the sum of $1,430 not yet accounted for. This is about three-eighths of the total gap of $3,790. Unless and until we can find other explanations for it, it must stand as an estimate of income discrimination. ... Specifically, it is the difference between Negro and white incomes that cannot be attributed to differential occupational levels, differential educational attainment ... differences in size of family of origin, or differences in the socio-economic status thereof.... (Duncan, 1969, p. 100)

Duncan has succeeded in applying sophisticated techniques to a problem of more than academic interest. It is of interest to those in government who make policy decisions vital to the socioeconomic advancement of blacks. Duncan's findings strongly suggest that the elimination of disadvantaged backgrounds is not a sufficient cure for poverty, that racial discrimination plays a powerful role in the transfer of poverty from generation to generation. He suggests that

> if there were remedies for all these forms of discrimination, so that only the handicap of family background remained, that handicap would be materially diminished in the next generation. It would be further attenuated in successive generations under these ideal conditions, and while some persisting differential in achievement due solely to initial background handicaps would be observed for several decades, it would tend to disappear of its own accord. (p. 102)

This is a model application of sophisticated analysis performed on sample data.

Inside the Equations Duncan draws an additional picture with the results of his multiple regression analysis, which we should look at both for its substantive interest and the practice it will give you in understanding the results of such analyses. He examines several regressions in order to pick apart the transfer of background advantages (and disadvantages) from generation to generation.

He starts by examining the effects of the respondent's family characteristics on the respondent's own educational achievement, both for blacks and for whites. Figure 9-1 is a picture of two regression equations, one for blacks and one for whites, using *standardized* regression coefficients (beta weights). The straight arrows indicate a causal link with the head of the arrow at the

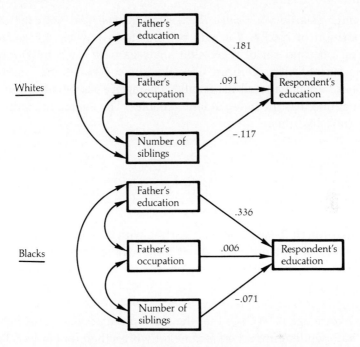

Figure 9-1 Regression of respondent's education on selected background characteristics by race.

dependent variable. The curved arrows show that there is a correlation between the two connected variables, but no causal link is being asserted in this analysis. If we put these diagrams in the form of equations, they would look like this:

$$\text{Whites: } \hat{E}_r = 4.04 + .181E_f + .091O_f - .117S$$
$$\text{Blacks: } \hat{E}_r = 3.07 + .336E_f + .006O_f - .071S$$

where $\hat{E}_r$ is the respondent's predicted education (the dependent variable), E_f is the respondent's father's education, O_f is the father's occupational status, and S is the number of siblings. (Compare these equations with Figure 9-1 and be sure you see that they describe the same thing.) The important thing to notice is that the impact of father's education on son's education is stronger for blacks than for whites (.336 versus .181), as reflected in the standardized regression coefficients. Thus, the regression analysis tells us that blacks are more successful than whites in passing on their educational level to their sons. This of course means that *low* achievement levels are passed on more consistently, too; nonetheless, if a black father gets a lot of schooling, he is more successful than a comparable white father in passing it on to his son.

This might startle you, since this seems inconsistent with the disadvantaged position of blacks. You will be less startled when you examine the results of a second set of regression analyses performed by Duncan.

In the second step, Duncan uses the son's occupational status as the dependent variable and all his family's characteristics plus the son's educational attainment as the independent variables. The equations, for blacks and whites, look like this:

Whites: $\hat{O}_r = .73 + .256E_f + .138O_f - .365S + 7.964E_r$

Blacks: $\hat{O}_r = 4.72 + .811E_f + .005O_f - .105S + 3.653E_r$

Again, go through these equations, starting with the dependent variable, and be sure you understand the parts and the whole before going on.

Take a look at the regression coefficients for the respondent's education. Clearly, the impact of additional years of schooling on occupational status is much greater for whites than for blacks (7.964 versus 3.653). Now look at Table 9-2 once again. We see that the son's occupational status has a much greater independent impact on income for whites than for blacks (.071 versus .021). The story is now complete. While black fathers have more success in passing their educational achievements on to their sons, the sons of white fathers are more able to convert the educational advantage into an occupational advantage and, in turn, the occupational advantage into a higher income. Thus, blacks are in a frustrating position: education does not pay off for them in better jobs to the extent it does for whites; and better jobs do not pay off in money terms as much as they do for whites. In each case, comparing the regression coefficients of blacks and whites allows us to compare the impacts of different advantages on such things as occupational prestige and income.

We have spent considerable space on this series of examples, but I hope you begin to see how much we can learn by carefully examining the results of complex and sophisticated statistical analysis. It allows us to address questions that are often fascinating and of considerable practical importance.

Path Analysis

Path analysis is a technique for evaluating entire causal models. It utilizes multiple regression techniques, but the results are presented in a way different from that discussed earlier in this chapter. Duncan, in his article on race and poverty, also uses path analysis. Recall that he was interested in the effects that father's education and occupation, the number of siblings in the respondent's family, and the respondent's education and occupation had on the respondent's income. You will also recall that Duncan performed several multiple regressions using these factors as independent variables, and that he performed separate analyses for blacks and for whites. If we drew a picture of the black regression equations, we would have the information shown in Figure 9-2.

Path analysis imposes a number of requirements on the relationships between the included variables. First, all the causal relationships work in one direction only (for example, father's education causes son's education, but son's education does not cause father's education), which is symbolized by a one-way arrow going from the cause to the effect; another way of putting this is that all relationships must be asymmetrical or **simple recursive** (but see Lieberson, 1985). Strictly speaking, path analysis does permit causation to run in both directions (racist values cause racist behavior which in turn reinforces racist values); but the calculation of coefficients is messy enough to justify sticking to simple recursive path analysis models here. Second, since path models are causal models, all the variables must have a definite time-ordering. Third, all variables that are not affected by other variables in

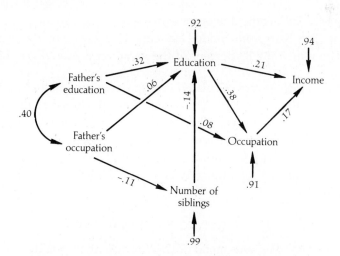

Figure 9-2 Path diagram representing a model of the socioeconomic life cycle in the black population, with path coefficients estimated for native men 25 to 64 years old with nonfarm backgrounds and in the experienced civilian labor force, March 1962.

SOURCE: Duncan, 1969, p. 90.

the model (father's education and occupation) are called **exogenous** variables and are joined by a curved arrow that represents a relationship that is not considered causal. In fact, the causal nature of the relationship is considered to be irrelevant for our purposes here.

There are two kinds of numbers represented in a path-analysis diagram. First, there are the numbers associated with the straight connecting arrows. These numbers are standardized regression coefficients (beta weights) and tell us how great an impact each variable has on another variable *independent of the other variables in the model*. Since the coefficients are standardized, we can compare their magnitudes: for example, among blacks, son's education has a greater impact on son's income (.21) than son's occupation (.17) has. These standardized regression coefficients, when used in a path analysis, are called **path coefficients**. (Be sure to look at the diagram and find the above numbers.)

You'll also notice that for four of the variables (all but father's education and father's occupation), there are arrows coming out of nowhere, with numbers attached. These numbers are coefficients of alienation (go back to Chapter 7 and refresh your memory if you need to) and reflect the amount of variation in these variables left *un*explained by the variables in the model (it is the square root of the unexplained variance). For example, the coefficient of alienation for income is .94. If we square this number, we get the proportion of the variation in income that is *not* explained by the variables in the model, or $(.94)^2 = .88$. Thus, the model explains 12 percent of the variation in income. The remaining 88 percent of the variation is due to variables *outside* the model.

The path coefficients are calculated through a *series* of multiple regressions. We begin with the independent variables that are closest to the dependent variable and work backwards. For example, we first do the regression of income on son's occupation and get the path coefficient (.17). We then perform a multiple regression of income on both son's occupation and son's education. This tells us how much of the effect of education is direct and how much is indirect.

INTERPRETING PATH DIAGRAMS

What are path diagrams good for? What do they tell us that a single multiple regression equation does not? First, multiple regression determines the direct (independent) effects of each independent variable on the dependent variable, as measured by the beta weights. Path analysis, however, allows us to determine the magnitude of the *indirect* effects as well. We do this by simply multiplying paths together. For example, education's direct effect on income is .21. In addition, education affects income indirectly through occupation. The path from education to occupation has a coefficient of .38, and that from occupation to income has a coefficient of .17. The indirect effect of education through occupation on income is measured by the *product* of these two paths, or (.38)(.17) = .06.

The effects of father's education on son's income provide a second example. Father's education has no direct impact on son's income (reflected by the absence of an arrow connecting the two), but this doesn't mean that father's education is unimportant. It *indirectly* affects son's income in three ways: through the son's education, (.32)(.21); through the son's education and then the son's occupation, (.32)(.38)(.17); and through the son's occupation, (.08)(.17). The sum of these compound paths is .10. Using path diagrams, we can see that while father's education has no direct effects on son's income (something that multiple regression equations would have told us—the beta weight would have been close to zero), it certainly plays an indirect role in determining the son's income (a finding that a multiple regression equation would not have told us).

Path analysis also allows us to spell out the relationships among the independent variables, something we cannot accomplish with a single multiple regression equation. For example, in Figure 9-2, we see that most of the impact of father's education on the son's income is exerted through the son's education [compound path of (.32)(.21) = .07], and then the son's occupation [compound path of (.32)(.38)(.17) = .02]; very little influence is exerted directly through the son's occupation [compound path of (.08)(.17) = .01]. Interestingly, father's occupation does not work directly through the son's occupation (there is no arrow between father's and son's occupation) but only through the son's education and *then* through the son's occupation [compound paths of (.06)(.21) = .01 and (.06)(.38)(.17) = .004] as well as a minor series of complex routings through number of siblings, son's education, and son's occupation. (You should go through the above paragraph, referring to the path diagram in Figure 9-2, and satisfy yourself that you understand where the compound paths come from and what they stand for.)

Path analysis is an important theoretical tool because it forces the researcher to specify explicitly *all* the relationships in a causal model. Multiple regression analysis by itself would describe only the relationships between the independent variables and the dependent variable. By working through the model with a series of multiple regressions, we see not only direct effects of independent variables on the dependent variable, but indirect effects and the patterns of interrelationships among all variables in the model as well. Because of this, the use of path analysis is growing as social scientists move toward more comprehensive explanatory models.

Summary

1. Multivariate analysis refers to relationships between dependent variables and entire sets of independent variables simultaneously. It allows us to explain variation in dependent variables with several variables at once, including the relative contributions of each variable independent of all the others. Using more than one independent variable also allows us to predict values for the dependent variable with greater accuracy.

2. In multiple regression, each slope is a partial slope that indicates the effect each independent variable has on the dependent variable over and above the effects of the other independent variables. Partial correlation coefficients have a similar interpretation except in terms of the strength of the relationship.

3. Standardized slopes are used to compare the relative magnitude of slopes for the independent variables in a multiple regression analysis.

4. Duncan's multiple regression analysis of race differences in income shows how multiple regression can be used to isolate the different factors that account for differences between groups. It also shows how the effects of different factors such as family background affect people differently depending on their race.

5. Unlike multiple regression, path analysis uses a series of multiple regressions that, together, enable us to measure both direct and indirect effects of each independent variable on the dependent variable, as well as to describe the relationships among the independent variables. Path analysis is an important analytical tool that forces researchers to specify all of the relationships in a causal model.

Key Terms

beta weight 230

correlation matrix 232

exogenous variable 240

multiple correlation coefficient 228

multiple regression 227

multivariate analysis 226

partial correlation coefficient 228

partial regression coefficient 228

partial slope 228

path analysis 238

path coefficient 240

simple recursive relationship 239

standardized partial slope (or regression coefficient) 230

unstandardized partial slope 230

Problems

1. Define and give an example of each of the key terms listed at the end of the chapter.

2. In an introductory sociology class, student semester grades (Y) are based on four exams (X_1, X_2, X_3, X_4), a final exam, and several short written assignments. Omitting the short written assignments and the final exam, the multiple regression of semester grades on the exam results is

$$\hat{Y} = a + .12X_1 + .12X_2 + .24X_3 + .31X_4$$
$$R^2 = .96$$

Variable	Standard deviation	Mean
Y	13	74
X_1	17	63
X_2	14	80
X_3	20	59
X_4	11	72

Given all of the above information, answer the following:
a. Complete the multiple regression equation above by computing the regression constant.
b. What does the equation as a whole mean?
c. What does each number and symbol in the equation stand for?
d. How well do the exam results explain variance in semester grades? Why is R^2 less than 1.00?
e. Compute the beta weights for each independent variable. What do they mean?
f. Which exam score has the greatest independent impact on semester grades? What evidence do you rely on for this answer?
g. Which exam score has the smallest independent impact on semester grades? What evidence do you rely on for this answer?
h. What would you predict as a semester grade for someone who got the following exam scores?

$$\text{1st exam} = 65$$
$$\text{2nd exam} = 74$$
$$\text{3rd exam} = 88$$
$$\text{final exam} = 92$$

3. In the 1984 GENSOC survey, respondents were asked to report their own years of schooling (Y) as well as that of their mothers (X_1) and fathers (X_2). The multiple regression of respondent's education on parents' education is

$$\hat{Y} = a + .26X_1 + .21X_2$$
$$R^2 = .28$$

Variable	Standard deviation	Mean
Y	3.43 years	12.4
X_1	4.24 years	10.3
X_2	3.18 years	10.1

Given all of the above information, answer the following:
   a. Complete the multiple regression equation above by computing the regression constant.
   b. What does the equation as a whole mean?
   c. What does each number and symbol in the equation stand for?
   d. How well does parents' education explain variance in respondents' education?
   e. Compute the beta weights for each independent variable. What do they mean?
   f. Which parent's education has the greatest independent impact on respondent's education? What evidence do you rely on for this answer?
   g. How many years of schooling would you predict for someone whose father has 11 years and whose mother had 16 years?
4. Use the path diagram in Figure 9-3 to answer the following:
   a. In terms of explained variance, which variable in the model is best explained by the other variables in the model?
   b. Which variable has the greatest direct effect on son's income?
   c. Compute the direct effect of number of siblings on son's income.
   d. Compute the indirect effect of father's occupation on son's occupation.
   e. Compute the indirect effects of father's education on son's income.
   f. Compute the indirect effects of father's occupation on son's income.
   g. Compute the *total* effect (direct and indirect) of son's education on son's income.
   h. What is the coefficient of alienation for son's income? How much of the variance in son's income is explained by this model?

i. What does the curved arrow connecting father's occupation and father's education mean?

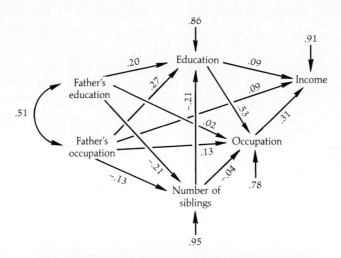

Figure 9-3 Path diagram representing a model of the socioeconomic life cycle in the white population, with path coefficients estimated for native men 25 to 64 years old with nonfarm backgrounds and in the experienced civilian labor force, March 1962.

SOURCE: Duncan, 1969, p. 90.

III

Statistical Inference

10

Sampling

Although we are usually unaware of it, each of us often relies on samples of information. We constantly select and interpret information from populations of groups, people, events, or objects in order to form impressions and make decisions. When we decide which candidate to vote for, whom to make friends with, which college to attend, or which job to apply for, we inevitably rely on only a small fraction of all the relevant information.

The problem with such everyday, informal samples is that there is no way to know how representative they are of the populations from which they are drawn. In most cases this is due to the fact that the entire population of information is not available to us to sample from. In short, we are unable to consider the entire population when we select our sample. The purpose of scientific sampling is to overcome these limitations by identifying and systematically considering all members of a population and making selections in such a way as to maximize the chances of obtaining a representative sample.

Suppose, for example, that you wanted to know how assembly-line workers feel about their jobs. How would you go about deciding whom to observe? What would you have to do in order to be confident that your results accurately represented the population of assembly-line workers without taking a census of them?

We would have to begin by precisely defining the population we are interested in. It might be something like "all people classified by management as workers at Famous Amos Chocolate Chip Cookie factories." If the number of workers were fairly small—a few hundred, perhaps—then we could easily conduct a census in which we gather information on everyone. If, however, there were thousands of workers, a census would be too costly in both time and money. In this kind of case, we would like to be able to ask our questions of only *some* workers and be able to use their answers to represent the feelings and experiences of *all* the workers accurately. In other words, we want to select a **representative sample**.

This chapter is about how scientific sampling is done and how you as a reader can evaluate sampling procedures in order to know how to evaluate the findings based on them. Keep in mind that we are rarely in a position to evaluate a specific sample, to say that this or that sample is representative. Most of the time we can evaluate only the procedure used to select a sample. If we have confidence in the procedure, then we infer that the resulting sample is probably a good one.

Sampling the Assembly Line: A "Simple" Sample

We have to begin by deciding how large a sample to select. One general rule is to take as many observations in your sample as you can afford. Suppose that it costs roughly $100 to gather and process information on each observation and that we have $50,000 to spend on our study. This would allow us to select around 500 workers for our sample.

SAMPLE SIZE

Intuitively, it probably makes sense to you that the larger the sample we select, the better off we are. But does it also matter what *proportion* of the population we include? If we select 1,000 people from a population of one million wouldn't that give us much greater accuracy than selecting 1,000 people from a population of one billion? While it may come as a surprise, in most cases where the population is large enough to force us to take a sample, the proportion we select does *not* significantly affect our accuracy. In such cases, the most important thing is the absolute size of the sample, not its relative size. For now you will have to take this on faith. Later on (Chapter 11) we will explore this issue in greater detail so that you can see why it is true.

MAKING A FRAME

Having settled on a sample size, we return to our definition of the population from which our sample will come: "All those classified by management as workers in Famous Amos Cookie factories." Our next step is to put this definition into a form that will allow us to make actual selections of people. To do this, we make a list of all the factories. From each factory we obtain a list of all workers. This list, which contains all members of the population, is called the **frame** of the sample.

In this case, the sampling frame consists of an actual list of the entire population, but in many studies the population is so large that this is impossible. It would be impossible, for example, to list the entire U.S. population. In such cases (as we will see later on in this chapter), it is sufficient to define the population precisely as a way of constructing a sampling frame. Thus, in order to select a sample, a frame must be constructed either as a list of all members of the population or as a precise definition.

THE THREE RULES OF SCIENTIFIC SAMPLING

Since we cannot know if a particular sample represents a population, our only source of confidence (or of doubt) lies in the procedures we use to select it. As in a lottery, the only way we can judge the fairness of a specific outcome (this person won instead of any of the others) is by judging the fairness of the procedure through which the outcome was generated.

The first and most important rule of scientific sampling is that every member of the population must have a chance to get into the sample, which is to say, the frame must be complete. If, for example, we want to select a sample of Illinois residents, and we use as a frame the list of subscribers to television cable services, our frame will be incomplete since many Illinois residents do not receive cable television. All those not on the list belong to the population, but since they are excluded from the frame they have no chance of being included in the sample. Such a procedure would violate the first rule of scientific sampling.

The importance of having a complete frame was made glaringly apparent in the early days of public opinion polling during the 1936 U.S. presidential election that pitted the Democratic incumbent Franklin D. Roosevelt against Republican challenger Alfred M. Landon. The *Literary Digest*, a popular magazine, selected a sample from telephone directories and automobile registration lists and asked people whom they intended to vote for. The result was a predicted landslide for the Republican candidate. Unfortunately for the *Digest* (and for Landon), those who were well off enough in 1936 to have telephones or own cars tended to be Republicans. Since the sampling frame tended to exclude Democratic voters, the sample result underestimated the support for Roosevelt and overestimated the support for Landon. To the shock of those who relied on the *Digest* poll, Roosevelt went on to a landslide victory of his own, winning in all but two states.

According to the second rule of sampling, the chance that each member of the population will be selected must be *known*. In other words, we have to be able to state precisely what the probability was that any member of the population would be picked. The only way to know this is, at each step in the sampling process, to know how many people are eligible to be chosen so that we can divide this number into the number of those who are actually picked and thereby calculate a simple probability.

It is generally desirable to have the probabilities of selection equal for all members of the population, but it is not a necessary condition of good sampling. Provided that the probabilities of selection are known, once the data have been gathered it is possible to make mathematical corrections (using what are known as sampling weights) in order to adjust for unequal selection probabilities.

The third rule of scientific sampling is that all selections must be *independent* of one another. As we saw in Chapter 5, in terms of probabilities, each selection is an event, and two events are independent of each other if the fact that one has occurred in no way affects the probability that the other will occur. In the case of sampling, this means that the selection of one member of the population should not increase or decrease the overall chance that someone else will be picked. For example, if we select a representative sample of 250 workers and then interview both them and their best friends at work, we cannot lump the two groups of 250 workers together in order to get our sample of 500. Why not? The answer is that the selection of the 250 best friends depended on the selection of the original 250 workers. Since people and their best friends tend to resemble each other on many characteristics (including their attitudes towards their jobs), our sample will tend to be more homogeneous than the population, and thus not representative. (We could use our sample to represent 250 *friendships*, however.)

Another example may help clarify the third rule. If we wanted to expose animal tissues to a chemical in order to see if they developed cancer, we might take pieces of tissue from many animals, expose them to the chemical, and compare the effects with unexposed tissues. Suppose we use 100 animals, taking one piece of tissue from each. We could, however, also obtain a sample of 100 pieces of tissue by taking 50 pieces from each of two animals or 100 pieces from a single animal. We would have 100 pieces of tissue in both cases, but would the results be as valid as before? No, because each large set of tissues is now drawn from only one or two animals, and each piece is not an independent example of the general population of animal tissue. Coming from the same host, the tissues will most likely have important similarities that would occur less frequently if drawn from many different hosts.

RANDOM SAMPLING

How do we go about satisfying the three rules of scientific sampling? In the case of our study of factory workers, the list of the population satisfies the first rule by providing us with a frame that includes every member of the population. There are several ways of satisfying the second and third rules of sampling, the most widely known of which is the **simple random sample** (or "random sample" for short).

This is how it is done. We begin by assigning a unique number to each person on the list. If there are 25,000 workers in our population, then we will have to number their names from 1 through 25,000 (no small task itself). Then we use a **random number table** to make our selections. We can also use random numbers generated by a computer.

What is a random number table? As you may recall, the word *random* has a precise scientific meaning: every possible outcome is equally likely. We could make a random number table ourselves by placing 10 balls numbered 0 through 9 in a pot, shuffling them well, and picking one ball. We write down its number, return it to the pot, reshuffle the balls, and repeat the procedure. If our mixing and picking were thorough and unbiased, the resulting list would consist of random digits. On each draw, each digit has an equal chance of being drawn regardless of what happened on the previous draws.

Generating a table large enough to do us any good would take a great deal of mixing and picking, so in most cases we have a computer do the job for us by substituting a random electronic process for the physical process of selecting numbered balls. (Although computer-generated random numbers are in fact only pseudorandom, they work effectively in most sampling problems.) There are also books that contain nothing but row upon row of truly random digits, sometimes as many as one million (Rand, 1955). See Appendix Table 4 for an example of a random number table.

Having decided to select a sample of 500 factory workers, we first select a place to begin in the random number table and then go through the table picking numbers between 1 and 25,000. Each time we find one, we go down our list of workers to that numbered person and check off him or her. We keep doing this until we have selected 500 different people. Every member of the population had an equal chance of being picked because no one's number was any more or less likely to occur than anyone else's.

We should note that the techniques of statistical inference assume that sampling is done *with replacement*, which means (as you may recall from Chapter 5), that each person could theoretically be selected more than once. In practice this is rarely done, and sampling is done *without* replacement: once a person is selected, he or she is removed from consideration on subsequent selections. This can have serious effects on the accuracy of results only if the sample amounts to a substantial proportion of the total population. The exploration of these possible effects as well as ways to correct for them is beyond the scope of this book, but you should be aware that field researchers are often pressed by the demands of practicality to violate theoretical assumptions that underlie inference techniques (see Kish, 1965).

With our list of 500 sampled names, we are ready to go into the field and gather our data.

SYSTEMATIC SAMPLING

Simple random sampling is the simplest kind of sampling from a theoretical point of view, but in practical terms it is often unwieldy: assigning numbers to 25,000 names and drawing 500 random numbers is a lot of work. In order to save time, energy, and money, researchers often make use of a technique called **systematic** sampling whenever they have a frame that consists of a complete list of the population.

There are 25,000 people on our list, and we want to select 500 of them. This means that we want to take one out of every 50 people on the list (25,000/500 = 50). To do this, all we need to do is select a random starting point among the first 50 names (by selecting a number from our random number table) and then select that name and every 50th name after it. In this case, 50 is called a **skip interval**, and we are in a sense "skipping" through the list with steps that are 50 names long, selecting each person on whose name we happen to land. If the random starting point was person number 25, then we would select people with the numbers 25, 75, 125, 175, 225, 275, and so on through person number 24,975.

From each of the 50 possible starting points (numbers 1 through 50), we could generate a unique sample of 500 people. Since each starting point is equally likely to be chosen (we used a random number table to determine it), each possible sample is also equally likely, and each person has an equal chance of having his or her name selected and thereby being included in the sample.

How do simple random and systematic samples differ? In a simple random sample, not only do all individuals have an equal chance of selection, but all *combinations* of individuals are equally likely as well. In systematic sampling, however, not all combinations are possible once the list is finalized. In our assembly-line study, for example, workers with numbers 125 and 126 cannot appear in the same sample since the skip interval is 50 names long and these two people are only one name apart. Only people whose names are separated by exactly 50 names can appear together.

If we assume, however, that the list is "well shuffled," then all combinations are equally probable. In practice, this is usually not a risky assumption to make, but it can be dangerous when the list is arranged in order from high to low on a characteristic that is relevant to the subject being studied. For example, if the list was arranged from high to low status within the factory, choosing a starting point near the top of the list would produce a sample with higher average status than would be the case if we chose a starting point nearer the bottom of the list. In the first case, the highest status person on the list would be person number 1, and the lowest status person would be number 24,925. In the second case, the highest status person would be person number 50, and the lowest status person would be number 24,975.

The problem of **periodicity** arises in systematic sampling when an important characteristic recurs at regular intervals throughout the list. Suppose, for example, that we were sampling apartments from a list of units in a large building. If the skip interval were ten and if every tenth apartment happened to be a corner apartment, we might get very biased results when studying patterns of social interaction in large apartment buildings. Why? Because people in corner apartments tend to be more socially isolated than those in interior apartments who have neighbors on both sides. For these reasons, it is important to check lists beforehand to ensure that they are well shuffled.

Does selecting a sample randomly or systematically *guarantee* that our sample will be representative of the population from which it is drawn? No. Our ideal sample would be a group that matches the population on every characteristic except size. We are very unlikely to draw such a sample, for there will be variation from sample to sample just by chance. The larger our sample, however, the more confidence we can have in our results. We will talk about sample size and accuracy in greater detail later on (Chapter 12).

Sampling Error

As was the case with measurement (Chapter 2), the possibility of random error and bias is an inevitable part of the sampling process. Random error, in fact, is unavoidable in even the most perfect sampling procedure, and as we will see in Chapters 11, 12, and 13, the estimation of such error lies at the heart of statistical inference.

Unlike random error, however, sampling bias is not something that can be simply estimated from sample data. Sampling bias usually results from an incomplete sampling frame that tends to exclude certain groups in a population, but it can also result from errors in the field. It is sometimes possible to detect bias in a sample if we know something about the way it should appear on certain variables. A sample of New York City residents, for example, that did not include a large percentage of minorities would be immediately suspect, as would a sample from Palm Beach, Florida, that did not include a relatively large percentage of elderly or a sample of Beverly Hills, California, residents that did not include a large proportion of wealthy people.

How do we know how large such percentages should be? One way is to consult recent U.S. census data and compare the sample with the population on key characteristics such as income, age, race, gender, and education. This kind of checking can at least give clues to possible bias. Beyond this, the only way to evaluate the possibility of bias is to inspect the sampling frame and field procedures carefully for any systematic exclusion of members of the population.

STRATIFICATION: MINIMIZING RANDOM ERROR

Although random error is an inevitable part of the sampling process, there are some ways to minimize it. In our study of assembly-line workers, even the most carefully drawn sample might wind up overrepresenting men and underrepresenting women, or overrepresenting younger workers and underrepresenting older workers. Random number tables are truly random only in the long run. What can we do?

One way to minimize random error is to increase the amount of detail in the sampling frame through a process known as **stratification**. If, for example, we had a list that included the workers' ages, we could guarantee that the age distribution in our sample would precisely match the age distribution in the population. All we have to do is divide the frame into different age groups (called **strata**) and draw separate samples *within* each age group (or **stratum**) in proportion to their relative numbers in the population. If, for example, 35 to 40-year-old workers make up 10 percent of the worker population, then we would select 50 workers from that age group for our sample (10 percent of 500). Following this procedure for each age group will result in a sample of 500 workers that exactly mirrors the age distribution in the population.

When we stratify a sampling frame, we narrow down the possible outcomes by making it impossible to draw samples that are in error on the characteristics we use to define the strata. Thus, a sample of 500 workers drawn from a stratified frame will have less error than a sample of 500 people drawn without stratification. Although the stratified sample is of the same size as the unstratified one, it will tend to be more accurate because of the way in which it is drawn. When a sample design tends to make samples of a given size more accurate than other samples of the same size, we say that the first sample has greater **efficiency**. In most cases, stratification improves the efficiency of a sample design. At the very worst, a stratified sample will yield the same level of accuracy as a simple random sample of the same size. In other words, stratification cannot hurt a sample. It can only help or have no effect at all.

Stratification can be used to ensure proper representation for as many characteristics as are available in the sampling frame. If, of course, we cannot identify members of the population according to the characteristics we want to stratify by, then we cannot use stratification in order to minimize random error.

Whether or not we stratify, it is important to remember an important rule: If we draw a sample that accurately represents a group, then that sample will also represent all *sub*groups as well, provided that the sample is large enough (for more on this, see the section on oversampling later in this chapter). If we select a representative sample of all U.S. elementary schools, then we also have a representative sample of all private schools, public schools, parochial schools, small schools, large schools, city schools, rural schools, and so on. As long as our selection procedures follow the three rules of sampling, our sample will go wrong only by chance; and as we will see in the next few chapters, by drawing a sufficiently large sample, we can make those chances of random error very, very small.

RESPONSE RATES: MINIMIZING BIAS

Bias can be introduced into a sample design if the frame is incomplete; but it can also occur later on in the sampling process if we fail to obtain information about all members of the sample. We select our sample of 500 assembly-line workers, and we send out our interviewers. What do we do if 50 of the people refuse to grant interviews? We now have a 90 percent **response rate** (450/500 × 100). Most people would say that a loss of only one in ten is not very serious, and they would probably be right; but what if we were refused by 100 workers, or 200? What then? Can we feel confident that those who did respond represent not only themselves but those who refused?

If everyone in a sample is equally likely to refuse to be interviewed, then there is random error and a loss of sample size, but no bias. It is generally the case, however, that people who refuse to be interviewed tend to differ from those who agree. High-income people, for example, are less likely to allow interviews (and are more adept at avoiding them) than are middle-income people. In Turkey there has been a long-standing conflict between the central government and Kurdish tribesmen who live in the eastern end of the country. Census takers sometimes enter such areas at considerable risk to their own safety with the result that Kurds are often underrepresented in studies of the Turkish population. While it is possible to estimate the extent of random error in a sample, all that we can know about this kind of bias is that its likelihood increases as the response rate goes down.

Even when we are able to gather information from all members of a sample, there are often specific variables for which replies are missing for a variety of reasons, including refusals, oversight, and coding errors. Whatever the cause, this results in some variables having response rates that are lower than the response rate for the study as a whole. Suppose, for example, that we conduct a survey and get a response rate of 80 percent. Further suppose that one of the questions is controversial and only 70 percent of the respondents reply. In spite of the fact that we have an 80 percent response rate for the rest of the study—a perfectly respectable level of response—in terms of representing the population on this question, the response rate is only (80% × 70%) = 56 percent, which is quite low. Thus, it is important when you read research reports that you pay attention to both the overall response rate and the rate for individual variables.

As a general guideline, a response rate of 75 percent or more is acceptable. Response rates of the National Opinion Research Center's annual General Social Surveys, for example, are typically just under 80 percent, with refusals running around 17 percent and unavailability for other reasons (such as illness) at around 4 percent (Davis, 1986). As response rates drop off below 70 percent, the chance of bias is quite serious, and anything below 50 percent is almost useless. I do not mean to suggest that information on 250 people out of an original sample of 500 is of no value or interest. Rather, the problem is that we cannot rely on the information to accurately represent the population it came from.

One way to see how disastrous low response rates can be is to consider voter preference polls that are conducted just before elections. Suppose that 30 percent of the population favors Wilson, 40 percent favors King, and 30 percent gives no response. Should King be confident of winning? The only way to be sure (and fair) is to assume the worst, which is to say that all of those who gave no response will vote for Wilson, giving her a solid 60 percent majority (a landslide by most political standards). If King receives anything less than a third of the votes from those who gave no response, he will lose the election. The only way either candidate can make a prediction based on such sample results is by making an *assumption* about the way the nonrespondents will actually vote, and there is little reason to expect such assumptions to be true.

This problem played an important part in the 1948 U.S. presidential contest between Thomas Dewey and Harry Truman. Just before the election there was a large group of people who were undecided and refused to declare for either candidate. The pollsters assumed these undecideds would eventually vote in the same proportions as those who had made a decision at the time of the poll. The pollsters thus predicted that Dewey would defeat Truman by a landslide. In fact, however, the undecideds voted overwhelmingly for Truman who went on to win the election (see Mosteller et al., 1949).

In spite of the difficulties inherent in small response rates, there are many instances in which rates as small as 5 percent or 10 percent are reported, although rarely would academic researchers do so. In commercial research such as that conducted by magazines, for example, low response rates are often erroneously considered acceptable. Whenever you find a published study that has a low response rate or does not report the rate at all, you have immediate reason to seriously doubt the representativeness of the sample involved.

OVERSAMPLING

As you may recall from the section on stratified sampling, if a sample represents a population, then it also represents all *sub*groups of that population *provided the sample is large enough*. Suppose, for example, that we want to conduct a study of people's attitudes towards various aspects of race relations in the United States, with a special focus on American Indians. Since American Indians make up only 0.5 percent of all Americans, in a sample of 1,500 cases that accurately reflected the U.S. population we would expect to find only seven or eight American Indians (1,500 × .5%). As we saw in Chapter 7, however, this number of cases is far too small to examine relationships between variables. A relatively small 3 × 4 table, for example, would contain more cells than the total number of American Indians in the sample. This means that we would have some empty cells even without controlling for additional variables. For purposes of analyzing group differences in race relations attitudes, such a sample would be worthless.

If we select more than eight American Indians, their proportion in the sample will exceed their proportion in the population, and the sample will not be representative of the entire U.S. population. In this kind of situation, however, this is precisely what we do: we **oversample** the members of the relatively small subgroup in order to ensure an adequate number for statistical analysis purposes. We might, for example, select a sample of 1,100 whites and 400 American Indians.

By oversampling American Indians, we are in a position to make comparisons between them and whites; but what happens when we perform an analysis on the entire sample? If we try to use our data to represent the entire U.S. population, we will be relying on a sample that is only 73 percent non-Indian (versus 95.5 percent in the population) and 27 percent American

Indian (versus .5 percent in the population). This presents no problem if Indians and non-Indians do not differ on the characteristics we are studying; but suppose we want to estimate something like the extent of unemployment? Since American Indians are far more likely than non-Indians to be unemployed, the oversampling among Indians will have inflated the number of unemployed people in the sample without changing the total size of the sample. Therefore, when we calculate the percentage of the sample that is unemployed, we will get a number that is too large.

There is, of course, a way out of this. We have included a number of American Indians that is 50 times as large as it would be if they were represented in true proportion to their numbers relative to the U.S. population (400/8 = 50). If we multiply each American Indian's unemployment score (1 if unemployed, 0 if employed) by the reciprocal of this—one-fiftieth (1/50)—then the total number of unemployed American Indians will be the same as what we would expect if we had only selected 8 in our sample rather than 400.

We would then have to correct each non-Indian's unemployment score by a similarly appropriate correction factor. Our sample included only 1,100 non-Indians instead of the proportionately correct figure of 1,492 (99.5 percent of 1,500). This means that the non-Indian subgroup is only 74 percent (1,100/1,492) of its correct value. To correct for this, we simply inflate each non-Indian unemployment score by multiplying it by the *reciprocal* of this number—1,492/1,100 = 1.36. This will give us the number of unemployed non-Indians that we would have found had we included the correct number of non-Indians in the sample to begin with.

In each of these cases, the correction factors are called **sampling weights** because they correct for unbalanced representation among subgroups caused by the sampling process. Sampling weights are also used in situations that do not involve deliberate oversampling (see the following section on cluster sampling) but in which we nonetheless need to correct for unequal probabilities in the selection process. In order to do this, of course, we must know what the true probabilities of selection were (as we did in the above case of deliberate oversampling), a vital bit of information that is available to us only if we follow the second rule of scientific sampling.

In this example we stratified by race, a relatively easy characteristic to observe. Suppose, however, that we wanted to do a study of people who have attempted suicide at some time in their lives. Since it would be impossible to identify such people before we drew a sample, and since they amount to a very small proportion of the population, we could not use oversampling in order to obtain a subsample that would be large enough to enable us to draw conclusions about people who have attempted suicide. One solution would be to draw an enormous sample. If, for example, attempted suicides amounted to 1 percent of the population and we wanted to have 200 such people in our sample, we would have to draw a sample of 20,000 people in order to get an expected value of 200 (1 percent of 20,000).

In practice, it is more effective to send interviewers to a large sample of households and conduct very short interviews designed to "screen" potential respondents to identify those who possess the rare characteristics we are interested in. It is difficult to design an interview that is both short and able to detect rare characteristics, and such studies are still very expensive—which is why studies of people with rare characteristics are themselves quite rare.

Complex Samples

Simple random and systematic samples work quite well with populations that are small enough to be listed. When you want to study a population as large as a city, state, or country, however, you are unlikely to have a list of everyone's name and address. What then?

The answer is that an adequate sampling frame does not have to consist of a list including every member of the population; the frame can also consist of a precise definition. In these situations, the sampling process can get quite complicated, but a few relatively simple examples can give you some idea of just how **complex samples** are drawn.

MULTISTAGE SAMPLES

Consider the problem of drawing a sample of individuals from a population the size of Mexico City, which contains roughly 20 million people spread out over many square miles. Obviously it is impossible to construct a frame that would allow us to select a simple random or systematic sample. The alternative is to draw the sample in *stages*, beginning with larger units and working our way down to individuals.

First, we get a map of Mexico City that shows each block. We then divide the city into districts (which correspond to the idea of census tracts in the U.S.). Using data from the most recent census, we can get a good idea of how many people live in each district of the city and, from this, of just how we will have to distribute our sample selections among the districts in order to represent the city accurately.

The second step is to draw a sample of blocks from each district (we might first stratify the districts according to any information we might have about them). To do this, we number each block within each district (we literally write a number on each block on the map, which may add up to several thousand blocks). We then use random numbers to select the blocks.

The third step is to send "listers" into the field to walk around each selected block and write down everything they find, from single-family dwellings to stores, apartment buildings, and vacant lots. At this point, we have a list of all the households on our selected blocks. We could string the lists together to make one huge list for each district. Then we could draw a systematic sample of households from that list. In this kind of complex multistage design, every household has an equal chance of being selected because every block has an equal chance and every house on every block has an equal chance. At each stage of the process, everyone in the defined area has a chance of being selected, we know what the probabilities of selection are, and all selections are independent of one another. Thus, even without a frame that lists each member of the population, we are able to satisfy all three rules for scientific sampling.

CLUSTER SAMPLES

With large multistage samples such as the one just described, each area from which samples of blocks are drawn may cover many square miles. Because of the economic constraints of large surveys, especially those done on a national level, it is usually preferable to select bunches of houses (known as **clusters**) from selected blocks. By going to five houses on a single block rather than to five houses spread out over five different blocks, interviewers save both time and money.

How is it done? Suppose we have a city with 100,000 dwelling units (DUs), from which we want to take a sample of 1,000 DUs (or 1 percent). We stratify the population into several geographic areas. For purposes of illustrating **cluster sampling**, we will focus on just one, which we will call the "West End."

Suppose the West End has 5,000 DUs. We want to select 1 percent of them into our sample, or $(.01)(5,000) = 50$ DUs. Since the West End is spread out over many blocks, we decide to select clusters of 5 households per block once we have selected a sample of blocks. This means we will divide our 50 West End DUs among 10 blocks (50/5). Note that because some blocks may be very small and some very large, we may wind up picking *no* clusters from some blocks and more than one cluster from the larger ones. Keep this in mind as you read on.

Suppose the West End has 500 blocks. We number the blocks and randomly select 10 of them. We send our listers out to record every DU on the 10 selected blocks and get the results shown in Table 10-1. Column I gives the

Table 10-1 Selecting Clusters

(I) Block Number	(II) Number of DUs	(III) DU Number	(IV) Number of Clusters	(V) Number of DUs
1	20	1–20	2	10
2	6	21–26	0	0
3	15	27–41	2	10
4	9	42–50	1	5
5	9	51–59	1	5
6	6	60–65	0	0
7	5	66–70	1	5
8	8	71–78	1	5
9	10	79–88	1	5
10	12	89–100	1	5
TOTAL	100		10	50

block number, and column II tells us how many DUs are on each block. If we think of the 10 separate lists as one big list with the DUs numbered consecutively from 1 to 100, column III tells us which DUs are located in which blocks. For example, DUs 1 through 20 are in Block 1 and DU 43 is in Block 4. For the moment, ignore the last two columns.

We want to select 50 DUs from the 100 DUs in our 10 selected blocks. As we saw earlier, 50 DUs are equivalent to 50/5 = 10 clusters of 5 households each. The next step is to determine how many clusters to take from each of the 10 blocks. To do this, we use systematic sampling with a skip interval of 100/10 = 10 (if you do not understand why the skip interval is 10, review the section on systematic sampling). Suppose we select a random starting point of 8. Our selected DUs will then be 8, 18, 28, 38, 48, 58, 68, 78, 88, and 98.

The 8th DU is on Block 1, as is the 18th DU. Therefore, we will select 2 clusters from Block 1, or 10 DUs (5 per cluster). Notice that this does not necessarily mean that we will wind up including the 8th DU in our sample. At this stage of sampling, we are "skipping" through the list of blocks, and each of our "skips" is 10 DUs long. Every time we "land" in a block, we select one cluster of 5 DUs from that block. The actual selection of those DUs, however, must wait for the final stage of sampling.

The 28th and 38th DUs are on Block 3. Therefore, we will select 2 clusters from Block 3. Notice that we did not land in Block 2 because it was too small: it is only 6 DUs long compared to our skip interval of 10 DUs. The chances of selecting any given block, then, are directly proportional to the number of

DUs on the block, which is as it should be if we are going to keep our probabilities of selection equal for all households. This is known as sampling with **probabilities proportional to size** (or *PPS* sampling, for short).

The 48th DU is on Block 4, the 58th on Block 5, the 68th on Block 7, the 78th on Block 8, the 88th on Block 9, and the 98th on Block 10. We will select one 5-household cluster each from Blocks 4, 5, 7, 8, 9, and 10 (see column IV in Table 10-1 for a summary). Block 1 has 20 DUs, of which we are going to pick 2 clusters of 5 households each for a total of 10 DUs. We can use systematic sampling within the DU list for Block 1, and use the same procedure to select the correct number of clusters from each of the other selected blocks.

In the first stage, in which we selected the original ten blocks, each household's block had an equal chance of being picked. In the second stage, we preserved the equal selection probabilities for households by selecting clusters within blocks according to the size of the blocks. If we had taken the same number of households from each block, it would not have been fair to those who lived on large blocks, since they would be up against more competition than those on small blocks. So, we took a larger number of households from the large blocks than from the small. In this way, we keep the chances of being selected equal throughout all stages of the sampling process.

There are some sample designs in which the chances of being selected are not equal for all members of the population. In the case of oversampling, for example, we do this deliberately and correct later with the use of sampling weights. In some cases, we do know how many DUs there are in an area and therefore may give people in that area a disproportionately large (or small) chance of selection in comparison with people in other areas. If the sampling is done carefully, however, with accurate records and an understanding of the mathematics of sampling, we can use sampling weights to correct for these unequal probabilities after the sampling is finished.

Cluster sampling is designed to save both time and money, but it does so at some costs in accuracy. People who live near each other tend to be more alike on a variety of characteristics than people who live far away from each other. For this reason, cluster samples tend to be more homogeneous than simple random samples of the same population would be. In other words, cluster samples tend to underestimate variance.

Cluster sampling is a good example of the kinds of compromises one must often make in scientific research. The ideal techniques dictated by statistical theory are often impractical if not impossible, and we must accept a trade-off between practicality and accuracy. In most cases, the losses that accompany cluster sampling are not serious ones, and researchers can make some corrections for them.

Let the Buyer Beware

Although the language of scientific sampling is often used in describing the seemingly countless polls conducted each year in the United States, in reality the language is often misused. Academic researchers and some commercial pollsters adhere rigidly to the rules of sampling, but in many other cases, sampling is "scientific" in name only.

Beware of the common misuse of the word "random," as in "shoppers selected at random." This rarely means what it says. The roving reporter who collars people on the street, for example, is not taking a representative sample of any group we can define. In spite of the occasional acknowledgement that such samples are not random, more often than not the impression left is that the replies of the people interviewed represent a much broader population than themselves. It is important not to take such sampling seriously as a representation of anything more than those who were actually selected.

Mass-circulation magazines often send out tens of thousands of questionnaires and then print stories based on 5,000 or even 10,000 replies. The sample sounds impressively large (and it is), but this often distracts the reader's attention from the fact that it may represent no more than a 5 percent response rate—virtually worthless from a scientific perspective. In general, it is a good idea never to trust the results of mailed questionnaires unless the authors provide a response rate that is acceptably high. The absence of a published response rate should make you immediately suspicious.

Some commercial pollsters use **quota samples**. This means that the interviewer is instructed to obtain respondents who have specified characteristics or combinations of characteristics. In locating 100 respondents, for example, an interviewer might be told to find 90 whites and 10 blacks or 5 upper-class, 70 middle-class, and 25 working-class people, and so on. The problem with this method is that the interviewers can select anyone they want by any means, so long as the final sample has the proper number of people with the specified characteristics. Although the sample may reflect the population on *those* characteristics, there is no way of knowing that it reflects characteristics for which there was no quota.

In practice, quota samples have a fairly good track record. Many research organizations have used them, and various checks have shown them to be surprisingly accurate. The problem is that *there is no theoretical reason to expect this to happen*. The fact that some number of quota samples in the past have been reasonably accurate does not mean that future quota samples will be accurate. Studies that use samples drawn according to the three rules of sampling not only work well in practice, but are theoretically sound as well. As you will see in the chapters to come, we can estimate the probability that sample estimates of population characteristics are in error, but we cannot do this with quota samples.

Whether to accept the results of quota samples is a difficult issue. If past performance is used as a criterion, then accepting quota sample results is not terribly risky. If, however, we are conservative, then quota samples should be regarded with considerable reservation.

You should keep in mind that although commercial pollsters can do work of high quality, they must often cut corners in order to remain competitive. Unfortunately, the corners are often cut in the sampling stage as well as in field work (by, for example, substituting a readily available neighbor for a selected respondent who is not home). It is important that you develop the ability to ask critical questions and evaluate the answers intelligently. If you encounter a poll of an urban northeastern state, for example, and only 1 percent of the respondents are black, you *know* something is wrong with the sample. The critical evaluation of information requires patience and effort, but you will find that it is worthwhile.

Summary

1. The purpose of scientific sampling is to select cases in such a way that we can then make statements about populations without gathering data on all of their members. In general, it is the size of the sample that is most important in determining accuracy, not the proportion of the population that is selected.

2. According to the three rules of scientific sampling, every member of the population must have a chance of being included (the sampling frame must be complete), the probability of being selected must be known, and selections must be made independently of one another.

3. There are many different designs that satisfy the three rules of sampling, including simple random, systematic, and complex multistage and cluster samples. Each design is an attempt to satisfy the scientific requirements in different practical situations.

4. Like measurement, the sampling process is always subject to both random error and bias. Stratification is a technique for minimizing random error. Getting a high response rate is a major way of minimizing sampling bias.

5. Oversampling is useful in studying small subgroups of populations. Correction for the distorting effects of oversampling is achieved through the use of sampling weights.

6. The language of scientific sampling ("random," for example) is often misused, and in consuming published data it is important to pay attention to the way in which a sample is selected. Quota samples are an example of a sample design that violates scientific principles, but which nonetheless seems to have a good track record. In the end, consumers must use informed judgment in deciding how to interpret and whether to believe results based on samples.

Key Terms

cluster 261
cluster sampling 261
complex sample 260
efficiency 255
frame 250
multistage sample 260
oversampling 258
periodicity 254
probabilities proportional to size (PPS) 263
quota sample 264

random number table 252
representative sample 249
response rate 256
sampling error 254
sampling weight 259
simple random sample 252
skip interval 253
strata (stratum) 255
stratification 255
systematic sample 253

Problems

1. Define and give an example of each of the key terms listed at the end of the chapter.
2. If we select a sample from a very large population, which will have a greater impact on the accuracy of our results, the size of our sample or the proportion of the population that we include? Why?
3. What are the two main ways of constructing a sampling frame? Give an example of each.
4. What are the three rules of scientific sampling? Give an example that illustrates the *violation* of each rule.
5. What does "random" mean, and in what way is the term often misused?

6. What is the difference between sampling with replacement and sampling without replacement? Under what circumstances can sampling without replacement negatively affect a sample? Under what circumstances can sampling with replacement negatively affect a sample?

7. How does systematic sampling differ from random sampling? Under what circumstances is it most likely to be used? What kinds of sampling frames cause problems with systematic sampling? How can they be solved?

8. Give examples of random error and bias in sampling. What is the connection between stratified sampling and random error? What is the connection between response rates and bias? Be specific and given examples that show what you mean. Under what circumstances can stratification hurt the accuracy of a sample?

9. When is oversampling most likely to be used? What kinds of problems does it introduce, and how are they solved? Be specific and use examples to show what you mean.

10. Under what circumstances are complex samples most likely to be used? Be specific for each type—multistage and cluster. What are the main advantages and disadvantages of cluster sampling?

11. Why should one be wary about quota samples even though they have a good "track record"?

12. We want to do a study of Vietnam veterans using a sample of 2,000 veterans drawn from Veterans' Administration lists. For this study:
    a. Define the population.
    b. What would be the sampling frame?
    c. What kind of sampling procedure would you use? Why?
    d. Would sampling with or without replacement make a difference in this case? Why or why not?
    e. How and why might you stratify this sample?
    f. How and why might you oversample some groups?

13. How might you design a sample of households in your community?

14. If we drew a sample of households from telephone directories in order to represent the entire population of households in your state, how would you evaluate the scientific validity of this design?

11

Statistical Inference

Statistical inference is the process of using sample results to estimate population characteristics. It is a process by which we use what we know (a sample result) in order to make statements about what we do not know (a population characteristic). We attach probabilities to these estimates that express the degree of confidence we have in them.

In order to use samples in this way, we must rely on special theoretical probability distributions that are derived mathematically to act as a bridge between what we know and what we do not know. To understand these theoretical distributions, you will first need to become familiar with some of the technical tools and language that describe them and their use. Some of the ideas will be familiar (probability distributions), but most will be new. Most important, we will describe these new ideas without telling you how they are used until we have covered them all. If you think of inference techniques as a jigsaw puzzle, then we will describe the pieces individually before we put the puzzle together.

To read this chapter, try to pick a time when you are rested and free from interruption. Concentrate on understanding each piece of the puzzle, and have faith that before too long you will see how they all fit together to enable us to use sample data to make statements about populations. I think that you will find it worth your patience.

Some New Ways of Looking at Distributions

As we saw in earlier chapters, distributions show how often events or characteristics occur in a set of data. If we have a variable with a set of possible

Table 11-1 Cigarettes Smoked per Day by People Who Smoke, United States, 1980

Cigarettes Smoked	(I) Frequency*	(III) Percent	(III) Probability
1–14	15,200	29.4%	.294
15–24	21,806	42.1	.421
25–34	6,800	13.1	.131
35–60	7,945	15.4	.154
TOTAL	51,751	100.0%	1.000

SOURCE: U.S Bureau of the Census, 1983, p. 124.
* in thousands

scores, a frequency distribution shows the number of observations that have each score or range of scores. A percentage or proportional distribution gives the relative number of observations that have each score.

Consider, for example, the data in Table 11-1 which show 1980 daily cigarette consumption for smokers in the United States. To simplify the example, I have assumed that no one smokes more than 60 cigarettes a day. For each level of smoking we have both the frequency column (I) and the percent column (II). Notice also, however, that column III shows the probability that a case drawn at random will fall in each of the four categories of smoking. The probability that a smoker consumes 1 to 14 cigarettes a day, for example, is .294, or, in percentage terms, 29.4 percent. Since column III lists the probability for each of the four possible outcomes, it constitutes a probability distribution. Keep in mind that probabilities and proportions are the same thing. This means that we can turn any frequency or percentage distribution into a probability distribution.

SOME NEW GRAPHICS: PROBABILITY DENSITY FUNCTIONS

There are, of course, many ways to describe distributions using tables and graphs of various kinds. In the graphs we have seen thus far, the height of the curve (or bar) was used to indicate the frequency or percentage of cases with a particular score or range of scores. Represented as a histogram, for example, Table 11-1 would look like Figure 11-1.

If we round off the corners and turn the histogram into a frequency polygon, the result is the line superimposed on the histogram. The polygon indicates that most smokers are concentrated in the 15 to 24-cigarettes-a-day

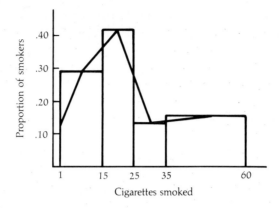

Figure 11-1 Histogram for cigarettes smoked per day by people who smoke, United States, 1980.

SOURCE: U.S. Bureau of the Census, 1983, p. 124.

category. The distribution is unimodal and slightly skewed to the right (look at the graph and be sure you see this).

The histogram and frequency polygon are not the only methods we could use to graphically display this kind of distribution. Instead of representing proportions with the *height* of the curve, for example, we could think in terms of the *area beneath the curve*. We could, for example, use calculus to measure the total area enclosed by the curve and the two axes in square inches. Suppose, then, that when we do this, the result for Table 11-1 looks like Figure 11-2.

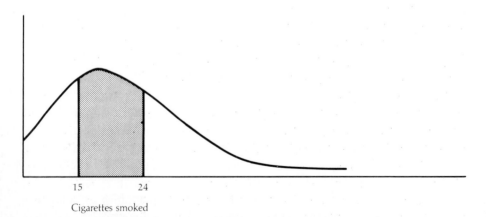

Figure 11-2 Figure 11-1 represented with areas beneath the curve (probability density function).

Unlike Figure 11-1, in Figure 11-2, the vertical axis no longer represents proportions. In this curve, it is the area beneath the curve that is important, not the height of the curve at any given point. In Figure 11-1, for example, we would find the proportion of smokers who smoke between 15 and 24 cigarettes per day by seeing how high the polygon line was above the point on the horizontal axis midway between 15 and 24 cigarettes. It rises to the level of .42 (42 percent). In Figure 11-2, however, the proportion of smokers in that same category is equivalent to the proportion of the total area beneath the curve that is enclosed by the two vertical lines rising from the "15" and "24" points on the horizontal axis. The shaded area in Figure 11-2, then, represents a new way of graphically expressing the proportion of cases that fall within a particular range of scores. Here we have made a direct translation from the proportion of cases in a distribution to the proportion of the area beneath a curve. If we have drawn the curve correctly, the proportion of the total area beneath the curve that is shaded should be .42.

In Figure 11-1, the polygon is drawn in such a way that we can use its height to represent the proportion of cases in a distribution with different ranges of scores. In Figure 11-2, we use a different kind of curve that is drawn in such a way that we can use *areas* to represent those same proportions in a distribution. As a mathematical line, the curve is an example of a *function* (just as a regression line is a mathematical function), but it goes by the special name of **probability density function**. It is called a *probability* function because by indicating the proportion of cases that lie within a particular range of scores, it is also indicating the probability that a case will fall within that range. It is called a probability *density* function because it uses areas to represent those probabilities. The less variation there is in the distribution, for example, the more densely packed the cases will be within a relatively narrow range.

Probability density functions have an additional characteristic that you should be aware of before we go on. With histograms and polygons, we can determine the probability for a single score: it simply corresponds to the height of the curve or bar at that point. With probability density functions, however, it is impossible to determine the probability for a single score. If you look at Figure 11-2, for example, how would you represent the probability of selecting a score of exactly 24.00? In this kind of graph, probabilities are represented by areas, but since we are dealing with an exact score instead of a range of scores, the area would correspond to a straight vertical line rising from the point "24" on the horizontal axis.

As you may remember from mathematics, however, a line has no area, and thus if we try to use the probability density function to show the probability of selecting someone with an exact score of 24.00 or any other exact score, the result will always be zero. This does not mean, of course, that the actual probabilities for all exact scores are zero, since we know that many people smoke 24 cigarettes per day and cigarette consumption is measured in units of whole cigarettes. What it does mean, however, is that the probability density function is set up in such a way that it can be used only to calculate the probabilities associated with *ranges* of scores rather than single scores.

If you feel that you understand the idea of a probability density function and using areas beneath curves to represent probabilities for ranges of scores, then you are ready to go on. If not, review the last section.

Theoretical Probability Distributions: Sampling Distributions

Statistical inference uses theoretical probability distributions in order to use sample results to estimate population characteristics. To see what these distributions are about, consider the problem of estimating the proportion of young people (18 to 25 years old) who are current users of marijuana (have smoked it within the past month).

If we interviewed the entire population of young adults and measured their marijuana use, the **population distribution** of current marijuana users would consist of the proportional distribution of some 33.6 million observations (Table 11-2). This distribution tells us that the probability that we will randomly select a current marijuana user from this population is .37.

Since measuring marijuana use for all 33.6 million young people would be prohibitively expensive, we would probably take a sample rather than a census, as the Census Bureau did in 1979. The **sample distribution** of the

Table 11-2 Current Users* of Marijuana among 18 to 25-Year-Olds, United States, 1979, Hypothetical

Marijuana Use	Proportion
User	.37
Nonuser	.63
TOTAL	1.00
(N)	(33,600,000)

* Current users are those who have used marijuana at least once during the previous month.

Table 11-3 Current Users*
of Marijuana among 18 to
25-Year-Olds, United States,
1979, National Sample

Marijuana Use	Proportion
User	.35
Nonuser	.65
TOTAL	1.00
(N)	(2,044)

SOURCE: U.S. Bureau of the Census, 1983, p. 123.
* Current users are those who have used marijuana at least once during the previous month.

results is shown in Table 11-3. Whereas Table 11-3 is the probability distribution for a single sample of observations, Table 11-2 is a hypothetical probability distribution for the population from which that sample was drawn. In actual practice, of course, when we draw a sample we do not know what the distribution for the population looks like (we do not have a Table 11-2). In fact, the purpose of inference is to use sample results like those found in Table 11-3 to make intelligent estimates of what Table 11-2 probably looks like.

For the purpose of illustration, suppose that Table 11-2 does in fact represent the population of young adults that our sample was drawn from. In our example, we have drawn a sample of 2,044 people, and for that sample we have calculated the proportion who are current marijuana users (.35). Notice that the sample proportion is different from the population proportion (.37). If we were to use the sample result as our best estimate of the population proportion, we would be off by .02 (or .37 − .35).

Although we would not do this in practice, suppose we drew a second sample of 2,044 young adults and computed the proportion who are current marijuana users. What might we get as a sample result then? We might have drawn a sample in which the proportion who are current users is .34 or .32 or .38 or even .37. Simply by chance, we would expect that results would vary from one sample to another.

Now imagine that we select not just one or two, but an enormous number of samples—a billion, perhaps—each with 2,044 young adults. For each sample we compute the proportion who are current marijuana users. We now have a billion statistics, each of which is a proportion that represents an

Table 11-4 Hypothetical Distribution of Sample Proportions of People Who Are Current Users of Marijuana, for Samples of $N = 2,044$

Sample Proportion	Number of Samples	Percentage of Samples	Probability
.34	50,000,000	5%	.05
.35	100,000,000	10	.10
.36	200,000,000	20	.20
.37	300,000,000	30	.30
.38	200,000,000	20	.20
.39	100,000,000	10	.10
.40	50,000,000	5	.05
TOTAL	1,000,000,000	100%	1.00

independently drawn sample of 2,044 observations. We could then construct a distribution of these sample statistics, showing how many *samples* had proportions of .34, .35, .36, .37, and so on, as in Table 11-4.

Before we go on, it is important to stop for a moment and be sure you know where we are. We start with a population of 33,600,000 young adults from which it is possible to draw an infinitely large number of samples of 2,044 people each. We are imagining that we actually selected a billion such samples, and for each we computed the proportion of young adults who are current marijuana users. Table 11-4 shows the possible outcomes for these billion samples, which I have arbitrarily limited to .34 to .40 in order to simplify things. This distribution, like any other, shows the number of cases that fall in each category. In 50,000,000 of the billion samples, for example, the sample proportion of people who were current users was .34. This constitutes 5 percent of all the samples we drew, which corresponds to a probability of .05. Thus, according to Table 11-4, the probability of drawing a sample in which the proportion of current users is .34 is .05.

Table 11-4, then, lists possible outcomes (sample proportions) and the probability associated with each. In other words, what we have here is a *probability distribution of possible sample results*. This kind of distribution is called a **sampling distribution**. Notice the "ing" in samp*ling*, which distinguishes it from a sample distribution. Do not confuse the two. A sam*ple* distribution shows the results for the observations that make up a single sample. Each of the 2,044 cases in the sample distribution in Table 11-3 is a young adult. A sam*pling* distribution, however, shows the results for all possible samples. Each of the billion cases in the sampling distribution in Table 11-4 is a sample statistic such as a mean or proportion. Thus the scores in a sample distribution are the scores for individual observations ("user" or "nonuser"). The "score" in a sampling distribution, however, is a *statistic—proportion who are users* that describes not an individual observation, but an entire sample of observations.

There are an infinite number of samples that could be drawn with replacement from a population (see Chapter 10 if you need to refresh your understanding of this). One billion samples is certainly a huge number of samples, but it does not exactly represent the *theoretical* sampling distribution because the latter is based on an *infinite* number of samples. No matter how many samples we drew, the resulting probability distribution of sample results would only approximate the theoretical sampling distribution. I have used the figure of one billion samples to introduce you to the idea of a probability distribution of sample statistics, but you should be aware that as theoretical distributions, sampling distributions are based on an infinite number of samples. They cannot be constructed by selecting one actual sample after another.

The Characteristics of Sampling Distributions

Like all distributions, sampling distributions can be described in terms of central tendency, variation, and shape. Since a proportion is a ratio-scale measurement (a proportion of .40 is twice as large as one of .20), sampling distributions for proportions have means and standard deviations. The sampling distribution in Table 11-4, for example, has a mean of .37 and a standard deviation of .014. It is unimodal and symmetrical. If we represented it with a probability density function, it would look something like Figure 11-3.

The mean of a sampling distribution is always equal to the population value we are trying to estimate. If we are measuring the proportion of young adults who are current marijuana users, for example, the average proportion in the sampling distribution will always equal the proportion for the population from which the samples are drawn. This general statement about sampling distributions does not hold for all of the population characteristics we might want to estimate, but it does for all those that we discuss in the chapters to follow.

Of course we do not know what the mean of the sampling distribution is. If we did, there would be no reason to conduct a study. Thus, you might think that it does us little good to know that the mean of the sampling distribution (which we do not know since we draw only one sample) equals the population value we are trying to estimate (which we also do not know). For the moment, you must be patient. You will see soon enough why this is important.

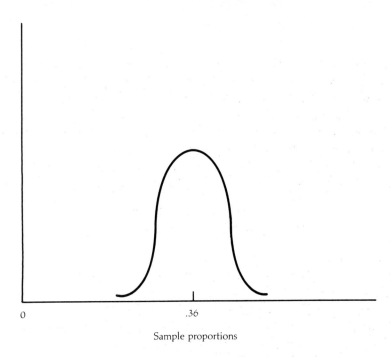

Figure 11-3 Probability density function showing sampling distribution in Table 11-4.

THE STANDARD ERROR

One of the most important characteristics of a sampling distribution is its standard deviation—so important, in fact, that it has its own name, the **standard error**. Standard errors are symbolized by the Greek letter *sigma* (σ) with a subscript that tells us which statistic the sampling distribution represents. The standard error for sample proportions would apply to Table 11-4, for example, and is symbolized by σ_{P_s} (the P stands for "proportion" and the s stands for "sample"). The standard error for sample means is represented by $\sigma_{\bar{x}}$.

What distinguishes the standard error from other standard deviations, of course, is the word "error." The idea of error is important in sampling distributions because the whole point of statistical inference is to use sample results to make accurate estimates of population characteristics. The mean of the sampling distribution in Table 11-4, for example, is .37, which is the same as the proportion of current users in the population from which these samples are drawn (Table 11-2). Of the billion samples, however, only 30 percent are within rounding error of the true value that each is trying to estimate. The rest are off by some amount. Twenty percent, for example, are .01 too high and another 20 percent are .01 too low (be sure you see this).

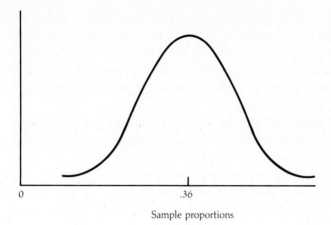

Figure 11-4 Sampling distribution for sample proportions with relatively large standard error.

A large standard error means that on average the sample results are widely spread out, which means that a relatively large proportion of the potential samples are in error by a wide margin. The sampling distribution describes the pool of possible samples we could draw, from which we draw just one. A large standard error (as in Figure 11-4) tells us that the pool contains a very large number of samples that differ from the population proportion by quite a lot. Any one of those samples is unlikely to be a very accurate estimate of the population. If the standard error is small, however (as in Figure 11-5), this means that the sample results are tightly clustered around the population

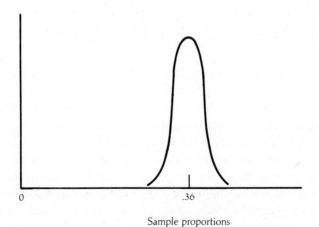

Figure 11-5 Sampling distribution for sample proportions with relatively small standard error.

proportion (which is what we are trying to estimate), which makes any one sample that we in fact draw relatively likely to be accurate. Thus, the size of the standard error reflects the relative amount of error involved in using any *one* sample from a sampling distribution as a basis for estimating population characteristics—which, of course, is exactly what we wind up doing. Hence the name, standard *error*.

It is important to keep the technical language straight here since it can be confusing. A sam*pling* distribution is a probability distribution in which each observation is a sample result, and the scores consist of sample statistics such as proportions and means. Since the sampling distribution has a mean, we can find ourselves talking about the mean of the sample means (the average mean in a sampling distribution of sample means) or the mean sample proportion (the average proportion in a sampling distribution of proportions). Sampling distributions also have standard errors. In this case, we are talking about the amount of variation in sample statistics—the fact that not all samples will have the same proportion or mean for a particular variable.

To minimize confusion, try to keep in mind just what statistic is represented in the distribution (means, proportions, etc.) and then remember that the mean of that distribution refers to whatever statistic the sampling distribution describes.

Calculating the Standard Error In general, the standard error for a sampling distribution of means, percentages, or proportions is calculated as the standard deviation in the population divided by the square root of the sample size. For sample means, for example, the formula for the standard error is

$$\sigma_{\bar{X}} = \frac{\sigma}{\sqrt{N}}$$

If we were trying to use a sample of 1,500 smokers to estimate the mean number of cigarettes smoked per day by smokers for the U.S. population, the standard error for the sampling distribution of sample means would be the population standard deviation for number of cigarettes smoked divided by the square root of the size of the sample (1,500).

As you may have noticed, however, we do not know what the population standard deviation is. If we did know it, we would also know the population mean for cigarette consumption since we cannot calculate a standard deviation without knowing the mean. And, of course, if we already knew the mean for the population, there would be no need to conduct a study to find it out. How, then, do we calculate a standard error?

The answer is that we use the *sample* standard deviation in order to estimate the population standard deviation in the numerator of the equation. With reasonably large samples (roughly 100 or more cases) this is a good estimate of the population standard deviation. In the long run, however, the average sample standard deviation tends to be slightly smaller than the population standard deviation—in other words, it is biased slightly downward. To compensate for this, we make the denominator of the standard error slightly smaller by subtracting one from the sample size. This makes the standard error slightly larger. The formula for the estimated standard error for sample means, for example, is

$$\hat{\sigma}_{\bar{X}} = \frac{s}{\sqrt{N-1}}$$

For the data in Table 11-1, the sample mean is 21.59 cigarettes per day, and the sample standard deviation is $s = 13.09$. Using the sample standard deviation as a substitute for the population standard deviation, the estimated standard error for the mean number of cigarettes smoked is

$$\hat{\sigma}_{\bar{X}} = \frac{13.09}{\sqrt{51{,}751-1}} = .06$$

As you can see, with samples as large as this one, the correction factor in the denominator is trivial, but with small samples, it can make a real difference.

The formula that we use to estimate the standard error depends on what we are trying to estimate. There is a different formula for estimating proportions, for example, just as there is a different formula for trying to estimate *differences* between proportions or differences between means. We will introduce these standard errors more formally in later chapters when we talk about how to use the sampling distributions that go with them.

THE CENTRAL-LIMIT THEOREM AND THE SHAPE OF SAMPLING DISTRIBUTIONS

In order to use a distribution in statistics, we have to know not only its central tendency and variation, but its shape as well. If we can know the shape of a sampling distribution, then we know everything we need to know in order to use it to make inferences about a population.

As before, we have a problem: since we do not actually draw all of the samples that make up a sampling distribution, how can we know its shape?

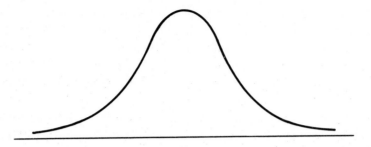

Figure 11-6   The normal distribution.

The answer lies in the **central-limit theorem**. According to this theorem, *the more cases we include in a sample, the more closely does the sampling distribution approximate a particular shape called the normal curve. This holds regardless of the shape of the population distribution.* The normal curve is a unimodal, symmetrical, "bell-shaped" curve (Figure 11-6). If the size of our sample is large enough, the central-limit theorem allows us to assume that the sampling distribution from which our sample was drawn is normally shaped.

An example will demonstrate the central-limit theorem. Instead of focusing on samples of people, we will focus on dice. To begin, imagine the population of all the dice in existence. If we could locate them all and record the numbers on the upturned face of each, the population distribution would be rectangular (a far cry from a normal curve) like that shown in Figure 11-7. Each number on a die has an equal chance of being face up. Since there are six faces, one-sixth of all dice should have a "1" showing, one-sixth should have a "2" showing, and so on. The mean face value for the dice population is $(1 + 2 + 3 + 4 + 5 + 6)/6 = 3.5$.

Suppose we want to draw a random sample from the dice population and use it to estimate the population mean. What would the sampling distribution from which that sample mean was drawn look like? How would its shape

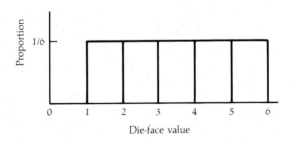

Figure 11-7   Histogram showing proportional distribution of the face values for the population of all dice.

Table 11-5 Sampling Distribution for Outcomes (Means) for Throws of Two Dice ($N = 2$)

Outcome (Mean)	Probability of Outcome
1.0	1/36 = .028
1.5	2/36 = .056
2.0	3/36 = .083
2.5	4/36 = .111
3.0	5/36 = .139
3.5	6/36 = .167
4.0	5/36 = .139
4.5	4/36 = .111
5.0	3/36 = .083
5.5	2/36 = .056
6.0	1/36 = .028
TOTAL	36/36 = 1.001

change as the sample size is increased? We start with a very small sample, $N = 2$. We roll two dice, add the two face values together, and divide by 2 to get a sample mean. There are 11 possible outcomes, for each of which there is a probability (see Table 11-5). Since each of the means in Table 11-5 is a sample result, Table 11-5 is a sampling distribution.

To see where the probabilities in Table 11-5 come from, look at Table 11-6. In this table, the row variable is the face value for the first die we roll; the

Table 11-6 Outcomes (Means) for Throws of Two Dice

First Die	Second Die					
	1	2	3	4	5	6
1	1.0	1.5	2.0	2.5	3.0	3.5
2	1.5	2.0	2.5	3.0	3.5	4.0
3	2.0	2.5	3.0	3.5	4.0	4.5
4	2.5	3.0	3.5	4.0	4.5	5.0
5	3.0	3.5	4.0	4.5	5.0	5.5
6	3.5	4.0	4.5	5.0	5.5	6.0

column variable is the face value of the second die; and the cells (where columns and rows intersect) are the means of each row-column combination. In the first row of the second column, for example, we see that if we roll a 1 on the first die and a 2 on the second die, the mean face value is $(1 + 2)/2 = 1.5$.

Notice that there are 36 cells, corresponding to each of the possible outcomes in Table 11-5. To calculate the probabilities in Table 11-5, all we have to do is count up the number of ways of getting each outcome and divide by 36. There is only one way to get a mean of 1.0, for example, so the probability of getting a mean of 1 with two dice is $1/36 = .028$.

If we drew a line graph of the sampling distribution in Table 11-5, it would look like Figure 11-8. Notice that this graph indicates the probability of a particular outcome by the height of the curve and not the area (it is not a probability density function). For the purposes of this demonstration, this does not matter. Also notice that the peak of the curve is above the point 3.5, which is the population mean. The mean of this sampling distribution is exactly equal to the population mean, just as we expected. The most important thing to notice is that even though the sample size is extremely small, the sampling distribution is symmetrical and is a lot closer to a normal curve than the rectangular population distribution in Figure 11-7.

What happens when we increase the sample size to $N = 3$? In tabular form, the sampling distribution is shown in Table 11-7. If we drew a line graph of this sampling distribution, it would look like Figure 11-9. If you compare Figure 11-9 with Figure 11-6, you can see that the sampling distribution for

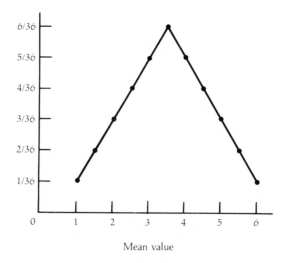

Figure 11-8 Sampling distribution for sample outcomes (means) for throws of two dice ($N = 2$).

Table 11-7 Sampling Distribution for Outcomes (Means) for Throws of Three Dice

Outcome (Mean)	Probability of Outcome
1.00	1/216 = .0046
1.33	3/216 = .0139
1.67	6/216 = .0278
2.00	10/216 = .0463
2.33	15/216 = .0694
2.67	21/216 = .0972
3.00	25/216 = .1158
3.33	27/216 = .1250
3.67	27/216 = .1250
4.00	25/216 = .1158
4.33	21/216 = .0972
4.67	15/216 = .0694
5.00	10/216 = .0463
5.33	6/216 = .0278
5.67	3/216 = .0139
6.00	1/216 = .0046
TOTAL	216/216 = 1.0000

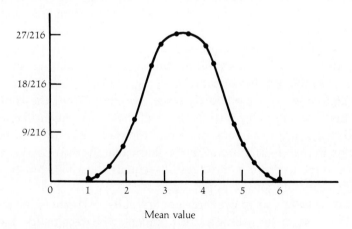

Figure 11-9 Sampling distribution for sample outcomes (means) for throws of three dice ($N = 3$).

sample means is moving closer to a normal shape even with only three cases. Although this example does not *prove* the central-limit theorem, it does show how sampling distributions rapidly approach normality even when the samples are drawn from populations that are decidedly not normal themselves. In practice, the sampling distribution will approach a normal shape very quickly as sample size increases. Even with samples of only 30 cases, the sampling distribution is so close to normal that the difference involved is trivial for most research purposes.

Technically, the central-limit theorem applies only to simple random samples drawn with replacement. As we saw in Chapter 10, however, samples are rarely drawn with pure randomness and are very rarely drawn with replacement. So long as cases are selected with independence and with known probabilities, however, the violation of the assumptions underlying the central-limit theorem does not have serious consequences. In addition, there are correction factors that can be used with small samples. There is error involved, of course, but not so great as to undermine the usefulness of most practical applications (see Kish, 1965).

Getting Some Perspective

It might be useful to stop for a moment and get some perspective on where this chapter is taking us. When we do a study, we focus on a population of observations. For each variable that we study, there is a population of observations that are distributed in some way. In practice, rather than gathering the entire population of observations, we usually draw a sample in a way that maximizes the likelihood that it represents the population with enough accuracy to justify relying on a sample rather than a census.

Once the data are gathered, we have only one sample. Whatever statistic we use to describe our sample (such as a mean), it is drawn from a pool—a sampling distribution—consisting of an infinite number of means that describe an infinite number of samples of a given size.

If our sample is even moderately large, we know from the central-limit theorem that the sampling distribution will look very much like a normal curve and that we can estimate the size of the standard error that measures the variation in that distribution. We also know that the mean of the sampling distribution—the average sample result—equals the characteristic that we are trying to estimate for the population.

All of this will allow us to use our sample results—whether it be a mean or a proportion—to make estimates of population characteristics and attach probabilities that express our level of confidence in the estimates. To do this, we need to use special tables of probabilities that are associated with the normal distribution. In the remainder of this chapter we will take a close look at the normal curve and the probability tables that describe it. In Chapters 12 and 13 we will introduce the uses of the curve to make actual estimates of population characteristics.

Before you continue, be sure you have a firm grasp of the idea of a probability distribution and how we use areas beneath a curve to represent a distribution. Focus for a moment on the idea of a sampling distribution and how it differs from a sample distribution. Imagine a variable such as income, and see if you can use it to explain the difference between a population distribution and a sample distribution; between a population mean, a sample mean, and a sampling distribution for sample means; and between the population standard deviation, the sample standard deviation, and the standard error for the sampling distribution of sample means. Focus on the central-limit theorem and what it tells us about sampling distributions. Getting the clearest grasp of these central concepts now will be more than worth your while later on.

The Normal Distribution

The purpose of statistical inference is to use sample results to estimate population characteristics and attach probabilities to those estimates. The key to this process lies in the sampling distribution from which our sample result is selected. In order to use a sampling distribution to calculate probabilities, we have to know three things about it: its mean, its standard error, and its shape. Although we do not know the actual value of the sampling distribution's mean, we do know that it equals the population characteristic we are trying to estimate. We can use the sample standard deviation and sample size to construct an estimate of the standard error. The central-limit theorem allows us to assume a normally shaped sampling distribution if our sample is moderately large. All that is left to understand is how to put all of this knowledge together in order to make statistical inferences. To do that, we need to know more about the normal curve.

The bell-shaped normal curve is one of the most widely used probability distributions in statistics. The upper and lower ends of the curve are called **tails** (Figure 11-10). Because the curve is symmetrical, half of the cases in the distribution lie above its mean and half lie below.

In order for a distribution to be normal, it must be more than bell-shaped. It must conform to a rather complicated mathematical formula. Some normal curves are flatter (a larger variance) or more peaked (a smaller variance) than others. In short, the word "normal" refers to a family of curves that conform to the same mathematical function but differ in their variance.

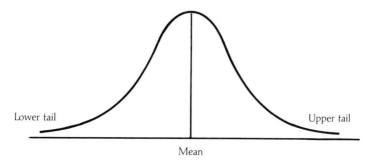

Figure 11-10 The normal distribution.

There are many bell-shaped curves that are not normal. They differ from the normal curve by being either too flat or too peaked. The degree to which a curve deviates from the normal shape is called **kurtosis** (from the Greek word for "curve"). If a curve is flatter than a normal curve, it is called **platykurtic** (or "flat curve"). If it is more peaked than a normal curve, it is called **leptokurtic** (or "thin curve"). You will rarely encounter these terms in the social and behavioral sciences (although economists do make use of them), but you should at least be familiar with them.

USING THE NORMAL DISTRIBUTION

In inferential statistics, the **normal distribution** is used to describe a sampling distribution. Because it is in the form of a probability density function, the height of the curve is not important to us; we are interested in the area beneath the curve. We use the curve to calculate the probability that any given case will fall within a specified range of scores, and the range generally includes the mean of the distribution. Therefore, we measure distances between the mean and some point above or below it. To measure the distance between the mean of a normal distribution and some point above or below it, the *units* of the horizontal axis must be in terms of standard deviations (standard errors when dealing with a sampling distribution), and this fact is crucial for our use of the curve.

If the variable is income, for example, with a mean of $20,000 and a standard deviation of $6,000, then instead of saying that a score of $32,000 is $12,000 above the mean, we say that it is $12,000/$6,000 = 2 standard deviations above the mean. This is simply a translation from one set of units to another, just as we might convert from temperature in degrees Fahrenheit to temperature in degrees centigrade or from value in dollars to value in Mexican pesos or French francs. If the standard deviation is $6,000, then each distance of $6,000 on the horizontal axis is equivalent to one standard deviation.

Consider, for example, a distribution of College Board scores which is normal with a mean of 500 and a standard deviation of 100 (Figure 11-11). This figure is laid out so that we can express distances from the mean in two ways. On the upper row, distances are measured in units of actual test scores, also known as **unstandardized** scores. A score of 600 is 100 points above the mean. Since each 100 points is equivalent to one standard deviation, then a score of 600 is one standard deviation above the mean (lower row of numbers). An unstandardized score of 600, then, is equivalent to a *standardized* score of $(600 - 500)/100 = +1$ standard deviation, or $+1\sigma$. Similarly, an unstandardized score of 300 is 200 test points below the mean which is equivalent to a standardized score of $(300 - 500)/100 = -2$ standard deviations, or -2σ.

Standard scores represent distances between a particular score and the mean of a distribution, expressed in units of standard deviations. In a normal distribution, standard scores are also known as **Z-scores**. Thus, in the normal distribution of College Board scores, an unstandardized score of 400 corresponds to a Z-score of $(400 - 500)/100 = -1.0\sigma$. The minus sign indicates that the score is below the mean, and the number tells us how many standard deviations below the mean the score is. A score of 700 corresponds to a Z-score of $(700 - 500)/100 = +2.0\sigma$, indicating a score that is 2 standard deviations above the mean.

To use the normal curve to calculate probabilities, we find the probability that a score lies within a certain distance of the mean—in other words, the probability that a score lies between the mean and a certain number of Z-scores above or below it. To make this easier, mathematicians have

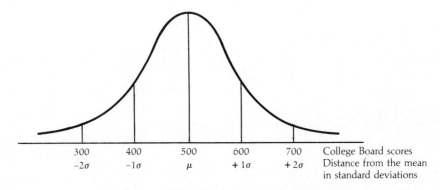

Figure 11-11 Distribution of College Board scores for high school seniors.

computed a table that shows the proportion of cases in a normal distribution that will lie between the mean of that distribution and any specified number of standard deviations above or below it (Table 11-8).

To use the table of normal curve probabilities, we first express the distance between the mean and a score in units of standard deviations accurate to the nearest hundredth (such as 1.00 or 2.34). To find the proportion of cases that

Table 11-8 Areas under the Normal Curve

Fractional parts of the total area (10,000) under the normal curve, corresponding to distances between the mean and scores which are Z standard-deviation units from the mean.

Z	.00	.01	.02	.03	.04	.05	.06	.07	.08	.09
0.0	0000	0040	0080	0120	0159	0199	0239	0279	0319	0359
0.1	0398	0438	0478	0517	0557	0596	0636	0675	0714	0753
0.2	0793	0832	0871	0910	0948	0987	1026	1064	1103	1141
0.3	1179	1217	1255	1293	1331	1368	1406	1443	1480	1517
0.4	1554	1591	1628	1664	1700	1736	1772	1808	1844	1879
0.5	1915	1950	1985	2019	2054	2088	2123	2157	2190	2224
0.6	2257	2291	2324	2357	2389	2422	2454	2486	2518	2549
0.7	2580	2612	2642	2673	2704	2734	2764	2794	2823	2852
0.8	2881	2910	2939	2967	2995	3023	3051	3078	3106	3133
0.9	3159	3186	3212	3238	3264	3289	3315	3340	3365	3389
1.0	3413	3438	3461	3485	3508	3531	3554	3577	3599	3621
1.1	3643	3665	3686	3718	3729	3749	3770	3790	3810	3830
1.2	3849	3869	3888	3907	3925	3944	3962	3980	3997	4015
1.3	4032	4049	4066	4083	4099	4115	4131	4147	4162	4177
1.4	4192	4207	4222	4236	4251	4265	4279	4292	4306	4319
1.5	4332	4345	4357	4370	4382	4394	4406	4418	4430	4441
1.6	4452	4463	4474	4485	4495	4505	4515	4525	4535	4545
1.7	4554	4564	4573	4582	4591	4599	4608	4616	4625	4633
1.8	4641	4649	4656	4664	4671	4678	4686	4693	4699	4706
1.9	4713	4719	4726	4732	4738	4744	4750	4758	4762	4767
2.0	4773	4778	4783	4788	4793	4798	4803	4808	4812	4817
2.1	4821	4826	4830	4834	4838	4842	4846	4850	4854	4857
2.2	4861	4865	4868	4871	4875	4878	4881	4884	4887	4890
2.3	4893	4896	4898	4901	4904	4906	4909	4911	4913	4916
2.4	4918	4920	4922	4925	4927	4929	4931	4932	4934	4936
2.5	4938	4940	4941	4943	4945	4946	4948	4949	4951	4952
2.6	4953	4955	4956	4957	4959	4960	4961	4962	4963	4964
2.7	4965	4966	4967	4968	4969	4970	4971	4972	4973	4974
2.8	4974	4975	4976	4977	4977	4978	4979	4980	4980	4981
2.9	4981	4982	4983	4984	4984	4984	4985	4985	4986	4986
3.0	4986.5	4987	4987	4988	4988	4988	4989	4989	4989	4990
3.1	4990.0	4991	4991	4991	4992	4992	4992	4992	4993	4993
3.2	4993.129									
3.3	4995.166									
3.4	4996.631									
3.5	4997.674									
3.6	4998.409									
3.7	4998.922									
3.8	4999.277									
3.9	4999.519									
4.0	4999.683									
4.5	4999.966									
5.0	4999.997133									

SOURCE: Rugg, 1917, Appendix Table III, pp. 389–90.

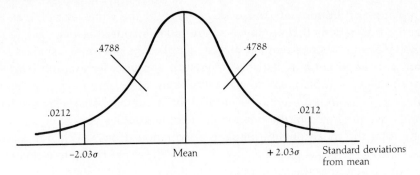

Figure 11-12 Sampling distribution for sample means.

lie between the mean and 2.03 standard errors above the mean, we go down the left-hand column to the number 2.0 (our Z-score to the nearest tenth) and then go across that row until we are under the column headed by ".03" (corresponding to 2.03 rather than just 2.0). The four-digit number found there (4788) is the proportion of cases (.4788) that lie between the mean and 2.03 standard deviations above the mean (the decimals have been left out of the table to save room). Since the curve is symmetrical, this is also the proportion of cases that lie between the mean and 2.03 standard deviations *below* the mean. If we add these two together (or simply multiply times two since the proportion is the same for both), we get the proportion of cases in a normal distribution that lie *within* 2.03 standard deviations of the mean: .4788 + .4788 = .9576. I use the word "within" to refer to those scores that lie between the mean and *either* 2.03 standard deviations above *or* 2.03 standard deviations below (see Figure 11-12).

Since half of the cases in a normal distribution lie above the mean and half lie below, if we subtract .4788 from .5000 the result (.0212) is the proportion of cases that will lie *at least* 2.03 standard deviations above the mean. This is also the proportion that will lie at least 2.03 standard deviations below the mean. If we double this proportion, the result (.0424) is the proportion of cases that will lie at least 2.03 standard deviations above *or* below the mean (see Figure 11-12).

THE NORMAL CURVE AND SAMPLING DISTRIBUTIONS

The normal curve table shows that in *any* normal distribution roughly two-thirds of the cases will lie within one standard deviation of the mean, just over 95 percent will lie within two standard deviations, and more than 99 percent will lie within three standard deviations of the mean. If we are talking about a sampling distribution, of course, everything is the same except that the standard deviations that are involved are called standard errors.

Being able to assume that a sampling distribution is normally shaped is of enormous importance in statistical inference. Suppose, for example, that we want to use a sample mean to estimate the mean income for a population. Our sample mean represents one of the infinite number of sample means we could have drawn from the pool of all possible samples of a given size. We know that the mean of the sampling distribution (from which our sample mean was selected) equals the population mean we are trying to estimate. If we can assume that the sampling distribution is normally shaped, however, we know something else of great value. We know, for example, that of all the sample means we could have drawn, the proportion of sample means that will lie between the mean of the sampling distribution and two standard errors above *or* below is just over 95 percent. This means that if we make the statement, "Our sample mean is no more than two standard errors away from the actual population mean," 95 percent of the time such a statement will be correct. In general, if we think of proportions of the area under the normal curve as probabilities, *even though we do not know what the population mean is*, we can calculate the probability that our sample mean lies within any given distance of the population mean. In Chapters 12 and 13 we will make great use of this key fact.

TRANSLATING BETWEEN STANDARD ERRORS AND UNSTANDARDIZED SCORES

Thus far we have measured distances from the mean of a sampling distribution in units of standard deviations and standard errors. Most of us find it hard to think in terms of standard deviations. The proportion of cases that lie *more* than 2.03 standard errors from the mean is $.0212 \times 2 = .0424$, for example, which means that the probability of drawing a sample mean that is more than 2.03 standard errors from the population mean is only .0424 (as in Figure 11-12). The statement would be easier to grasp, however, if it were expressed in terms of actual dollars of income, such as, "There is only a .0424 probability that our sample mean will be more than X dollars away from the population mean we want to estimate."

We can translate from standard scores to unstandardized scores by recalling what a standard error is: it is the square root of the average squared deviation of the sample means about the mean of the sampling distribution, and its units are the same as those of the unstandardized scores. In other words, in the case of income, each standard error is equivalent to a certain number of dollars. If we gathered income data from 401 people and found a standard deviation of $6,000, the estimated standard error would be:

$$\hat{\sigma}_{\bar{X}} = \frac{\$6,000}{\sqrt{401-1}} = \$300$$

This tells us that each distance of one standard error on the horizontal axis of the sampling distribution is equivalent to $300 of income. A distance of 2.03 standard errors from the mean of the sampling distribution is therefore equal to 2.03 × $300 = $609. We can now say that with 401 cases, a normal sampling distribution, and an estimated standard error of $300, there is only a .0424 probability of drawing a sample mean that will be more than $609 above or below the mean of the sampling distribution, which happens to be equal to the population mean.

If we want to find the probability that a sample mean will lie within a certain number of dollars of the population mean, we reverse what we have just done by dividing the distance in dollars by the size of the standard error. This translates the unstandardized distance in dollars into a standardized distance in terms of standard errors:

$$Z = \frac{\$450}{\$300} = 1.5 \text{ standard errors}$$

With the Z-score, we then go into the normal table and find the area that corresponds to 1.50 standard errors, or .4332. If the sample size is 401 and the sampling distribution is normal, then the probability of drawing a sample whose mean is *within* $450 of the mean of the sampling distribution is 2 × .4332 = .8664. The probability of selecting a sample whose mean is *more* than $450 from the mean of the sampling distribution is 1.0000 − .8664 = .1336.

In summary, to translate a distance in standard errors into one in unstandardized units, we *multiply* the number of standard errors times the size of the standard error (such as 1.50 × $300 = $450). To translate a distance in unstandardized units into standard errors, we *divide* the distance by the size of the standard error (such as $450/$300 = 1.5 standard errors). Now, think of some examples of your own and practice this back-and-forth translation between standard errors and unstandardized scores. Playing around with actual problems helps immeasurably in learning about the normal curve and how it works. Work hard at the normal curve problems at the end of this chapter—*now* is the time to work out any aspects of it that you do not understand.

Summary

1. Statistical inference is the process of using sample results to make statements about a population with a level of confidence expressed as a probability. The purpose of this chapter has been to introduce the concept of sampling distributions; the ways of describing them in terms of central tendency, variation, and shape; and the use of the normal distribution to calculate probabilities for different sample outcomes.

2. Inference relies on special probability distributions that take the form of probability density functions, graphs that represent proportional distributions in terms of the area beneath a curve rather than the height of a curve at any given point. Unlike other graphic displays, these functions represent the proportion of scores that lie within a given range and cannot be used to determine the probability associated with any specific score.

3. In order to use sample distributions to make statements about population distributions, we make use of a special theoretical distribution known as a sampling distribution. A sampling distribution is a probability distribution for all the possible sample results we could have drawn from the infinite number of possible samples of a given size.

4. Like all distributions, sampling distributions can be described in terms of central tendency, variation, and shape. The mean of the sampling distribution (the average sample result) is always equal to the population value we are trying to estimate. Variation is measured by a standard deviation called the standard error. The central-limit theorem tells us that when the sample is large enough, the shape of a sampling distribution for means or proportions can be assumed to be normal.

5. Although many bell-shaped curves appear to be normal, only those that conform to the formula for a normal curve are in fact normal. Curves that are flatter than normal are called platykurtic, while those that are more peaked than normal are leptokurtic.

6. The normal distribution is set up in such a way that we can determine the probability that a score in that distribution will lie between the mean and any given number of standard deviations above or below the mean. Distances from the mean must always be measured in units of standard deviations (or, with sampling distributions, in standard errors). This means that it is always necessary to translate between raw, unstandardized scores and standard scores. In a sampling distribution, standardized scores are known as Z-scores.

Key Terms

central-limit theorem 280
kurtosis 286
   leptokurtic 286
   platykurtic 286
normal distribution 286
population distribution 272
probability density function 271
sample distribution 272

sampling distribution 274
standard error 276
standard (Z) score 287
statistical inference 268
tail 285
unstandardized score 287
Z-score 287

Problems

1. Define and give an example of each of the key terms listed at the end of the chapter.

2. How does a probability density function differ from other ways of graphically displaying a probability distribution?

3. Suppose we are interested in the mean weight of 12 to 16-year-old U.S. adolescents. In terms of this variable, what is the difference between the population mean, the sample mean, and the sampling distribution for mean weight? How are they related?

4. Under what circumstances is it necessary to estimate the value of the standard error for means or proportions? In the case of means, how is the standard error estimated?

5. If the population standard deviation for Y is 3.0, and we select a sample of 100 observations, compute the standard error for the mean of Y.

6. If we do not know the population standard deviation for Y, and we select a sample of 900 observations that has a sample standard deviation of 10.8, how would you compute the estimated standard error for the mean of Y? What is your result?

7. What is the central-limit theorem, and why is it important?

8. Suppose we have a sample of occupational prestige scores that is normally distributed with a mean of 80 and a standard deviation of 20. Given this information, compute the proportion of cases with

occupational prestige scores that are:
a. higher than 80
b. less than 80
c. between 60 and 80
d. between 80 and 90
e. 75 or higher
f. 50 or lower
g. within 20 points of the mean
h. at least 20 points above *or* below the mean
i. within 30 points of the mean
j. at least 40 points above *or* below the mean
k. exactly 72
l. exactly 45
m. between 70 and 90
n. between 60 and 85

9. If the mean of a normal distribution is 35, what is the value of the median?

10. Suppose we draw a representative sample of 401 married people and ask them at what age they married their current spouse. The sample mean is 28 years and the sample standard deviation is 10 years.
   a. Estimate the standard error for the sampling distribution of sample means.
   What is the probability of selecting a sample whose mean:
   b. falls between the unknown population mean and 1.0 standard error above the mean?
   c. falls between the unknown population mean and 1.0 standard error below the mean?
   d. falls *within* 1.5 standard errors of the unknown population mean?
   e. falls within 2.5 standard errors of the unknown population mean?
   f. is at *least* 2.25 standard errors above the unknown population mean?
   g. is at *least* 2.75 standard errors below the unknown population mean?
   h. is at least 1.96 standard errors above *or* below the unknown population mean?
   i. is at least $\frac{1}{2}$ year above the population mean?
   j. is at least 1 year below the population mean?
   k. is within $1\frac{1}{2}$ years of the population mean?
   l. is at *least* $1\frac{1}{2}$ years above *or* below the population mean?
   m. is *no more than* 1 year above or below the population mean?
   n. is greater than the population mean?
   o. is less than the population mean?
   p. equals 26.2 exactly?
   q. equals 31.7 exactly?

🙏 12 🙏

Estimating Population Means and Proportions

Statistical inference techniques can be used to make statements about populations in two basic ways. The first is the estimation of population characteristics such as proportions or means. Using sample means, for example, we might want to infer the mean income in a population, the mean number of children people intend to have, the number of years of schooling they have, the number of hours they work per week, the number of employed adults in a household, or the average frequency of church attendance.

The second kind of statement about populations rests on **hypothesis testing**. With hypothesis testing, we make a specific assumption about a population characteristic (such as a population mean) and then see if our sample data justify the *rejection* of that assumption. Although most of the literature emphasizes hypothesis testing as the method of choice in scientific research, we are going to start with estimation techniques for two reasons. First, I share the belief of many researchers that hypothesis testing is less useful than estimation and more conducive to misinterpretation. Second, statistical inference is easier to understand if we start with estimation. Hypothesis testing involves a reverse logic that can be quite confusing when you encounter it for the first time.

In this chapter we will describe techniques for estimating population proportions and means as well as *differences* between means and proportions for two populations. With this as a thorough grounding in basic inference procedures, in Chapter 13 we will turn to hypothesis testing. To understand the material in this chapter, it is vital that you understand Chapter 11. If you need to, take some time to review that chapter before reading on.

Point Estimates

To estimate a population characteristic such as a mean or proportion, one obvious and relatively simple strategy is to use the comparable statistic for a sample. To estimate a population mean, for example, we might simply use the mean for a sample drawn from that population. As you may recall, this is exactly what we did in Chapter 11 to estimate the standard error for sample means: since we did not know the population standard deviation, we used the sample standard deviation instead in the numerator of the standard error formula. This kind of estimate, that consists of a single value, is called a **point estimate**.

As convenient as this may be, it is of course very unlikely that a sample statistic such as a mean will exactly equal the corresponding value for the population it comes from; but if we want to use a single specific value as an estimate, the sample statistic is certainly our best guess.

As we saw in Chapter 11, in the long run the average sample mean will be the same as the corresponding population characteristic—which is to say, the mean of the sampling distribution of sample means will equal the population mean. In the same way, the average sample proportion in a sampling distribution will equal the population proportion. Because of this, means and proportions are called **unbiased estimators**. They are called "unbiased" because in the long run sample means will be no more likely to underestimate the population mean than they will be to overestimate it.

As we also saw in Chapter 11, however, this is not true for the standard deviation, which is a **biased estimator**. In the long run, sample standard deviations tend to underestimate the population standard deviation. For this reason, when we substituted the sample standard deviation for the population standard deviation in the formula for the standard error, we compensated by reducing the denominator slightly, dividing by $\sqrt{N-1}$ instead of $\sqrt{N}$:

$$\hat{\sigma}_{\bar{X}} = \frac{s}{\sqrt{N-1}}$$

The advantage of point estimates is their precision and convenience: when we think of the average family size in a society, it is easier to think in terms of a single number (2.3 children) than a range of numbers. Unfortunately, this is offset by the fact that it is extremely unlikely that any point estimate is correct. As is often the case in statistics, what we gain in precision and convenience we then must pay for in lost confidence and uncertainty.

As you may have begun to see in Chapter 11, however, when we think of *ranges* of outcomes rather than specific outcomes, we are in a position to calculate various kinds of probabilities. Probabilities, of course, are very useful tools for measuring the likelihood of error and, therefore, the degree of confidence we can have in an inference made from a sample to a population.

Confidence Intervals: Estimating Population Means

The number of children that people consider to be ideal has long been of interest to demographers and others examining the relationship between cultural values and population growth. Suppose, then, that we want to estimate the population mean for the variable, "ideal number of children." Because of the enormous cost involved in a census, we would probably take a sample of adults, such as that found in the GENSOC surveys. We ask each respondent, "What do you think is the ideal number of children for a family to have?" and record the answers. The result is the distribution found in Table 12-1.

As with every ratio-scale distribution, this one has a mean and a standard deviation. The mean ideal number of children is 2.60, the standard deviation

Table 12-1 Number of Children
Considered Ideal by U.S. Adults, 1986

Number of Children	Frequency	Percent
0	17	1.2
1	24	1.8
2	742	54.2
3	359	26.2
4	185	13.5
5	24	1.8
6	14	1.0
7	5	0.4
TOTAL*	1370	100.1

SOURCE: Davis, 1986, p. 232.
* There are 100 cases with inadequate or missing data from the original sample of 1,470.

is 0.97, and the sample size is 1,370. The estimated standard error is

$$\hat{\sigma}_{\bar{X}} = \frac{s}{\sqrt{N-1}} = \frac{.97}{\sqrt{1369}} = .03$$

Statistical inference is the process of estimating a population characteristic (such as a mean) and then attaching a probability which expresses a level of confidence in what we have done. Ideally, we would like to be able to make a precise estimate in which we have a great deal of confidence, such as "We are 99 percent sure that the mean ideal number of children is 2.60." There are, however, two major problems with this kind of statement. First, as we saw earlier, the probability that a point estimate actually equals the population mean is infinitesimally small since there are an infinite number of possible sample results. We could never be remotely sure that our sample mean equalled the population mean exactly. Second, even if this were not a problem, our sample result either equals the population mean or it does not. We cannot legitimately attach a probability statement to a specific sample outcome.

One way to solve both of these problems is to use a range of values—an **interval estimate**—such as 2.52 to 2.68. This would solve the problem of not being able to have confidence in a single value used to estimate the population mean. We solve the problem of not being able to attach a probability to a specific estimate (even with an interval, it either includes the population value or it does not) by expressing our confidence in the procedures used to generate the estimate rather than the estimate itself. Instead of saying, for example, that we are 99 percent sure that our interval includes the population value, we can say that 99 percent of the time, the procedures we have used will result in an interval that includes the population value. Our confidence, then, is based on the procedures we use—from the sampling process to the mathematics of statistical inference—rather than the estimate itself. The result is called a **confidence interval**.

Our sample mean in this case is 2.60 children. We are going to use an interval rather than a single value to estimate the population mean. Since the sample mean is the best single estimate we have of the population mean, it makes sense to include that in the interval; and since the sample mean is as likely to be too high an estimate as it is to be too low, it also makes sense to place the sample mean at the *center* of the interval. We might, for example, add and subtract 1.00 child to the sample mean and get an interval that runs from 1.60 to 3.60. If we wanted to be more precise, we could narrow the interval to, perhaps, 2.52 to 2.68 (as a range of only ±0.08 in contrast with a range of ±1.00 in the first interval). Suppose that we want to be fairly precise and we select as an interval the sample mean plus and minus 0.08 child, or 2.52 to 2.68. It is conventional to write such an interval as

$$\bar{X} \pm 0.08$$

In case you are wondering about the basis for selecting what may seem to be a rather arbitrary interval, I should point out that in practice, the width of the interval is dictated by the level of confidence we want to have in it, something which is usually decided *before* constructing the interval. We are doing things backwards here because in the long run it makes it easier to understand what confidence intervals are about.

Having decided on an interval that we hope includes the population mean, we are left with the job of finding the probability that our procedures have resulted in an accurate interval. For this we return to the normal distribution first introduced in Chapter 11.

The sample mean is based on a sample that satisfies the three conditions of scientific sampling described in Chapter 10 (we base this on our justifiable confidence in the National Opinion Research Center's sampling techniques as described in the appendixes to its statistical reports). Since the sample is large, we may use the central-limit theorem to assume that the sampling distribution for sample means is normally shaped. As we saw in Chapter 11, the normal distribution table is set up in terms of standard deviations or, in the case of sampling distributions, standard errors. Before we can use the normal table to find the appropriate probability, then, we must convert the length of our interval from unstandardized scores to Z-scores.

We constructed the interval by adding and subtracting a distance of 0.08 child to the sample mean. We can convert this to units of standard errors by dividing the length of each segment of the interval (.08) by the number of children in each estimated standard error (.03). This gives us an interval whose length is expressed in standardized units:

$$\text{Confidence interval} = \bar{X} \pm \frac{.08 \text{ child}}{.03 \text{ child per standard error}}$$

$$= \bar{X} \pm 2.67 \text{ standard errors}$$

We began with an interval stated as "the sample mean plus and minus 0.08 child." By standardizing, we now have an *equivalent* interval stated as "the sample mean plus and minus 2.67 standard errors." Stop for a moment and satisfy yourself that if we take this new interval and multiply the Z-score (2.67) times the number of children in each standard error (.03), we wind up with the original interval (within rounding error).

If we think of our interval as a straight line with the sample mean at its center, it would look like this:

$$2.67\sigma_{\bar{X}} \qquad 2.67\sigma_{\bar{X}}$$

$$2.52 \qquad\qquad 2.60 \qquad\qquad 2.68$$

The sample mean that lies at the center of our interval is one of an infinite number of sample means that could have been chosen from the sampling distribution for sample means. Figure 12-1 shows the sampling distribution from which this single sample result was selected. Notice that it is normally shaped and has as its mean the population mean we are trying to estimate. Also notice that the length of the horizontal axis is measured in Z-scores— units of standard errors—which is necessary for using the normal distribution table. The point marked "-1$\sigma_{\bar{X}}$," for example, represents a distance of one standard error below the population mean. As we saw in Chapter 11, the normal table tells us what proportion of cases will lie between the mean of a normal distribution and a given number of standard errors above or below. Just over two-thirds of all possible samples (.6826), for example, will have means between one standard error below and one standard error above the population mean (the shaded area in Figure 12-1).

Our interval is also in units of standard errors, and we could draw it as a horizontal line below the sampling distribution as in Figure 12-2. In this figure

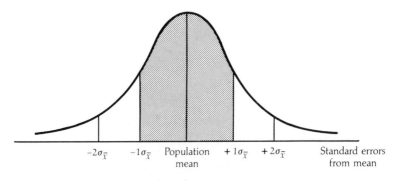

Figure 12-1 Sampling distribution for the mean ideal number of children.

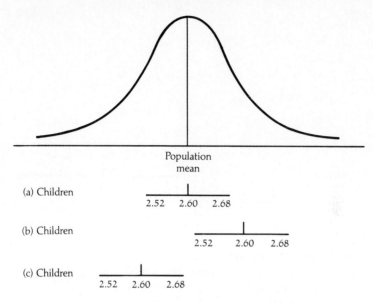

Figure 12-2 Sampling distribution for mean ideal number of children with possible locations for a confidence interval based on a single sample.

we have placed the interval at three possible locations. In (a), for example, the sample mean equals the population mean exactly; in (b), the sample mean is higher than the population mean; and in (c), the sample mean is lower than the population mean. The point is that since we have constructed an interval that has the sample mean at its center, the interval could be located in an infinite number of possible positions *depending on how far above or below the population mean our sample mean actually is*.

Since we do not know the value of the population mean, however, how do we know where our interval actually is in relation to it and, therefore, how accurate our interval is? The answer might be clearer if we rephrase the question: "How far off does our sample mean have to be in order for the *interval* to miss the population mean?" If you look at Figure 12-3, you can see that in order for the interval to miss the population mean, the sample mean must be so far above the population mean [as in (a)] that the lower boundary of the interval misses the population mean which lies at the center of the sampling distribution. The interval will also miss the population mean if the sample mean is so far below the population mean [as in (b)] that the upper boundary of the interval misses the population mean. How far above or below the population mean does the sample mean have to be in order for this to happen? The answer for (a) is *just over the length of the lower leg away*. In (b),

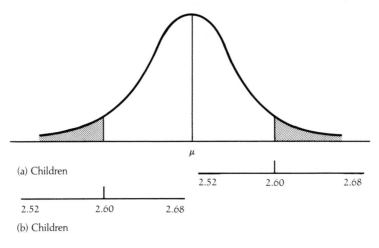

Figure 12-3 Sampling distribution for sample means, showing how much in error the confidence interval must be in order to miss the population mean.

the sample mean is just over the length of the upper leg away. As we know, the lower and upper legs of the interval are the same length, 0.08 child or 2.67 standard errors.

As you may have sensed already, we are close to the answer. The interval we have constructed will miss the population mean if the sample mean is more than 2.67 standard errors above *or* below the population mean. Since the population mean is also the mean of the sampling distribution, the question becomes, "What is the probability of drawing a sample mean that is more than 2.67 standard errors from the mean of a normal sampling distribution?" We are looking for the probability of drawing a sample mean that will fall in *either* of the two shaded regions in Figure 12-3. To find this, we look in the normal table (Appendix Table 1), down the left-hand column to 2.60 and over to the column headed by ".07" to get the probability that corresponds to a Z-score of 2.67. The result—.4962—is the proportion of cases in a normal sampling distribution that lie between the mean and plus or minus 2.67 standard errors. This corresponds to each *un*shaded area on either side of the mean in Figure 12-3. To get the proportion of cases that lie *beyond* these areas, we subtract .4962 from .5000. The result (.0038) is the probability that our sample mean will fall in *one* of the shaded regions. Since our sample estimate is as likely to be too high as it is to be too low, if we double .0038, we get the probability (.0076) that our sample result will fall in *either* shaded region (this is a good place to stop and go over this paragraph to be sure you understand it before reading on).

We have calculated the probability (.0076) of drawing a sample of 1,370 cases and getting a mean ideal number of children that is more than 2.67 standard errors above or below the mean ideal number of children in the population. This is the probability that in the long run, an interval constructed in this way will *miss* the population mean. Our estimate is: The population mean lies somewhere between 2.52 and 2.68 children or, expressed as an interval, the population mean is estimated to be

$$\bar{X} \pm 0.08 \text{ child}$$

There is a probability of .0076 (less than 1 percent) that we have drawn a sample whose mean is so far from the population mean that the interval we constructed with the sample mean at its center fails to include the population mean. Therefore, there is a $(1 - .0076) = .9924$ probability that intervals constructed in this way *would* include the population mean. Our level of confidence in what we have done is thus .9924, or, expressed as a percentage, 99.24 percent. The interval would be known as "a 99.24-percent confidence interval," or, somewhat less precisely (and more modestly), as a "99-percent confidence interval."

There are an infinite number of possible samples that we could have drawn from the population. Thus, there are an infinite number of sample means and an infinite number of possible confidence intervals with those means at their center. From the normal distribution we found that 99 percent of those means will fall within 2.67 standard errors of the unknown population mean, and, therefore, that 99 percent of all possible confidence intervals (constructed by adding and subtracting 2.67 standard errors to a sample mean) will include the unknown population mean. Our confidence level of 99 percent reflects the long-run chances of drawing one of those 99 percent of all possible sample results that will include the population mean. As Davis (1971, p. 57) put it, "When we say that our confidence level is .95, we mean that we are using a crystal ball that is known to work 95 times out of a hundred, although on any given trial it either works or fails."

In this way we have used statistical inference to estimate an unknown population mean with a known degree of confidence. Notice that we never needed to know the population mean, only that the sample was properly drawn and that the sampling distribution was normally shaped. *We have taken something known (the sample mean) and used a theoretical distribution (the sampling distribution) to estimate something unknown (the population mean), and we have calculated the probability that we are correct.*

The ability to estimate population means is a powerful statistical tool that we can use to describe aspects of the world that would otherwise be beyond our reach. Using sample data, for example, sociologists can estimate the average number of weeks that unemployed people are out of work, the average size of families and households, the average family income, the average number of hours that people work each week, or the average level of educational attainment. Public health and medical researchers can estimate the average level of blood pressure in the U.S. population, the average blood cholesterol level, or the mean number of years that cancer patients survive after their initial diagnosis. Educational researchers can estimate average test scores for entire populations of millions of students and see how those scores change over time and differ among population subgroups defined by such characteristics as gender, age, race, ethnicity, and region of residence. All of these estimates can be made with known levels of confidence for even the largest populations.

How Confidence Intervals Are Actually Constructed

In the previous section we began with an interval and then found the confidence level that went along with it. In practice, however, we begin with an idea of how confident we want to be and then construct the appropriate interval. The procedure is the same, except in reverse order.

With a 99.24-percent confidence interval, there is a $(1 - 99.24) = .76$ percent chance that the interval we construct will *not* include the population mean. Since our estimate is just as likely to be too high as it is to be too low, there is a $.76\%/2 = .38$ percent chance that it will be too low and a .38 percent chance that our interval will be too high. This means that we have made an interval by adding and subtracting a leg that is so long that we will miss the population mean only if the sample mean is among the upper or lower .38 percent of all possible sample means (see Figure 12-4).

If the sample mean falls in either of the shaded areas in Figure 12-4, the interval will miss the population mean. We have decided that we want this to happen only .76 percent of the time—that sample means and the confidence intervals constructed around them will be too high only .38 percent of the time and too low only .38 percent of the time. How wide must the interval be in order for this to be true?

As before, it is easier to answer this question if we rephrase it: How many standard errors do we have to add and subtract from the sample mean in order to cut off the upper and lower .38 percent of all the means in a normal sampling distribution? To find out, we must go back to the normal table, but now instead of going from standard errors to probabilities, we first locate

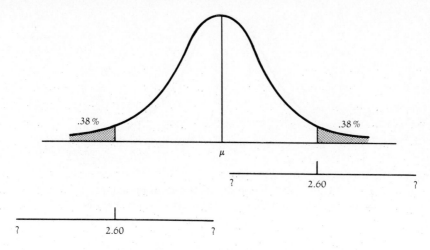

Figure 12-4 Sampling distribution for sample means, showing the upper and lower .38 percent of all sample means.

the appropriate probability and then see how many standard errors correspond to it. To find the Z-score that cuts off the upper and lower .38 percent of the cases, we must find the Z-score that cuts off (.5000 − .0038) = .4962 of the cases between itself and the mean (each of the *unshaded* areas to either side of the mean in Figure 12-4). We look in the body of the table for .4962 (or the number that is closest to it). This corresponds to a Z-score of 2.67, just what we had before (be sure to go into the normal table, Appendix Table 1, and see this for yourself).

This tells us that if we add and subtract 2.67 standard errors to sample means, we will get confidence intervals that will miss the population mean only 2(.38%) = .76 percent of the time. To convert the interval into units of children rather than standard errors, we simply multiply the 2.67 standard errors times the number of children per standard error (.03), which gives us an interval of

$$\bar{X} \pm 0.08 \text{ child}$$

which is just what we had before.

Suppose that we were willing to tolerate a lower level of confidence in our estimate. Suppose we want to be only 95 percent sure of our crystal ball. What would the interval look like then? The procedures are exactly the same as

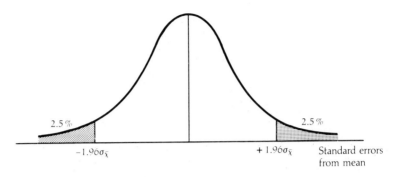

Figure 12-5 Sampling distribution for sample means, showing the upper and lower 2.5 percent of all sample means.

before, except with different numbers (stop here and try to figure it out yourself before going on).

A 95-percent confidence interval will miss the population mean when the sample mean is in one of the shaded areas of Figure 12-5. This will occur only 5 percent of the time, which means that each tail of the sampling distribution contains $5\%/2 = 2.5$ percent of the possible sample means. How many standard errors do we have to go away from the population mean to cut off a tail that contains 2.5 percent of the cases? To find this we go into the normal table and look for a proportion that is closest to $(.5000 - .0250) = .4750$, which corresponds to each of the unshaded areas on either side of the mean in Figure 12-5. In the normal table you will find that .4750 corresponds to a Z-score of 1.96 standard errors.

If we go 1.96 standard errors above and below the unknown population mean, we will cut off the upper and lower 2.5 percent of all the possible sample means (for a total of 5 percent). Intervals that consist of the sample mean plus or minus 1.96 standard errors will miss the population mean *only* when the sample mean is one that is more than 1.96 standard errors above *or* below the population mean, and according to the normal table, this can happen only 5 percent of the time. Our confidence interval is thus

$$\bar{X} \pm 1.96 \text{ standard errors}$$
$$\bar{X} \pm 0.06 \text{ child}$$

Our estimate of the mean ideal number of children runs from 2.0 to 3.2 children, and our level of confidence is 95 percent.

Notice that our new interval is considerably narrower than the 99-percent confidence interval, which means that our estimate is more precise. This illustrates an important principle of statistical inference that we will discuss at greater length later on: the more *confident* we want to be, the less *precise* we can be.

STUDENT'S *t*-DISTRIBUTION: Z FOR SMALL SAMPLES

You may recall that according to the central-limit theorem we may assume that the sampling distribution is normal if the sample size is large enough. What do we do, then, with small samples? In general, we cannot use the normal table to calculate probabilities and construct confidence intervals when the sample size falls to roughly 100 cases or fewer, because the sampling distribution is not normally shaped. In the case of means, however, there is an alternative.

Student's *t*-distribution (which is known by the pseudonym used by the great statistician William S. Gosset) is a sampling distribution that is used in exactly the same way as the normal table. The meaning of a *t*-score is the same as that of a Z-score. The difference is that the distribution of Z-scores is normal while the distribution of *t*-scores is more spread out than a normal curve (it has a greater variance and is therefore platykurtic). We estimate means with the *t*-distribution in order to compensate for small samples and the random error involved in using the sample standard deviation to estimate the population standard deviation in the estimation of the standard error.

The *t*-table (Table 12-2) is set up in a slightly different way than the normal table. While the body of the normal table is full of probabilities, the *t*-table uses only selected probabilities found in the two top rows of numbers. For confidence intervals, we use both tails of the sampling distribution (since the interval can be wrong by being either too high or too low), so we will use the second row (under the heading "Level of Significance for Two-tailed Test"). Since we want a 95-percent confidence interval, we use the column headed by ".05" (the middle column) which indicates the probability of error $(1.00 - .95 = .05)$.

To find the appropriate *t*-score for this level of confidence, we must go down that column until we find a row that corresponds to the size of our sample minus one (the *df* at the head of the first column represents **degrees of freedom**, which is simply the sample size minus one, found in the denominator of the estimated standard error for sample means that we saw earlier). Thus, with 120 degrees of freedom (121 cases), a 95-percent confidence interval is constructed by adding and subtracting 1.98 standard errors to the

Table 12-2   Distribution of t

	\multicolumn{6}{c}{Level of Significance for One-tailed Test}
	.10	.05	.025	.01	.005	.0005
	\multicolumn{6}{c}{Level of Significance for Two-tailed Test}
df	.20	.10	.05	.02	.01	.001
1	3.078	6.314	12.706	31.821	63.657	636.619
2	1.886	2.920	4.303	6.965	9.925	31.598
3	1.638	2.353	3.182	4.541	5.841	12.941
4	1.533	2.132	2.776	3.747	4.604	8.610
5	1.476	2.015	2.571	3.365	4.032	6.859
6	1.440	1.943	2.447	3.143	3.707	5.959
7	1.415	1.895	2.365	2.998	3.499	5.405
8	1.397	1.860	2.306	2.896	3.355	5.041
9	1.383	1.833	2.262	2.821	3.250	4.781
10	1.372	1.812	2.228	2.764	3.169	4.587
11	1.363	1.796	2.201	2.718	3.106	4.437
12	1.356	1.782	2.179	2.681	3.055	4.318
13	1.350	1.771	2.160	2.650	3.012	4.221
14	1.345	1.761	2.145	2.624	2.977	4.140
15	1.341	1.753	2.131	2.602	2.947	4.073
16	1.337	1.746	2.120	2.583	2.921	4.015
17	1.333	1.740	2.110	2.567	2.898	3.965
18	1.330	1.734	2.101	2.552	2.878	3.922
19	1.328	1.729	2.093	2.539	2.861	3.883
20	1.325	1.725	2.086	2.528	2.845	3.850
21	1.323	1.721	2.080	2.518	2.831	3.819
22	1.321	1.717	2.074	2.508	2.819	3.792
23	1.319	1.714	2.069	2.500	2.807	3.767
24	1.318	1.711	2.064	2.492	2.797	3.745
25	1.316	1.708	2.060	2.485	2.787	3.725
26	1.315	1.706	2.056	2.479	2.779	3.707
27	1.314	1.703	2.052	2.473	2.771	3.690
28	1.313	1.701	2.048	2.467	2.763	3.674
29	1.311	1.699	2.045	2.462	2.756	3.659
30	1.310	1.697	2.042	2.457	2.750	3.646
40	1.303	1.684	2.021	2.423	2.704	3.551
60	1.296	1.671	2.000	2.390	2.660	3.460
120	1.289	1.658	1.980	2.358	2.617	3.373
∞	1.282	1.645	1.960	2.326	2.576	3.291

SOURCE: Fisher and Yates, 1974, Table III.

sample mean. With only 30 degrees of freedom, we must add and subtract 2.042 standard errors, and so on.

Notice that these values are larger than the 1.96 standard errors we would get from the normal table. With small samples, a value of t is larger than a comparable value of z, thereby making the confidence interval wider. The larger the sample is, the more closely the t-distribution approximates the normal distribution. With samples of more than 500 cases, the two distributions are so similar that t-scores and Z-scores yield virtually identical results. The t-distribution should be used only when estimating means and when the sample size is small.

Estimation and Sample Size

Confidence intervals can be used to demonstrate the effects of a general principle that is, perhaps, intuitively obvious: the larger a sample is, the more precise our estimates can be.

In our example, we gathered data from 1,370 adults. What if we had a sample of only 200? How would this affect our estimates of the mean ideal number of children? The difference is felt through the size of the estimated standard error. Since we divide the sample standard deviation by the square root of the sample size minus one, the larger the sample is, the *smaller* the standard error will be. With 200 people, for example, the estimated standard error for mean ideal number of children would be

$$\hat{\sigma}_{\bar{x}} = \frac{.97}{\sqrt{199}} = .07$$

instead of .03. A 95-percent confidence interval consists of the sample mean plus and minus 1.96 standard errors. With 1,370 cases, the resulting interval is

$$\bar{X} \pm 1.96(.03)$$

or 2.54 to 2.66 children. But with only 200 cases, the standard error and, therefore, the interval are more than twice as large

$$\bar{X} \pm 1.96(.07)$$

or 2.46 to 2.74 children. Since the standard error is larger with the smaller sample, the legs that comprise the confidence interval must be longer.

Notice that the level of confidence in both estimates is the same (95 percent), but our precision is greater with the larger sample. Thus, for any given level of confidence, the larger the sample, the more precise our estimates will be.

Estimating Population Proportions

To estimate a population proportion, we follow exactly the same procedures that we use to estimate a population mean, except that we use a different formula to estimate the standard error:

$$\hat{\sigma}_{P_s} = \sqrt{\frac{PQ}{N}}$$

where P is the proportion of cases that have a characteristic and Q is the proportion that do not—or $(1 - P)$. Since we do not know the value of P (that is what we want to find out), we usually use a value of .5. Why? If we set P at .5, then Q is also .5 since P and Q must add up to 1.0. This means that the product PQ is $(.5)(.5) = .25$. If we use any number higher or lower than .5 for the value of P, then the product PQ will be lower than .25 (try some different values for P and you will see that this is true).

What we have done by setting P at .5 is to make the numerator of the standard error as large as it can possibly be. By inflating the size of the standard error, we also tend to inflate the size of the confidence interval, which increases the chance that it includes the population value we are trying to estimate. By setting P at .5, we are being conservative—bending over backwards to be sure our estimate is accurate.

If our sample proportion differs greatly from .5 (either higher or lower), then we might adjust the value of P in the standard error accordingly. If the sample proportion is only .1, for example, it is hard to maintain the assumption that P is in fact in the range of .5. In this case, we would most likely use a value of P that is smaller than .5—perhaps splitting the difference and using .3 or even a value as low as .2. In this way, we acknowledge the evidence of our sample while still deliberately overestimating the standard error. In no case are we using the absolutely correct value of P since, as with means, we are estimating the standard error.

To see how this works in practice, suppose we want to construct a 95-percent confidence interval that estimates the proportion of U.S. adults who expect that the United States will fight another world war within the next ten years. In 1986, 47.5 percent of a GENSOC survey sample of 1,396 U.S. adults said they shared this expectation.

As we saw previously, to construct a 95-percent confidence interval we must take the sample result and add and subtract 1.96 standard errors:

$$P_s \pm 1.96 \text{ standard errors}$$

where P_s is the sample proportion of adults who expect war within the next ten years. The estimated standard error in this case is

$$\hat{\sigma}_{P_s} = \sqrt{\frac{PQ}{N}} = \sqrt{\frac{(.5)(.5)}{1,396}} = .013$$

This results in a confidence interval of

$$P_s \pm 1.96 \text{ standard errors} = .475 \pm 1.96(.013) = .475 \pm .025$$

Box 12-1 One Step Further: *PQ* as a Variance

As we have just seen, the formula for the estimated standard error for sample proportions is

$$\hat{\sigma}_{P_s} = \sqrt{\frac{PQ}{N}}$$

where P is the proportion of cases that have a characteristic and Q is the proportion that do not. You might be wondering at this point just where this formula comes from. The denominator should look familiar, since all standard errors are divided by the sample size; but why the product PQ in the numerator?

When we first encountered a standard error formula in Chapter 11, it was for sample means, and it had in its numerator the standard deviation for the population (or, if we included it under the square root sign, we would have to use the variance). The standard error for sample proportions is simply a variation on this same formula. With proportions, each case has one of two scores: it either has the characteristic or it does not. Suppose we give each case that has the characteristic a score of "1" and each case that does not have the characteristic a score of "0." If we then go ahead and compute a standard deviation, the result will be identical to the square root of PQ, where P is the proportion of cases that have the characteristic, and Q is the proportion that do not.

In short, the product PQ is simply the variance for any dichotomy scored as "1" or "0."

which runs between .45 and .50, or, in percentage terms, between 45 and 50 percent. We estimate that in 1986 between 45 and 50 percent of the U.S. adult population expected another world war within ten years. There is a 95-percent probability that the procedures we have used have resulted in an interval that includes the actual population proportion.

With large samples, these kinds of inference techniques can be particularly powerful. If we drew a sample of 10,000 observations from *any* population—*no matter how large*—we could construct a 95-percent confidence interval for proportions that would be only 2 percentage points wide (±1 percent). In the long run, samples of 10,000 will be this accurate whether they are drawn from the state of Florida with its ten million people, China with its more than one billion people, or the entire world with its five billion people. The size of the population is for the most part irrelevant—only the size of the standard error and the level of confidence we require are important. The only circumstance in which the size of the population is relevant is when the sample constitutes a substantial portion of the population—say, 10 percent or more. The effect of this, however, is to *increase* our precision by giving us a narrower interval estimate for a given level of confidence.

Most national surveys use samples of around 1,500 cases. One reason that this size is preferred is that it is large enough to represent most subgroups of interest to social researchers. More important, however, is the fact that samples of this size represent a compromise between cost and precision. The larger the sample, the smaller the standard error is, and the smaller the standard error is, the narrower (and therefore more precise) the confidence interval will be. However, the standard error is decreased in size by the *square root* of the increase in sample size (since we calculate the standard error by dividing the population standard deviation by the square root of the sample size). To double our precision we have to cut the standard error in half, and to do this, we must take a sample that is *four* times as large, not twice as large. Thus to cut the length of the confidence interval in half and double our precision, we must quadruple the sample size. Given the high cost of each interview in national surveys, there comes a point of diminishing returns at which the additional cost of increasing the sample size is not worth the relatively small gains in precision.

SOME INSTANT CONFIDENCE INTERVALS FOR PROPORTIONS

When you come across poll returns in your newspaper that include findings such as "52 percent of the population approves of the government's foreign policy in Latin America," you might like to see if the sample result in fact indicates that a majority of the population approves or is simply the result of chance fluctuations in sample results.

An effective way to do this is to construct a confidence interval and see if it includes values above 50 percent. If our interval ran from 44 to 48 percent, for example, we would have to conclude that less than a majority approve, since the entire interval is below 50 percent. If, however, the interval ran from 50 to 55 percent, we could be confident that the population percentage who approve is somewhere above 50 percent. If it straddled the 50 percent mark—running, say, from 48 to 52 percent, we would not be in a position to estimate confidently that a majority approved *or* disapproved.

To make this easy to do, I have constructed Table 12-3 which can be used to evaluate many of the poll results that report percentages (or proportions). To use these figures, of course, we have to be confident that the sample on which the percentages are based was drawn according to correct scientific procedures. We could not use them with quota samples or with samples with low response rates.

For each sample size in the left-hand column of Table 12-3, the corresponding figure on the right completes the following statement: "The 99-percent confidence interval is the sample percentage plus or minus _____ percent." If the sample has 1,000 cases, and the percentage who approve of the government's policies is 45 percent, then our estimate of the population proportion who approve is 45 ± 4 percent, or from 41 to 49 percent (with proportions, the comparable figures are .45 ± .04, or from .41 to .49). If the level of approval was 48 percent, the 99-percent confidence interval would run from 44 to 52 percent (what would we conclude then?).

Table 12-3 "Instant" 99-Percent Confidence Intervals for Sample Proportions

Sample Size	The 99-Percent Confidence Interval Is the Sample Percentage Plus or Minus
50	18%
100	13%
300	7%
500	6%
800	5%
1,000	4%
1,500	3%
2,000	3%
3,000	2%
5,000	2%
10,000	1%
50,000	$\frac{1}{2}$%

Estimating Differences between Sample Proportions

Although there are research questions that can be answered by estimating a single mean or proportion, scientific questions usually involve comparisons between groups. Do those who attend church often express more or less racial prejudice than those who attend less often? How much of a difference does a drug make in the recovery rate of patients? How much more money do college graduates earn in comparison with those who never went to college? How does the size of the average poor family compare with that of the average middle-class family? How does the average number of sick days taken by male workers compare with the average for female workers?

Because most research focuses on the effects of one variable on another, it also necessarily focuses on questions about differences between means and proportions. For example, how large is the age difference in support for the right of unwed mothers to have abortions? In 1984, 49 percent of adults under the age of 40 favored abortion rights in this circumstance compared with 40 percent of those 40 years old or older (Table 12-4). This results in a sample age difference of (49% − 40%) = 9 percent, or, in terms of proportions, .09. How large is the difference in the population?

Since we are estimating the *difference* between the proportions of older and younger people who favor abortion rights, the sample difference of .09 will lie at the center of our interval. In order to use the normal table to determine the

Table 12-4 Position on Abortion Rights for Unwed Mothers by Age, United States, 1984

	Age		
Position	Less Than 40	40 or Older	All
For	49%	40%	45%
Against	51	60	55
TOTAL	100%	100%	100%
$(N)^*$	(712)	(703)	(1,415)

SOURCE: Computed from 1984 General Social Survey raw data.
* There are 55 cases with missing data from the total sample of 1,470.

length of the confidence interval "legs" and the associated probabilities, we need a normally shaped sampling distribution for the *difference* between two sample proportions.

What *is* such a distribution? With single samples we imagined drawing all possible samples and constructing a probability distribution of single means or proportions. In this case, however, we imagine that we draw all possible *pairs* of samples (older people and younger people), and for each pair we are calculating the *difference* between the two sample proportions in favor of abortion rights. The probability distribution of those differences is what constitutes the sampling distribution.

As with single sample means and proportions, the central-limit theorem is important here. With larger and larger samples, the sampling distribution for differences between samples closely approximates a normal curve. In addition, the mean of this sampling distribution (the average difference between sample proportions) equals the actual difference between the populations of older and younger people. Finally, like all sampling distributions, this one has a standard error. The estimated standard error for differences between sample proportions is

$$\hat{\sigma}_{P_{s_1} - P_{s_2}} = \sqrt{\frac{P_1 Q_1}{N_1} + \frac{P_2 Q_2}{N_2}}$$

where P_1 is the proportion of younger people who favor abortion rights, Q_1 is the proportion who oppose abortion rights, and P_2 and Q_2 are the corresponding proportions for older people. The respective sample sizes are represented by N_1 and N_2. Since our sample proportions are not terribly far from .5, we will use this value for P and Q in both cases, giving us an estimated standard error of

$$\hat{\sigma}_{P_{s_1} - P_{s_2}} = \sqrt{\frac{(.5)(.5)}{712} + \frac{(.5)(.5)}{703}} = .027$$

To give you some idea of how the use of .5 for the value of P and Q affects the size of the standard error, suppose we had instead used the sample values for each P and Q as estimates of the population proportions. In this case, the estimated standard error would then have been

$$\hat{\sigma}_{P_{s_1} - P_{s_2}} = \sqrt{\frac{(.49)(.51)}{712} + \frac{(.40)(.60)}{703}} = .026$$

a difference of only one-tenth of a percent. Had we used even smaller values (such as .3 for P and .7 for Q), the standard error would have been smaller (.024) but only slightly. As you can see, arbitrarily setting P and Q at a value of .5 for the sake of maximizing the standard error introduces a very small effect.

From this point on, we are on familiar ground. We have a sample result—the difference between two sample proportions. This difference was drawn from a normally shaped sampling distribution of such differences, with an estimated standard error of .027. If we want to estimate the age difference in the population with a 99-percent confidence interval, we need legs on the interval that are 2.575 standard errors long (look in the normal table, Appendix Table 1, and be sure you see why this is true):

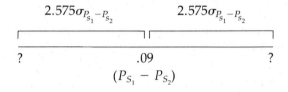

The 99-percent confidence interval is $(P_{S_1} - P_{S_2}) \pm 2.575$ standard errors. Since each standard error is equivalent to .027, the confidence interval can also be expressed as $09 \pm (2.575)(.027)$, or $.09 \pm .07$. The interval runs between .02 and .16, or, in percentage terms, from 2 to 16 percent. We estimate that the age difference in support for abortion rights for unwed mothers is somewhere between 2 and 16 percent, with a 99-percent confidence level.

OPEN-ENDED CONFIDENCE INTERVALS

If we compare the responses of men and women to the abortion question discussed above, we find a much smaller difference. Among men, 47 percent favor abortion rights compared with 42 percent among women, a difference of only 5 percent. The estimated standard error for differences in this case is also .027, but the 99-percent confidence interval is $(.47 - .42) \pm (2.575)(.027)$, or $.05 \pm .09$, which runs from $-.04$ to $+.14$. At this level of confidence, we cannot conclude that one gender or the other is more likely to favor abortion rights since the interval includes both possibilities as well as the possibility of no difference at all (a difference of zero).

Although the male sample has a higher proportion in favor than the female sample, we cannot conclude that the male population has a higher proportion—at least not with 99-percent confidence. How confident, then, *can* we be that men are more likely to favor abortion rights in this circumstance?

To answer this question, we are looking for an interval that corresponds to the statement, "when we subtract the female proportion from the male proportion, the result will be something greater than zero." To find this interval we add together two different intervals. The first piece runs upwards from zero, and we construct it by placing the sample result at the center of the interval and adding and subtracting legs that are long enough to reach zero, which is .05 ± .05 (or 0 to .10). Each leg in this interval has a length in proportions of .05. To calculate the confidence level associated with this interval, we must convert the length of each leg to standard errors: .05/.027 = 1.85. From the normal table we can see that the area contained between the mean and 1.85 standard errors above the mean is .4678. If we double this, the confidence level for this interval is .9356.

Our first interval (0 to .10) has a 93.56-percent confidence level. This means that the probability of drawing a sample leading to an *in*correct interval is (1 − .9356) = .0644. The probability that we have constructed an interval that is too low (i.e., the difference is greater than .10) is half of this, or .0322. So, if we construct an **open-ended confidence interval** consisting of two pieces (0 to .10 and .10 and above), the confidence level is .9356 + .0322 = .9678, or 97 percent. This is the confidence level for the estimate "men are more likely than women to favor abortion rights for unwed mothers."

Estimating Differences between Sample Means

The procedures for estimating differences between sample means are the same as those for estimating differences between sample proportions, except that, once again, we must use a different formula for the standard error. The estimated standard error for differences between sample means is:

$$\hat{\sigma}_{\bar{X}_1 - \bar{X}_2} = \sqrt{\frac{s_1^2}{N_1 - 1} + \frac{s_2^2}{N_2 - 1}}$$

where s_1^2 and s_2^2 are the variances in the two samples, and N_1 and N_2 are the respective sample sizes. Notice that since we are using the sample variances to estimate the population variances, we subtract one from each of the denominators.

Suppose we are interested in the effects of college education on the amount of television people watch. In 1982, for example, the GENSOC survey showed that those without any college education watched an average of 3.3

Table 12-5 Daily Television Viewing (Hours) by Education, United States, 1984

Television Viewing	Educational Attainment	
	Less Than College	At Least Some College
Mean	3.3	2.2
Variance	5.3	2.6
N*	777	379

SOURCE: Computed from 1984 General Social Survey raw data.
* There are 314 cases with no response from the total sample of 1,470.

hours of television a day, while those with at least some college education watched an average of 2.2 hours (Table 12-5), for a difference of 1.1 hours per day.

The estimated standard error for the difference between sample means is

$$\hat{\sigma}_{\bar{X}_1-\bar{X}_2} = \sqrt{\frac{s_1^2}{N_1 - 1} + \frac{s_2^2}{N_2 - 1}} = \sqrt{\frac{5.3}{776} + \frac{2.6}{378}} = .12$$

Suppose we want to estimate the difference with a 95-percent confidence interval. This means there will be a $(1 - .95) = .05$ overall probability of error or a $.05/2 = .025$ probability of error in either direction. We look in the normal table for the Z-score that corresponds most closely with $(.5000 - .0250) = .4750$, which is 1.96. Ninety-five percent of the possible differences between sample means will lie within 1.96 standard errors of the actual difference between the two population means. Each leg of the 95-percent confidence interval will be 1.96 standard errors long, or, in units of hours, $(1.96)(.12) = .24$. The confidence interval is thus

$$(\bar{X}_1 - \bar{X}_2) \pm .24 \text{ hours, or } 1.1 \pm .24 \text{ hours}$$

Our estimate is that those without at least some college education watch between .86 and 1.34 more hours of television per day than those with at least some college. Our level of confidence is 95 percent. (Notice that in constructing the interval we did not use the *t*-table even though we were estimating means. Why not?)

The Perpetual Trade-off: Confidence, Precision, and Sample Size

Ideally, we would like to estimate the characteristics of populations with both great precision and high levels of confidence; but in practice we usually compromise between the two.

The length of a confidence interval is calculated as a given number of standard errors multiplied by the size of the standard error in the units of the variable involved. In estimating the average ideal number of children, for example, each leg of the confidence interval was equal to a Z-score multiplied times the size of the standard error:

$$\text{Length of interval} = (\text{Z-score}) \frac{s}{\sqrt{N-1}}$$

precision   *confidence*   *sample size*

The first part of this equation is the length of the interval, which indicates the **precision** of the estimate: the longer the interval, the less precise the estimate is. The second part of the equation is the Z-score, which depends on the confidence level and, therefore, the **accuracy** of the estimate: the more confident we want to be, the farther out on the normal distribution we have to go in order to cut off the higher proportion of cases. For a 95-percent confidence interval, we use a Z-score of 1.96, but for a 99-percent confidence interval, we must use a Z-score of 2.575. Hence, the greater the confidence, the higher the Z-score has to be.

The third piece of the equation is the standard error. The size of the standard error depends on the sample size. Since sample size is in the denominator, the larger the sample, the smaller is the standard error. Keep in mind that the standard error is reduced by the *square root* of the sample size, which means that in order to cut the standard error in half we must quadruple the sample size.

As you can see from the relationship between the length of the interval, the confidence level, and sample size, we cannot increase precision (a narrower interval) without lowering confidence or taking a larger sample. If we make the interval narrower, in order to balance the equation we either must use a smaller Z-score (and, therefore, a lower confidence level) or achieve a smaller standard error. The only way to lower the standard error is to select a larger sample.

In the same way, if we want to increase our level of confidence (a larger Z-score) we can balance the equation only by increasing the width of the interval (which means less precision) or by making the standard error smaller by once again increasing sample size. This inevitable trade-off shows why it is important to select an appropriately large sample in the first place, for once the data are gathered and the sample size is fixed, confidence and precision cannot *both* be increased. An increase in one must be at the expense of the other.

INTERPRETING CONFIDENCE INTERVALS: SOME CAUTIONS

When you encounter confidence intervals in research reports, there are two important things you should keep in mind. First, what we have done here assumes that there is only random sampling error in the data, not bias. Statistical inference cannot make up for poor measurement.

Second, inference techniques assume that the data are based on samples that were drawn in ways that satisfy the requirements of scientific sampling (see Chapter 10). Without this, we cannot use the central-limit theorem to assume that the sampling distribution is normally shaped, and without that key assumption, we cannot construct confidence intervals.

It is also important to keep in mind that all interval estimates based on sample results have probabilities attached to them. Thus, part of the decision of what to make of a particular set of results hinges on the importance we attach to the probabilities of error that are involved. Ultimately the acceptance of a given level of confidence as adequate is a matter of judgment which rests not only with the personal preferences of the researcher but on some evaluation of the costs that are likely to result from making a mistake.

Summary

1. Statistical inference techniques are used to estimate the characteristics of populations and to test hypotheses about populations. Estimates can take the form of single values (point estimates) or intervals.

2. A confidence interval is an estimate that consists of a range of scores to which a probability of confidence is attached. The sample result is always placed at the center of the interval, to which is attached legs measured in units of standard errors. The level of confidence refers not to the specific

estimate but to the statistical procedures through which the estimate is obtained.

3. In practice, we construct confidence intervals by taking a sample result such as a mean and deciding what level of confidence we want to have. From the normal table, we then determine how many standard errors to add and subtract from the sample mean to create the interval with the desired level of confidence.

4. When using a small sample to construct confidence intervals for means, it is necessary to use Student's *t*-distribution rather than the normal table in order to compensate for the fact that the sampling distribution for sample means cannot be assumed to be normally shaped when samples are small. There is a separate *t*-distribution for each sample size.

5. In general, the larger the sample, the smaller the standard error is and the more precise the estimate can be (the narrower the interval).

6. Confidence intervals can be used to estimate any population characteristic for which it is possible to estimate a standard error, including proportions, means, differences between means, and differences between proportions.

7. An open-ended confidence interval allows us to attach a level of confidence to estimates that take the form of "the population value is equal to or greater than" a specified value.

8. The larger the sample, the smaller the standard error. The smaller the standard error, the narrower (more precise) an interval can be for a given level of confidence. The wider the interval, the greater is our confidence in it. This means that once a sample is selected, the only way to increase the confidence level for an estimate is to make the interval wider and, therefore, less precise. The only way to make the interval narrower and more precise is to accept a lower level of confidence.

9. The techniques for constructing confidence intervals assume that samples satisfy the requirements for scientific sampling and that there is no bias arising either from measurement or the sampling process.

Key Terms

accuracy 319
confidence interval 298
degrees of freedom 307
estimate 296
    biased 296
    interval 298

point 296
unbiased 296
hypothesis testing 295
open-ended confidence interval 316
precision 319
Student's *t*-distribution 307

Problems

1. Define and give an example of each of the key terms listed at the end of the chapter.

2. Suppose we want to estimate the average number of hours that workers spend at their jobs each week. We gather data from a sample of 1,601 people. The sample mean is 42.5 hours and the sample standard deviation is 12 hours.
   a. Estimate the standard error for sample means.
   b. Use a 95-percent confidence interval to estimate the population mean number of hours worked.
   c. Interpret your result in a sentence or two.

3. Suppose we want to estimate the average number of hours people spend listening to the radio each day. We gather data from a sample of 401 listeners. The sample mean is 3.7 hours and the sample standard deviation is 2.0 hours.
   a. Estimate the standard error for sample means.
   b. Use a 99-percent confidence interval to estimate the population mean number of hours spent listening to the radio.
   c. Interpret your result in a sentence or two.
   d. Suppose our sample had included only 41 respondents. How would you construct a 99-percent confidence interval now?

4. Suppose we manufacture computer chips, and we want to estimate the proportion of chips produced in a week that are defective. We take a random sample of 200 chips and subject them to rigorous testing. In our sample, we find that the proportion of chips that are defective is .40.
   a. Estimate the standard error for sample proportions.
   b. Use a 98-percent confidence interval to estimate the proportion of the week's output that is defective.
   c. Interpret your result in a sentence or two.

5. We want to estimate the proportion of U.S. adults who approve of birth control methods being made available to teenagers between 14 and 16 years old whose parents do not approve. In the 1986 GENSOC survey sample of 1,470 adults, the proportion approving was .56.
   a. Estimate the standard error for sample proportions.
   b. Use a 95-percent confidence interval to estimate the proportion of adults who approve.
   c. Interpret your result in a sentence or two.

6. We want to estimate the gender difference in support for capital punishment in the United States. In 1984, the GENSOC survey found that 80 percent of men approved ($N = 569$) and 71 percent of women approved $N = 807$).
   a. Estimate the standard error for the difference between sample proportions.
   b. Use a 99-percent confidence interval to estimate the difference in support for capital punishment between the male and female adult populations.
   c. Interpret your result in a sentence or two.

7. We want to estimate the difference between the percentage of Republicans and the percentage of Democrats and Independents who believe that welfare programs make people not want to work. In the 1984 GENSOC survey, 43 percent of Democrats and Independents agreed ($N = 1,030$) compared with 55 percent of Republicans ($N = 366$).
   a. Estimate the standard error for the difference between sample proportions.
   b. Use a 95-percent confidence interval to estimate the difference in the proportions who agree between Democrats and Independents on the one hand and Republicans on the other.
   c. Interpret your result in a sentence or two.

8. We want to estimate the difference between the mean number of hours of television watched by people who identify themselves as lower or working class on the one hand ("Low Class") and people who identify themselves as middle or upper class ("High Class") on the other. In the 1984 GENSOC survey, people in the "Low Class" category ($N = 587$) watched an average of 3.1 hours a day with a standard deviation of 2.2 hours. Those in the "High Class" category ($N = 569$) watched an average of 2.8 hours with a standard deviation of 2.1 hours.
   a. Estimate the standard error for the difference between two sample means.
   b. Use a 98-percent confidence interval to estimate the difference between means for the "Low Class" and "High Class" populations.
   c. Interpret your result in a sentence or two.

9. We want to estimate the difference between the mean number of children born to parents who earn less than $15,000 a year ("Low Income") and those who earn $15,000 a year or more ("High Income"). In the 1984 GENSOC survey, the mean number of children born to "Low Income" adults ($N = 565$) was 1.66 with a standard deviation of 1.68 children. The

mean number of children born to "High Income" adults ($N = 314$) was 1.91 children with a standard deviation of 1.70 children.
  a. Estimate the standard error for the difference between two sample means.
  b. Use a 95-percent confidence interval to estimate the difference between means for the "Low Income" and "High Income" populations.
  c. Interpret your result in a sentence or two.

10. If we quadrupled the size of a sample, how would this affect the width of a 99-percent confidence interval? How would it affect the width of a 95-percent interval? Why?

11. Suppose we have a quota sample. How would you apply the inference techniques described in this chapter to data from such a sample? Explain your answer.

13

Hypothesis Testing

In general scientific usage, a **hypothesis** is an unproven and yet testable assertion about the way things are or how things work. "Cancer is caused by a breakdown in the system of cell reproduction," "discrimination accounts for part of the earnings difference between men and women," "if voter registration was made easier in the U.S. a larger percentage of adults would vote in elections," and "as spending on public schools declines, so does the quality of education" are all examples of hypotheses that can be tested using scientifically gathered data.

In this chapter we are going to focus on a particular type of **hypothesis testing** that involves *testable assertions about the relationship between samples and populations*. Specifically, we are going to look at situations in which we have sample results that already suggest certain things about the world and the way it works, and use statistical inference to test the idea that such results are in fact chance occurrences that reflect nothing more than sampling error. If we can *reject* this kind of hypothesis, then we can conclude that the sample results reflect more than a chance occurrence and in fact indicate something about the population.

We might find a sample in which middle-class people tend to watch less television than working-class people, for example. The question we then must confront is whether this sample difference indicates that the *population* of middle-class people watches television less than the population of working-class people. We test this hypothesis by assuming its *opposite* to be true—that there is *no* class difference in television watching—and then see how inconsistent our evidence is with this assumption. This may sound like a rather backwards way of testing hypotheses, and it is; but it works, and by the end of this chapter you will see how.

Hypothesis Testing as an Everyday Activity

Although you may not be aware of it, we all use a form of hypothesis testing in our everyday lives. Consider a situation, for example, that involves someone we will call George, who has always acted like a friend to you. Today you have a date to meet him for lunch, but he never shows up. Is this a reflection of his feelings toward you? Can you still assume that he is your friend? To make such a decision, you would probably talk to him and find out what happened. Suppose he says he forgot. Now what?

We want to test the hypothesis that he is your friend. We do this by testing the *opposite*, that George is *not* your friend. Why? The answer to this will make more sense as we go along, but basically it comes down to this: If we make a specific hypothesis about something, and reject that hypothesis, we can use sampling distributions to calculate the probability that we have made an error in rejecting the hypothesis. If we cannot reject the hypothesis, we cannot calculate a corresponding probability that *that* is a mistake.

We suspect—on the basis of evidence—that he may not be your friend. In order to support this idea *with a known probability of error*, we assume the opposite to be true and see if the data justify rejecting that assumption. In general, we confirm a hypothesis by rejecting its opposite, since probabilities of error can be calculated only for the act of rejecting a hypothesis. We test the hypothesis that he is not your friend by assuming that he is. The hypothesis that represents what we actually think is going on (he is not your friend) is called the **substantive hypothesis** (symbolized as H_1). We test it by testing its opposite—he is your friend—which is called the **null hypothesis** (symbolized as H_0). The only way we can support the substantive hypothesis is by rejecting the null hypothesis with a known probability of error.

TESTING THE NULL HYPOTHESIS

George, like everyone else, is complicated. His behavior arises from many different aspects of his personality and the shifting social situations he finds himself in. We have only a sample of his behavior (he stood you up), and on the basis of that we are going to test a hypothesis about his relationship to you.

We begin by assuming that the null hypothesis—he is your friend—is true. Given this, you would most likely ask, "Would a true friend forget to meet you for lunch?" to which the most likely answer is, "It happens." That, however, does not help us very much. We could phrase the question in a more precise way that allows us to make a clearer decision: How *likely* is it that a true friend would forget a lunch date? Or, in terms of probabilities, what is the conditional probability that someone would stand you up *given* that he is a true friend?

Chances are that you would conclude that although forgetting lunch is not usual behavior for friends, it is not all that unusual, either. Many things can make the best of friends forget each other from time to time. You probably would not reject the null hypothesis that George is your friend *because the evidence of his behavior is not very inconsistent with the assumption that the null hypothesis makes about him*. Since the evidence does not justify a rejection of the null hypothesis, then we cannot support the substantive hypothesis that he is not your friend.

Now consider a more extreme example. George meets you for lunch, you eat together, and he abruptly walks off, not only leaving you to pay the check, but stealing your car and driving 300 miles before selling it and hopping a plane for Brazil. Once again the substantive hypothesis is that he is not your friend, and the null hypothesis is that he is your friend. What is the probability that he would behave *that* way given the assumption that he is your friend?

You might well conclude that the probability of someone acting that way given that he is your friend is very low, so low in fact that it is totally inconsistent with the assumption that he is your friend to begin with. Somewhere in your thinking process there is a cutoff point: if his behavior is *too* inconsistent with the null hypothesis that he is your friend, then you will reject that assumption in favor of the substantive hypothesis that he is not. The only alternative would be to conclude that something extraordinarily unlikely has happened.

Notice what we have done here. We began with evidence that pointed to a particular hypothesis about what was going on (George is not your friend). We assumed that the opposite was true (George is your friend) and then considered the degree of inconsistency between the assumption and the evidence by asking how likely the behavior was *given* the assumption contained in the null hypothesis. The more *un*likely the evidence is (the lower the conditional probability), the more likely we are to reject the null hypothesis that he is your friend. If, however, his behavior is not all that unusual for friends, it would provide us with no reason to reject the null hypothesis assumption that he is your friend.

Which conclusion you arrive at depends on the cutoff point you use to decide just how unlikely the evidence is before you reject the null hypothesis. No matter what you decide, there is always a probability of error. A friend is very unlikely to leave you with the bill, steal and sell your car, and flee to Brazil, but it is possible (he might have been a witness against organized crime figures who was pursued by paid assassins, and lacked the time or opportunity to tell you about it). On the one hand, if you conclude that he is not your friend, you might be rejecting a true null hypothesis (called a "type I error"). On the other hand, if you fail to reject a false null hypothesis by thinking George is your friend when he is not, you are also making a mistake (called a "type II error"). Unfortunately, there is no way of calculating the probability of error in this latter case (see Loether and McTavish, 1980, pp. 513–15, for some exceptions).

Rejecting the Null Hypothesis: The Probability of Error Imagine that you could actually calculate the probability that a friend would behave as George did in our extreme case. Suppose the probability is .0001 or 1 in 10,000: "Even among friends, this kind of thing would nonetheless happen 1 time in 10,000."

If you reject the null hypothesis and conclude that he is not your friend, there is a .0001 chance that you have misjudged the evidence and made a mistake. In other words, this could have been that 1-in-10,000 rare occurrence between true friends. This is the crucial point in hypothesis testing: if we can calculate the conditional probability that an event would happen *given* the null hypothesis assumption, *then we also know the probability that we will make a mistake if we reject it.* This is called the **significance level** of a hypothesis test. In everyday life we usually cannot calculate such probabilities, but with data based on well-drawn samples, we can.

ROLLING DICE: A SECOND EXAMPLE

Suppose you are rolling dice with Susan and she rolls two 7s in a row. Would you suspect loaded dice? Probably not, because although two 7s in a row is unusual with fair dice, it is not all that unusual. It is not terribly inconsistent with the idea of a "lucky throw." Suppose she rolls five more 7s in a row, what then? Chances are that you would be getting increasingly suspicious because the probability is *very* low that she would roll seven 7s in a row just by chance with a fair pair of dice (in fact, it is .0000036 or just under four chances in a million).

To test the hypothesis that the dice are unfair, we assume the opposite—that they are fair—and evaluate that assumption in light of the evidence. If the result for our sample of seven rolls is very unlikely given the null hypothesis assumption that the dice are fair, then we reject the null hypothesis and conclude that the dice are not fair. Since we know that the probability of getting seven 7s in a row by chance is extremely small, we can reject the null hypothesis with a known probability that the decision is a mistake (.0000036). In rejecting the null hypothesis, we are taking the risk that what has happened is that four-in-a-million rare occurrence when a fair pair of dice turns up seven 7s in a row. Given how unlikely that is, we reject the null without much worry. The only alternative is to assume that the dice are in fact fair and that an extraordinarily lucky thing has happened.

Suppose we had used as our null hypothesis the assumption that the dice were *not* fair. What is the probability of throwing seven 7s in a row with *un*fair dice? Since we have not specified just how unfair the dice are, it is impossible to calculate such a probability. The only kinds of hypotheses we can test directly are those that assume something specific about the population. When we assume that the dice are fair, we assume that all six numbers are equally likely to come up, which means that the probability for each of the six faces is $1/6 = .1667$. Having made this assumption, we can calculate the conditional probability of getting seven 7s in a row *given* that assumption. The assumption that the dice are *not* fair, however, does not tell us the exact chances of any total coming up on the dice and therefore does not allow us to calculate the exact probabilities necessary to test a null hypothesis.

The purpose of these two examples has been to give you some idea of the logic involved in hypothesis testing. In the sections that follow we will describe some specific applications of this general approach.

Single-Sample Hypothesis Tests: Means

You have been hired by your state department of education to find out if students in your state score above the national average on a standardized achievement test. The national mean is 70 on a scale of 0 to 100. To find out how well your state's students perform, you select a sample of 400 students using sampling techniques that satisfy the criteria for scientifically valid samples. For your sample, the mean test score is 73 and the sample standard deviation is 25 points.

Your problem now is to test the substantive hypothesis (H_1) that the mean score for your state is above the national average—in other words, greater than 70:

$$H_1: \mu > 70$$

where the symbol ">" means "is greater than." You test this hypothesis indirectly by testing its opposite, the null hypothesis (H_0) which states that the mean for your state is exactly equal to the national mean of 70:

$$H_0: \mu = 70$$

You might be wondering why the null hypothesis is not that the state mean is 70 or less rather than 70 exactly. As you will soon see, the only kind of null hypothesis we can test is one that specifies a single value for the mean of the sampling distribution, which in this case is the population mean. This means that the null hypothesis must specify a single number such as 70 rather than a range of numbers such as "70 or less."

As before, we have drawn only one of many possible sample means. The question is this: If we assume that the state mean is actually 70, what is the probability of drawing a *sample* mean of 73? Put differently, how inconsistent is our sample mean of 73 with the assumption that the population from which it was drawn has a mean of 70? If a sample mean of 73 is extremely unlikely to be drawn from a population with a mean of 70, then we would reject the null hypothesis in favor of the substantive hypothesis that the state population mean is in fact greater than 70.

In testing the hypothesis about George's friendship, we relied on a very imprecise cutoff point to indicate the degree of inconsistency between the evidence and the null hypothesis that would lead us to reject the null hypothesis. How inconsistent is "very inconsistent"? How low is a "very low" probability? The decision involves the same kind of judgment as the selection of an appropriate level of confidence for a confidence interval. In the end we must decide how large a probability of error is tolerable. In most scientific literature, 5 percent is considered to be the largest acceptable probability of error. Suppose we choose a 1-percent probability. This probability—the probability that rejecting the null hypothesis is a mistake—is the significance level of the test.

Having set the significance level in advance, we are now ready to proceed with the test. We need to calculate the probability of drawing a sample mean of 73 from a population that has a mean of 70. If the probability is .01 or less, this means that our sample result is very unlikely to have been drawn from such a population. This would leave us with one of two choices. On the one hand we could leave the null hypothesis unshaken and conclude that our sample mean simply represents one of those unusual events that sometimes occur by chance. On the other hand, we could conclude that while our sample mean is unlikely to have been drawn from a population with a mean of 70, it is *not* so unlikely to have been drawn from a population with a mean that is *greater* than 70. In this case we would reject the null hypothesis in favor of the substantive hypothesis.

To make a decision, we must calculate the appropriate probability, and for this we need to go back to basics. We have a sampling distribution of all possible sample means. To use it to calculate probabilities, we must (1) be able to assume that it is normally shaped; (2) know something about its mean; and (3) estimate its standard error.

With 400 cases in our sample, we are able to use the central-limit theorem to assume a normally shaped sampling distribution. According to the null hypothesis, we are assuming that the population mean is exactly 70. Therefore, since the mean of the sampling distribution always equals the population mean, this is also the value that we should assign to the sampling distribution mean. As you will see, this makes things considerably easier.

Estimating the standard error for sample means is quite straightforward:

$$\hat{\sigma}_{\bar{x}} = \frac{s}{\sqrt{N-1}} = \frac{25}{\sqrt{399}} = 1.25$$

It might help to see what we are dealing with if we drew a picture of the sampling distribution (Figure 13-1). We have assumed a normal sampling distribution with a mean of 70. Our sample mean is 73. What is the probability of getting a mean like ours (73) from this kind of sampling distribution (with a mean of 70)?

As we saw in Chapter 11, we cannot use the normal table to calculate the probability of any single score (such as 73), only ranges of scores. We can, therefore, calculate the probability of getting a score of 73 *or higher*, which would answer the question, "How likely are we to select a sample mean that is *at least* this inconsistent with the assumption that the population mean is 70?" To do this, we first convert the distance between the assumed sampling

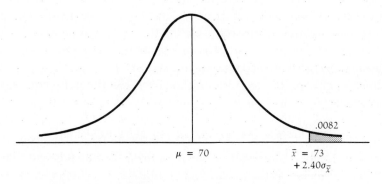

Figure 13-1 Sampling distribution for sample means.

distribution mean and our sample mean to standard errors (a Z-score):

$$Z = \frac{73 - 70}{1.25} = 2.40$$

We are looking for the probability of selecting a sample mean that is at least 2.40 standard errors above the mean of a normally shaped sampling distribution (the right-hand tail shaded in Figure 13-1). To find this we go into the normal table (Appendix Table 1), down the left-hand column to the row marked "2.4" and stay in the first column (headed by ".00"). The proportion of the area contained between the mean of the sampling distribution and a point 2.40 standard errors above it is .4918. Since we want the probability of falling at *least* 2.40 standard errors above the mean, we must subtract this from .5000 to get .0082.

What does it all mean? We began by assuming (H_0) that the population mean (and, therefore, the mean of the sampling distribution) was 70. How inconsistent is our evidence (a sample mean of 73) with our assumption about the population mean? The answer is that there is a less than 1-percent (.0082) chance of selecting a sample mean such as ours from a population with a mean of 70. Given this, we can draw one of two conclusions. The first is that our sample result is not so unusual as to make us doubt the null hypothesis. The second is that our sample result is so unusual that the null hypothesis is untenable and must be rejected.

If we decide to reject H_0, there is, of course, a chance that we are making a mistake. After all, even with a population mean of 70, one out of every hundred sample means will nonetheless be at least 3 points above that just by chance. Our sample mean could have been one of those unusual occurrences. How likely is it that just such an unusual occurrence is leading us to make a mistake? The answer is .82 percent. Since we decided in advance that we would tolerate a probability of error as large as 1 percent (the significance level), our result would justify our rejecting the null hypothesis and concluding that the state mean test score is greater than the national mean of 70. There is a .82-percent chance that our procedures have resulted in a sample mean that, by chance, has led us to mistakenly reject the null hypothesis that the state mean is in fact no greater than 70.

INTERPRETING HYPOTHESIS TEST RESULTS

We might describe the result of our hypothesis test as follows: "The null hypothesis is rejected at the .01 level of significance" (which means that the probability of error is .01 or less). We could be more precise by stating the actual probability of error as the significance level, as in "The null hypothesis is rejected at the .0082 level." We could also say that "the state mean is significantly greater than the national mean ($p < .01$)." The expression in parentheses reads "probability of error less than .01."

Suppose our sample mean had been only 72 rather than 73? In this case, the sample mean would have been only 1.60 standard errors above the assumed mean of the sampling distribution, and the probability would have been $(.5000 - .4452) = .0548$, which is considerably larger than the .01 probability of error we decided upon earlier. Under the strict rules of hypothesis testing, we would not have been able to reject H_0 without an unacceptably high probability of error. This would *not* mean that the null hypothesis was true; it would only mean that the evidence did not justify rejecting it.

Why are the rules of hypothesis testing so strict? Why is it necessary to establish in advance the level of significance? The problem is that we may allow our desire to reject a null hypothesis to influence our judgment about how much error to tolerate after the fact. To help maintain scientific objectivity, we decide beforehand under what conditions we will reject H_0 and then proceed. It is somewhat arbitrary, but many researchers feel it is a necessary precaution against subjective bias.

There are many other researchers, however, who object strongly to this approach. One of the reasons for their objection is that rejecting a null hypothesis depends entirely on the probability of error we select as a cutoff point (the significance level). If we had set the significance level at .001 in our example above, we would *not* have been able to reject the null hypothesis. It is all a matter of judgment, and where that level is set makes all the difference between rejecting and not rejecting the null hypothesis. Here lies one of the major weaknesses of hypothesis testing: on the one hand, the setting of the significance level is somewhat arbitrary. On the other hand, the decisions are black and white: we either reject H_0 or we do not. For this reason, many researchers prefer to report the actual probabilities of error associated with rejecting null hypotheses and then evaluate the findings in light of them. This is an increasingly popular alternative to the use of rigid significance levels. We will have more to say about this controversial issue later on in this chapter.

Notice that one way to express the results of our hypothesis test was "the state mean is significantly greater than the national mean ($p < .01$)." In reading scientific literature that relies on statistical inference, you are going to encounter the word "significant" over and over again. In popular usage, "significant" means "important" or "interesting," but in statistics it has a precise meaning that has nothing to do with importance. *Significant* in this case merely means that we are confident that the state mean is greater than 70. It could in fact be 70.00001. To say that the state mean is significantly greater than the national mean only indicates that we are sure it is not exactly 70 or less. In most research problems, knowing that the mean is above a certain value is of little use without an estimate of just how much above it is. This is where confidence intervals are more useful than hypothesis tests.

Before moving on to another example of single-sample hypothesis testing, there is an important situation that we should mention. What if, in testing the hypothesis about the state mean test score, we had drawn a sample in which the mean was *below* 70? What then? Since our substantive hypothesis is that the state mean is *greater* than 70, the sample result directly contradicts the substantive hypothesis. There is no way that such a result could be used to reject H_0 in favor of the hypothesis that the state mean is greater than 70. Therefore, there would be no need even to perform the hypothesis test—no need to calculate probabilities. Under those circumstances, we might then have tested a different substantive hypothesis—that the state mean is *below* the national mean.

The kind of hypothesis test we have just described is called a **one-tailed test** because we used only one tail of the sampling distribution to calculate probabilities (as in Figure 13-1). If the substantive hypothesis had been "the state mean does not equal the national mean," the null hypothesis would still have been "the state mean equals the national mean." Unlike before, however, we would have used both tails of the sampling distribution to calculate probabilities because our sample mean could be either above *or* below the population mean and still support the substantive hypothesis. Since substantive hypotheses almost always predict the direction of the sample outcome, two-tailed tests are rarely appropriate.

Single-Sample Hypothesis Tests: Proportions

You are the campaign manager for a political candidate who is running in a very close race for state governor. In allocating campaign funds, you need to know in which districts of the state your candidate is ahead and in which your candidate is behind so that you can concentrate your limited resources in communities where you are in danger of losing.

To find out where you are in trouble, you commission an opinion poll in each election district. You hope that your candidate is ahead in each district, which is to say, that more than 50 percent of the voters are on your side. This is the substantive hypothesis: that the proportion of voters intending to vote for your candidate in the population is greater than .50:

$$H_1: P > .50$$

To test this hypothesis you must make a specific assumption about the mean of the sampling distribution—the null hypothesis. In this case, the null hypothesis is that $P = .50$, which is to say, the best you will achieve is a tie vote:

$$H_0: P = .50$$

Suppose you draw a sample of 300 voters from the first district and the sample proportion in favor of your candidate (p_s) is .55. Can you conclude from this sample proportion that you will get a majority of votes in the district? With 300 cases in the sample, the central-limit theorem allows us to assume a normally shaped sampling distribution. According to the null hypothesis, we are assuming that the proportion of voters who will vote for your candidate is exactly .50, which means that the mean of the sampling distribution (the average sample proportion) is also .50. All we need now is an estimated standard error:

$$\hat{\sigma}_{P_s} = \sqrt{\frac{PQ}{N}} = \sqrt{\frac{(.5)(.5)}{300}} = .029$$

To test the null hypothesis, we must select a significance level. Suppose you want to be very sure that the null hypothesis is false before you reject it, which means selecting a very small significance level, say, .001.

We have assumed that the population proportion (P) is .50. Our sample proportion (p_s) is .55. In a normally shaped sampling distribution, our sample proportion is $(.55 - .50) = .05$ above the mean of the sampling distribution. What is the probability of drawing a sample proportion that is at least .05 above the mean of the sampling distribution simply by chance? To calculate this probability, we must convert this distance to a Z-score:

$$Z = \frac{.55 - .50}{.029} = 1.72$$

To find the probability of getting a sample proportion that is at least 1.72 standard errors above the mean of the sampling distribution, we go into the normal table (Appendix Table 1) and look for 1.70 in the left-hand column. We then go across that row to the column headed by ".02." The probability of selecting a sample result that is *between* the mean of the sampling distribution and a point 1.72 standard errors above the mean is .4573. The probability of

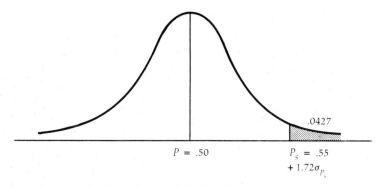

Figure 13-2 Sampling distribution for sample proportions.

selecting a sample proportion that is *at least* this far from the mean of the sampling distribution is (.5000 − .4573) = .0427, or just over 4 percent (see Figure 13-2).

Can we reject the null hypothesis? We assumed that P was exactly .50. How inconsistent is our sample result with our assumption about the population proportion? The answer is that there is a 4 percent chance of drawing a sample like ours from a population in which P = .50. This means that if we reject H_0, the probability of error is 4 percent. In setting the significance level earlier, however, we decided that we would not reject the null hypothesis unless the probability of error was no greater than 0.1 percent. In this case, however, if we reject H_0 the probability of error will be 4 percent, which is greater than the significance level we established earlier. Therefore, we cannot reject the null hypothesis. This is directly comparable to our deciding that George's standing you up was unusual behavior for a friend, but not unusual enough to justify rejecting the assumption of friendship *without an intolerably high probability of error.*

What conclusion would you reach? You might say that "the sample results do not show that you are doing significantly better than a tie vote in this district" or "the race is too close to call one way or the other." In either case, this is one district that you cannot assume you are ahead in without running what you have already established as an unacceptable risk of error.

Two-Sample Tests: Testing for Differences between Populations

The most common use of hypothesis testing focuses on the question of whether two populations differ on some characteristic. Suppose, for example, that we are interested in how Democrats and Republicans differ in their basic views of the causes of social inequality in the U.S. In the 1984 GENSOC survey, respondents were asked if they agreed with the idea that "Differences in social standing between people are acceptable because they basically reflect what people made out of the opportunities they had." A "yes" answer could be interpreted as assigning responsibility for inequality to individuals rather than the social system. A "no" answer, however, suggests that inequality of outcomes is at least in part generated by inequality of opportunity, which is the responsibility of society, not individuals. Given the historical differences between the Republican and Democratic parties, we would expect that Republicans would be more likely than Democrats to support this position by answering yes. Our substantive hypothesis (H_1), then, is that Republicans are more likely than Democrats to agree with this proposition, or that the difference between the two population proportions is greater than zero:

$$H_1: (P_r - P_d) > 0$$

where P_r and P_d are the proportions for Republicans and Democrats, respectively. The null hypothesis (H_0) is that there is *no* difference between the two populations, or that the difference between the two proportions equals zero:

$$H_0: (P_r - P_d) = 0$$

In the 1984 GENSOC survey, 81 percent of 358 Republicans and 73 percent of 519 Democrats agreed that inequality was due to what people made of their opportunities, for a difference of (81% − 73%) = 8 percent. Given that these two sample percentages—one for Republicans and one for Democrats—differ by 8 percent, can we conclude from this that the two populations from which these samples were drawn also differ in this same way? Or, in terms of our substantive hypothesis, does the fact that the *sample* differences support our hypothesis mean that our hypothesis holds true for the two *populations*?

As with single-sample hypothesis tests, to answer this question we must (1) be able to assume that the sampling distribution is normally shaped; (2) assume a value for the mean of the sampling distribution (the average difference between sample proportions); and (3) estimate the standard error for the corresponding sampling distribution. Given a sample of almost 900 cases, the central-limit theorem allows us to assume a normally shaped sampling distribution. The null hypothesis specifies that for the purposes of this test we are assuming that the difference between the populations (and thus the mean of the sampling distribution of sample differences) equals zero.

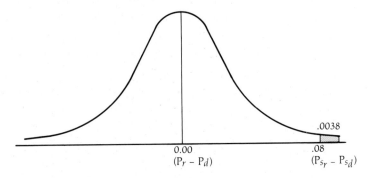

Figure 13-3 Sampling distribution for differences between sample proportions.

The final step is the estimation of the standard error:

$$\hat{\sigma}_{P_{s_1} - P_{s_2}} = \sqrt{\frac{P_r Q_r}{N_r} + \frac{P_d Q_d}{N_d}} = \sqrt{\frac{(.5)(.5)}{358} + \frac{(.5)(.5)}{519}} = .034$$

Note that we could have used the actual sample proportions (.81 and .19 for Republicans and .73 and .27 for Democrats) in the formula for the estimated standard error, but the difference in the result would have been slight (an estimated standard error of .028 rather than .034, amounting to no difference at all when we round to the nearest 100th). Any overestimation of the standard error that comes about by using .5 in each case only minimizes the probability of error attached to our final result, which is all to the good.

Once again, a graph may help summarize what we are doing (Figure 13-3). Shown on the sampling distribution are the assumed difference between the two populations (0.00) and the observed difference between the two samples (.81 − .73) = .08. Before going into the normal table we must decide on the significance level: how unlikely must our observed sample difference be—given the null hypothesis—in order for us to conclude that the null hypothesis is false? Suppose we select a .01 significance level.

The question now is this: If we assume that we are drawing samples from two populations that are *identical* on this variable, what is the probability of drawing two sample proportions that differ by .08 or more? Or, given that the assumed mean of the sampling distribution is zero, what is the probability of getting a sample difference that is .08 or more? To find out, we first express the distance between the sample outcome (.08) and the assumed mean of the sampling distribution (0.00) in standard error units:

$$Z = \frac{.08 - 0}{.03} = 2.67 \text{ standard errors}$$

Our observed sample difference is 2.67 standard errors above the value assumed by the null hypothesis. In a world in which these two populations (Republicans and Democrats) are identical on this issue, how likely are we to get a sample difference of 2.67 standard errors or more just by chance? This corresponds to a probability represented by the shaded region in Figure 13-3. To find it, we look in the normal table (Appendix Table 1), down the left-hand column to the row labeled "2.6" and across to the column headed by ".07." This corresponds to an area beneath the curve of .4962. Subtracting this from .5000 we get a probability of .0038, or just over a third of 1 percent.

The difference between the two sample proportions is very inconsistent with the null hypothesis. In fact, if the difference between the Republican and Democratic populations was actually zero, then we would expect to get sample results like ours less than 1 percent of the time by chance. If we reject the null hypothesis in favor of the substantive hypothesis, our probability of error will be only .0038, or .38 percent, which is well below our significance level of .01, or 1 percent. We are, therefore, quite certain that Republicans are more likely than Democrats to support this view of the world.

Our results could be summarized as "The difference between Republicans and Democrats on this issue is significant at the .0038 level," or "Republicans are significantly more likely than Democrats to attribute social inequality to what individuals make of their opportunities." Remember, however, that the word "significant" must be interpreted within its strict scientific meaning: the results of a hypothesis test tell us nothing about how great the difference between the two populations is, only that the difference between them is something greater than zero.

THE CONFIDENCE INTERVAL ALTERNATIVE

It is important to realize that we could accomplish all that a hypothesis test accomplishes and more if we use a confidence interval. If we construct a confidence interval in which the probability of error *in each direction* is 1 percent, we would have

$$(P_r - P_d) \pm 2.33(.034) \quad \text{or} \quad .08 \pm .08 \quad \text{or} \quad \text{from .00 to .16}$$

With this confidence interval, we would be 98 percent sure that the actual difference between the two populations was somewhere between a low of 0 percent and a high of 15 percent. In addition, we would be 99 percent sure that the difference was greater than zero.

We can use confidence intervals to test all of the hypotheses discussed earlier in this chapter. We could, for example, use a 98-percent confidence interval to estimate the mean achievement score for the state:

$$(73) \pm 2.33(1.25) \quad \text{or} \quad 73 \pm 2.9 \quad \text{or} \quad \text{from 70.1 to 75.9}$$

and conclude that the mean is no smaller than 70.1 but no larger than 75.9 (and, therefore, higher than the national mean of 70).

In a similar way, we could use a 95-percent confidence interval to estimate the proportion of people who are going to vote for your candidate:

$$(.55) \pm 1.96(.029) \quad \text{or} \quad .55 \pm .06 \quad \text{or} \quad \text{from 49 to 61 percent}$$

and conclude that your candidate will get no less than 49 percent of the vote and no more than 61 percent (an estimate, of course, that includes the possibility of losing). As you can see from these examples, by using a confidence interval, we not only get a test of the null hypothesis, but an estimate as well.

ABOUT "SIGNIFICANT DIFFERENCES"

The word "significant" is perhaps one of the most misunderstood and abused terms in the technical language of statistics. You will often find such statements as "The two groups are significantly different" or "People who used our brand of toothpaste had significantly fewer cavities." If you do not appreciate the limited technical meaning of "significance" you are likely to attach greater importance to it and the findings it describes than either deserves.

"People who use our toothpaste have significantly fewer cavities" means *only* that there is a small chance of making a mistake if we conclude that the cavity rate for those who use different brands of toothpaste are not *exactly equal. Two populations can be significantly different even though the actual difference is extremely small.* Suppose, for example, that our Republican and Democratic samples had been ten times as large (3,580 and 5,190, respectively). This would have given us an estimated standard error of .01 rather than .03, and a difference of less than 3 percentage points would have been significant at the .0038 level. A difference of less than 2 points would have been significant at the .05 level. While 2 percentage points does constitute a difference, it hardly qualifies as important.

The significance level attached to the rejection of a null hypothesis (or, for that matter, to a confidence interval) only expresses our degree of confidence in a statistical finding. Being confident of a finding, however, is no substitute for the finding itself. By itself, the significance level tells us very little. *Always look to see just what the finding is that the researchers are confident about.*

Chi-Square: Hypotheses about Relationships between Variables

While much of statistical inference is concerned with testing ideas about differences between two populations, many research questions focus on relationships between variables whose categories may define a large number of groups that are compared all at once.

Suppose, for example, that we examine the relationship between educational attainment and child-rearing values. Sociologists and developmental psychologists have long been interested in class differences in the goals parents have for their children, and one of the most important variables used to define social class is educational attainment. In the 1984 GENSOC survey, respondents were asked how important they thought it was that children obey their parents well. If we divide the results into two categories ("extremely" and "very" important in the "high" category and all lower rankings in the "low" category) and cross-tabulate this variable with educational attainment, we get the result shown in Table 13-1. As you can see, there is a moderately

Table 13-1 Valuing Obedience in Children by Educational Attainment, U.S. Adults, 1984

Importance of Obedience	Educational Attainment			
	At Least Some College	High School Graduate	Less Than High School	All
High	20.0%	35.1%	47.0%	32.5%
Low	80.0	64.9	53.0	67.5
TOTAL	100.0%	100.0%	100.0%	100.0%
(N)*	(559)	(476)	(400)	(1,435)

SOURCE: Computed from 1984 General Social Survey raw data.
* There are 38 cases of nonresponse out of a total sample size of 1,473.

strong negative relationship between education and support for this value (gamma = $-.40$): the higher one's education is, the *less* likely one is to value obedience in children highly.

As interesting as these sample findings are, our ultimate interest is not in the 1,435 people in the GENSOC survey but in the entire U.S. adult population from which it was drawn. Can we infer from a relationship between two variables in a sample that a similar relationship exists in the population? We know that differences between groups can occur in samples even when the populations from which the samples were drawn are identical. The same is true for relationships between variables.

We have found a relationship between these two variables in a sample, but can we infer from this that there is a negative relationship in the population? We approach this inference problem with the same logic as before. The substantive hypothesis (H_1) is that there is a negative relationship between the two variables in the population. The null hypothesis (H_0) is that the two variables are independent of each other in the population. To test the null hypothesis, *we must find the probability of observing a relationship this strong or stronger in a sample that is drawn from a population in which the variables are in fact independent of each other*. If we find that the probability is small enough, then we can reject the null hypothesis and support the substantive hypothesis with an acceptably small, known chance of error.

TESTING THE NULL HYPOTHESIS

In previous applications of hypothesis-testing techniques, we compared a sample result with an assumption about the population. We took the difference, for example, between a sample mean test score of 73 and the national mean of 70. According to the null hypothesis, we assumed that the state mean test score was the same as the national mean and then used the sample result to evaluate the credibility of that assumption.

We do the same thing with relationships between variables. We begin by putting Table 13-1 in the form of frequencies as in Table 13-2. Under the null hypothesis, we assume that the variables are independent of each other in the population. To see what this might look like, we keep the marginals in Table 13-2 the same but make the *cells* look as they would if the variables were stochastically independent. You may recall from Chapters 5 and 7 that when two variables are stochastically independent, the distribution of the dependent variable is the same for each category of the independent variable. To accomplish this, we would have to distribute the cases in each column so that the percentage distributions are the same as that for the "All" column in Table 13-1 (just as we did in Table 6-2 on page 126).

Table 13-2 Valuing Obedience in Children by Educational Attainment, U.S. Adults, 1984, Observed Frequencies

	Educational Attainment			
Importance of Obedience	At Least Some College	High School Graduate	Less Than High School	All
High	112	167	188	467
Low	447	309	212	968
TOTAL*	559	476	400	1,435

SOURCE: Computed from 1984 General Social Survey raw data.
* There are 38 cases of nonresponse out of a total sample size of 1,473.

In Table 13-3, the marginals are the same as in the previous tables, but the cells have been rearranged so that there is no relationship between education and the valuing of obedience in children. If we express this table in terms of frequencies, the result is Table 13-4. The frequencies in Table 13-2 are called **observed frequencies** because they are what we in fact observed in our data. The frequencies in Table 13-4, however, are called **expected frequencies** because they represent *what we would expect to find if the null hypothesis were true and the variables were stochastically independent.*

We now have two tables of frequencies—Table 13-2 and Table 13-4—with identical marginals. In previous examples of hypothesis testing, we measured the difference between our sample result and the null hypothesis assumption

Table 13-3 Valuing Obedience in Children by Educational Attainment, U.S. Adults, 1984, Assuming Independence

	Educational Attainment			
Importance of Obedience	At Least Some College	High School Graduate	Less Than High School	All
High	32.5%	32.5%	32.5%	32.5%
Low	67.5	67.5	67.5	67.5
TOTAL	100.0%	100.0%	100.0%	100.0%
(N)*	(559)	(476)	(400)	(1,435)

SOURCE: Computed from 1984 General Social Survey raw data.
* There are 38 cases of nonresponse out of a total sample size of 1,473.

Table 13-4 Valuing Obedience in Children by Educational Attainment, U.S. Adults, 1984, Assuming Independence, Expected Frequencies

Importance of Obedience	Educational Attainment			All
	At Least Some College	High School Graduate	Less Than High School	
High	182	155	130	467
Low	377	321	270	968
TOTAL*	559	476	400	1,435

SOURCE: Computed from 1984 General Social Survey raw data.
* There are 38 cases of nonresponse out of a total sample size of 1,473.

about the population. We then used that measured difference in a sampling distribution to find the conditional probability of getting a result such as ours given the assumption that the null hypothesis is true. If the sample result and the null hypothesis assumption were found to be too inconsistent with each other, we rejected the null hypothesis in favor of the substantive hypothesis.

With cross-tabulations, the logic is the same although the procedures are somewhat different since cross-tabulations are usually more complicated than a pair of proportions or means. To measure the difference between the sample results in Table 13-2 and the null hypothesis assumption in Table 13-4, *we take the difference between each corresponding pair of cells in the two tables.* We do this by subtracting each of the six expected frequencies in Table 13-4 from each corresponding observed frequency in Table 13-2: (112 − 182), (167 − 155), (188 − 130), and so on. Since, as with variances, the total of all these differences will always be zero, we square them. Finally, since we are interested in our results *relative* to what we would expect to find under the null hypothesis of independence, we divide this squared difference by the frequency we would expect to find under independence. The result is a statistic called **chi-square** (represented by the symbol χ^2). As a mathematical formula, the calculation of chi-square looks like this:

$$\chi^2 = \sum \frac{(f_o - f_e)^2}{f_e}$$

where f_o represents the observed frequencies in Table 13-2 and f_e represents the corresponding expected frequencies in Table 13-4. In our example, chi-square would be calculated as follows:

$$\chi^2 = \frac{(112-182)^2}{182} + \frac{(167-155)^2}{155} + \frac{(188-130)^2}{130} + \frac{(447-377)^2}{377}$$

$$+ \frac{(309-321)^2}{321} + \frac{(212-270)^2}{270} = 79.63$$

Before going on, it might be useful to review what we have done and why. We began with a table based on sample data showing a relationship between two variables. We then constructed a second table that represented the same variables with the same marginals but showing no relationship between the variables. This second table represents the null hypothesis of stochastic independence between the variables.

We then compared the sample table with the null hypothesis table—cell by cell—to see how inconsistent the sample result was with the assumption of independence expressed by the null hypothesis. We squared each difference for each pair of cells in order to avoid having the sum of the differences add to zero. We then divided each squared difference by the frequency expected under the condition of independence in order to get the difference *relative* to independence between the variables. By adding the six squared deviations from independence, we arrived at a summary measure—chi-square—of how much our sample results deviate from independence between educational attainment and the valuing of obedience in children.

What is the point of all this? We do it in order to evaluate the credibility of the null hypothesis by comparing it with the evidence from our sample. The question we want to answer is, "What is the probability of observing a *sample* relationship this strong or stronger assuming that the sample was drawn from a population in which the variables are independent of each other?" In more technical terms, the question is, "What is the probability of getting a value of chi-square this large or larger in a sample drawn from a population in which the variables are independent of each other?"

To answer this question, we use the value of chi-square just as we used values of Z and t earlier, except that in this case we use a different sampling distribution to locate the corresponding probability. Just as the normal table shows the distribution of Z-scores, so too does the chi-square table show the distribution of chi-square—the probability of getting various values of chi-square in samples drawn from populations in which two variables are independent of each other (Table 13-5).

Unlike the normal distribution, the chi-square distribution must compensate for the fact that tables with large numbers of cells will tend to have

Table 13-5 The Distribution of Chi-Square

df*	.99	.98	.95	.90	.80	.70	.50
				Level of Significance			
1	$.0^2157$	$.0^2628$	.00393	.0158	.0642	.148	.455
2	.0201	.0404	.103	.211	.446	.713	1.386
3	.115	.185	.352	.584	1.005	1.424	2.366
4	.297	.429	.711	1.064	1.649	2.195	3.357
5	.554	.752	1.145	1.610	2.343	3.000	4.351
6	.872	1.134	1.635	2.204	3.070	3.828	5.348
7	1.239	1.564	2.167	2.833	3.822	4.671	6.346
8	1.646	2.032	2.733	3.490	4.594	5.527	7.344
9	2.088	2.532	3.325	4.168	5.380	6.393	8.343
10	2.558	3.059	3.940	4.865	6.179	7.267	9.342
11	3.053	3.609	4.575	5.578	6.989	8.148	10.341
12	3.571	4.178	5.226	6.304	7.807	9.034	11.340
13	4.107	4.765	5.892	7.042	8.634	9.926	12.340
14	4.660	5.368	6.571	7.790	9.467	10.821	13.339
15	5.229	5.985	7.261	8.547	10.307	11.721	14.339
16	5.812	6.614	7.962	9.312	11.152	12.624	15.338
17	6.408	7.255	8.672	10.085	12.002	13.531	16.338
18	7.015	7.906	9.390	10.865	12.857	14.440	17.338
19	7.633	8.567	10.117	11.651	13.716	15.352	18.338
20	8.260	9.237	10.851	12.443	14.578	16.266	19.337
21	8.897	9.915	11.591	13.240	15.445	17.182	20.337
22	9.542	10.600	12.338	14.041	16.314	18.101	21.337
23	10.196	11.293	13.091	14.848	17.187	19.021	22.337
24	10.856	11.992	13.848	15.659	18.062	19.943	23.337
25	11.524	12.697	14.611	16.473	18.940	20.867	24.337
26	12.198	13.409	15.379	17.292	19.820	21.792	25.336
27	12.879	14.125	16.151	18.114	20.703	22.719	26.336
28	13.565	14.847	16.928	18.939	21.588	23.647	27.336
29	14.256	15.574	17.708	19.768	22.475	24.577	28.336
30	14.953	16.306	18.493	20.599	23.364	25.508	29.336

SOURCE: Fisher and Yates, 1974, Table IV.
* df = degrees of freedom

Table 13-5 *(continued)*

			Level of Significance			
.30	.20	.10	.05	.02	.01	.001
1.074	1.642	2.706	3.841	5.412	6.635	10.827
2.408	3.219	4.605	5.991	7.824	9.210	13.815
3.665	4.642	6.251	7.815	9.837	11.341	16.268
4.878	5.089	7.779	9.488	11.668	13.277	18.465
6.064	7.289	9.236	11.070	13.388	15.086	20.517
7.231	8.558	10.645	12.592	15.033	16.812	22.457
8.383	9.803	12.017	14.067	16.622	18.475	24.322
9.524	11.030	13.362	15.507	18.168	20.090	26.125
10.656	12.242	14.684	16.919	19.679	21.666	27.877
11.781	13.442	15.987	18.307	21.161	23.209	29.588
12.899	14.631	17.275	19.675	22.618	24.725	31.264
14.011	15.812	18.549	21.026	24.054	26.217	32.909
15.119	16.985	19.812	22.362	25.472	27.688	34.528
16.222	18.151	21.064	23.685	26.873	29.141	36.123
17.322	19.311	22.307	24.996	28.259	30.578	37.697
18.418	20.465	23.542	26.296	29.633	32.000	39.252
19.511	21.615	24.769	27.587	30.995	33.409	40.790
20.601	22.760	25.989	28.869	32.346	34.805	42.312
21.689	23.900	27.204	30.144	33.687	36.191	43.820
22.775	25.038	28.412	31.410	35.020	37.566	45.315
23.858	26.171	29.615	32.671	36.343	38.932	46.797
24.939	27.301	30.813	33.924	37.659	40.289	48.268
26.018	28.429	32.007	35.172	38.968	41.638	49.728
27.096	29.553	33.196	36.415	40.270	42.980	51.179
28.172	30.675	34.382	37.652	41.566	44.314	52.620
29.246	31.795	35.563	38.885	42.856	45.642	54.052
30.319	32.912	36.741	40.113	44.140	46.963	55.476
31.391	34.027	37.916	41.337	45.419	48.278	56.893
32.461	35.139	39.087	42.557	46.693	49.588	58.302
33.530	36.250	40.256	43.773	47.962	50.892	59.703

relatively large values of chi-square regardless of the strength of the relationship between the two variables. With a 2 × 3 table, for example, we add up six fractions just as we did above. In a 10 × 10 table, however, we will add up 100 fractions. Even with equal sample sizes, it is much easier to get a large value of chi-square with 100 fractions than it is with six. We compensate for this by requiring that the value of chi-square be much larger in a 10 × 10 table than in a 2 × 3 table in order to reject the null hypothesis with the same probability of error. This means that there is a separate distribution of chi-square for each table size.

In the chi-square distribution, table size is measured by a quantity called **degrees of freedom** (just as sample size was measured in degrees of freedom in the t-distribution table in Chapter 12). Degrees of freedom is found by multiplying the number of rows in the table minus one *times* the number of columns minus one, or $(r - 1)(c - 1)$. In our case, there are two rows and three columns; therefore, $(2 - 1)(3 - 1) = 2$ degrees of freedom.

To use the chi-square table (Table 13-5) to determine probabilities, we go down the left-hand column to the proper number of degrees of freedom—in our case, the second row. The numbers in the body of the table are values of chi-square such as the value of 79.63 that we computed earlier. The corresponding probabilities are found in the top row, ranging from a high of .99 on the left to a low of .001 on the right. As an example, the first value of chi-square in the second row is .0201. It is in the column headed by ".99." This indicates that with two degrees of freedom, the probability is .99 that we would draw a sample with a value of chi-square of .0201 or larger simply by chance *from a population in which the two variables are independent of each other*.

Our value of chi-square was 79.63. We go across the second row and look for a value that is as close as possible to ours. By the time we get to the end of the row, the value of chi-square has reached only 13.815, still far below 79.63. If we had obtained a value of 13.815 in our sample, this would have meant that our result would occur by chance with a probability of only .001. Since our value of chi-square is much larger than 13.815, this means that the probability associated with our result is much smaller than .001.

Thus, a relationship as strong as ours is very unlikely to occur by chance in a sample drawn from a population in which the two variables are independent. If we then reject the null hypothesis, the probability that we are making a mistake will be much less than .001. We would conclude that the negative relationship between education and valuing obedience in children is "significant at less than the .001 level," which is to say, we are confident that the two variables are negatively related in the population. This is often written as "$\chi^2 = 79.63, p < .001$."

If in our sample we had computed a value of chi-square equal to 6.0, then we would have concluded that the relationship was significant at the .05 level since 6.0 falls between 5.991 and 7.824 in the second row of Table 13-5 (be sure you see this). If our computed value of chi-square had been only 1.7, then the relationship would have significance between the .50 and .30 levels, roughly .44 (because the table is abbreviated, we must interpolate between points such as .50 and .30). This would mean that this kind of sample result would be very *likely* to occur in a population in which the null hypothesis is *true*. As a result, we would not reject the null hypothesis since the probability of error—roughly .44—is much too large.

Notice in Table 13-5 that for any given level of significance, the required value of chi-square varies directly with the table size as measured by degrees of freedom. With two degrees of freedom, a value of chi-square of 13.815 or larger is enough to reject the null hypothesis at the .001 level; but with six degrees of freedom (a 3 × 4 table) we need a value of chi-square of 22.457 or more.

While appreciating how table size affects the value of chi-square, it is also important to be aware of the effect of sample size. Sample size has a direct and proportional effect on chi-square: if we double the sample size, we also double the value of chi-square. If you go through the example used in this section and double all of the frequencies, you will see very quickly that this has the effect of doubling the final result as well. This is important, because given a large enough sample, *any* relationship, no matter how weak, will show up as statistically significant and justify rejecting the null hypothesis.

This creates problems only if we forget the precise and limited meaning of the concept of significance. The value of chi-square only leads us to a certain level of confidence that a relationship does exist in the population, which is to say, that the variables are not stochastically independent. Since there is less than a .001 chance that we would make an error in rejecting the null hypothesis, then we could attach a probability of greater than $(1 - .001) = .999$ to the substantive hypothesis that there *is* a relationship. When you see a value of chi-square for a relationship between variables, always look for the measure of association to see just how strong the relationship is. By itself, the significance level tells only part of the story.

Chi-Square and Goodness-of-Fit Tests

To calculate chi-square in the previous section, we compared sample frequencies in a cross-tabulation with those we would expect under the assumption that the two variables were independent of each other. This is a specific application of the general idea behind chi-square—that chi-square is a statistic that evaluates the differences between actual frequencies found in a sample and frequencies that are expected according to an assumption about the population from which the sample was drawn. As with other hypothesis tests, the "evaluation" of the differences takes the form of computing the probability that we would be making a mistake if we rejected the null hypothesis from which the expected frequencies are derived.

With **goodness-of-fit** tests, we use this same logic for a slightly different purpose. Suppose, for example, that we gather data from a national sample of adults in 1980. Before analyzing our data, we want to reassure ourselves that our sample is representative of the adult population it came from.

The best way to do this is to compare the sample distribution with the population distribution for those variables on which we have national data based on a complete census. We could, for example, compare the age distribution of our sample of adults with the age distribution for the entire country as recorded in the 1980 census of population. Column I of Table 13-6 shows the age distribution for the U.S. adult population based on the 1980 census. Suppose that the percentage age distribution in our sample looks like column II, with the corresponding observed frequencies (f_o) shown in column III.

We want to compare the sample age distribution with the known population distribution to see if there is any bias in the sample. There are 1,540 respondents in the sample. We are going to test the null hypothesis that states that there is no difference between the age distribution of the population our sample represents and the U.S. population age distribution. To test the null hypothesis, we compare the frequencies found in each age group in

Table 13-6 Using Chi-Square for a Goodness-of-Fit Test between Sample and Population Age Distributions, United States, 1980

	Age Distribution					
Age Group	(I) Census	(II) Sample	(III) f_o	(IV) f_e	(V) $(f_o - f_e)^2$	(VI) $(f_o - f_e)^2/f_e$
18–24	18.8%	14.2%	218	289	5,041	17.4
25–34	23.3	21.3	328	360	1,024	2.8
35–44	16.1	22.2	343	248	9,025	36.4
45–54	14.3	18.8	289	220	4,761	21.6
55–64	13.6	12.9	199	209	100	0.5
65+	13.9	10.6	163	214	2,601	12.2
TOTAL	100.0%	100.0%	1,540	1,540		90.9 = χ^2

SOURCE: U.S. Bureau of the Census, 1983, p. 33.

our sample with the frequencies we would *expect* to find if the null hypothesis were true.

How do we determine the expected frequencies? If our sample was distributed by age in the same way as the U.S. population is distributed in column I, then 18.8 percent of our 1,540 cases would fall in the 18 to 24-year-old category. In frequencies, this amounts to 18.8 percent of 1,540, or 289 cases. Similarly, we would expect to find 23.3 percent of our 1,540 cases in the 25 to 34-year-old category, or a total of 360 cases. The expected frequencies for each age category are found in column IV of Table 13-6.

From here on, we calculate chi-square just as before. We have a set of observed sample frequencies in column III and a set of expected frequencies in column IV. To compute chi-square we simply take the difference between each pair of observed and expected frequencies and square it, as in column V. We then divide the squared difference by the expected frequency (column VI):

$$\chi^2 = \sum \frac{(f_o - f_e)^2}{f_e}$$

The resulting sum of column VI—90.9—is the value of chi-square.

To find the probability that corresponds to our value of chi-square, we must first determine the number of degrees of freedom. With goodness-of-fit tests, degrees of freedom amounts to the number of categories minus one. In this case, for example, there are (6 − 1) = 5 degrees of freedom. We look down the left-hand column of the chi-square table (Table 13-5) to the row corresponding to 5 degrees of freedom, then across until we find a value of chi-square that comes the closest to the one we computed for our sample. As you can see, the largest value in this row—20.517—is far smaller than our value of 90.9 and corresponds to a probability of .001. This means that the probability that a properly drawn sample of the U.S. adult population would differ in its age distribution by this much or more is far less than .001, or 0.1 percent. We can, with virtual certainty, reject the null hypothesis and conclude that the age distribution of our sample does not represent the U.S. adult population.

Notice once again that the value of chi-square depends on the size of the sample. If our sample had included only 200 cases instead of 1,540, the value of chi-square would have been reduced proportionately to 11.81, a value that would have been barely significant at the .05 level. With a large sample, even a relatively small difference between two distributions will be statistically significant.

What does this result tell us about the representativeness of our sample? We have rejected the null hypothesis that there is no difference between the population and sample age distributions, with a probability of error that is almost zero. We are certain that the sample is biased on age. In what way is it biased? If we compare columns I and II, we can see that our sample underrepresents the younger and older age categories and overrepresents the middle two categories.

THE INDEX OF DISSIMILARITY

How serious is this degree of bias? There is no easy answer to this question, no statistical rule that classifies bias as "serious" or "trivial." There is, however, a useful measure that we can use to tell us how dissimilar two distributions are. To compute it, we compare the two distributions in percentage form by taking the difference between each pair of corresponding percentages (Table 13-7).

In column I of Table 13-7 we have the percentaged age distribution for the 1980 census, and in column II we have the corresponding distribution for our sample. Column III is simply the difference between the two. Notice that the sum of these differences is zero, which will always be the case since both distributions add to 100 percent. Also notice, however, that if we add up the absolute value of the positive differences and the absolute value of the negative differences, we get the same answer in both cases (10.6 percent). This quantity, found by adding up the positive or negative differences, is called the **index of dissimilarity**, and it has a precise and useful interpretation.

The index of dissimilarity tells us *what percentage of the cases in either distribution would have to be rearranged in order for the two percentage distributions*

Table 13-7 Computation of the Index of Dissimilarity for Table 13-6

(I) Census Age Distribution	(II) Sample Age Distribution	(III) (Census − Sample)
18.8%	14.2%	+4.6
23.3	21.3	+2.0
16.1	22.2	−6.1
14.3	18.8	−4.5
13.6	12.9	+0.7
13.9	10.6	+3.3
	Sum of positive differences	+10.6
	Sum of negative differences	−10.6
	Index of dissimilarity	10.6

to be identical. As such, it is a direct and precise measure of how different the two distributions are. In our case, in order to make our sample age distribution identical to that for the U.S. adult population, we would have to rearrange 10.6 percent of the cases, which is to say, we would have to reassign 10.6 percent of the cases to different age categories. We would, of course, never do that in practice since it would require us to change the ages of actual respondents; but the measure does provide a useful indicator of how dissimilar the two distributions are. In this case, the degree of dissimilarity indicates the degree of bias in the sample.

The index of dissimilarity can be used to compare any two percentage distributions so long as they have the same categories. Although it is not a widely used measure, some of its applications have been quite dramatic. If we examine the residential patterns of whites and nonwhites in the U.S., for example, we can use the index of dissimilarity to determine the degree to which housing is segregated by race (or any other measurable characteristic). We do this by comparing the actual distribution of nonwhites with the distribution we would expect to find if the only factor governing their representation in different neighborhoods was their relative numbers in the population as a whole. In this case, the index of dissimilarity tells us what percentage of nonwhites or whites would have to change their residence in order for the racial composition of neighborhoods to accurately reflect the racial composition of the population as a whole—in other words, in order to end residential segregation.

The results for most large U.S. cities are indexes of dissimilarity that range from the high 70s to the mid-90s, indicating that most members of one race or the other would have to move in order to bring about racial balance in urban neighborhoods (Taeuber, 1965; Sorensen et al., 1974). The index has been used to demonstrate that blacks constitute one of the most highly segregated racial categories in the U.S., second only to American Indians. In general, Asian and Hispanic Americans are much less segregated from whites than are blacks, with the exception of Puerto Ricans whose degree of segregation is similar to that for blacks (Massey, 1981).

Hypothesis Tests for Relationships between Variables

Earlier in this chapter we described the use of chi-square to test the null hypothesis that two cross-tabulated variables are independent of each other. In this section, we will briefly describe methods for testing null hypotheses about relationships between variables that take a form other than cross-tabulations.

For the most part, sampling distributions for nominal-scale measures of association are either quite complex or unknown (see Goodman and Kruskal, 1954, pp. 723–64), although Davis (1971, pp. 51–58) does provide a method for constructing confidence intervals for Yule's Q. For this reason, a common practice is to use chi-square as a test for independence between the variables and, if the relationship is found to be statistically significant, to report a measure of association (such as phi or Goodman and Kruskal's tau).

For ordinal variables, Goodman and Kruskal (1963) provide a means for using the normal curve to test the null hypothesis that gamma equals zero (see also Loether and McTavish, 1980, pp. 588–92). For relationships between rank-ordered variables, there are several significance tests that can be used to test the null hypothesis that no relationship exists between two sets of ranks. These are known as **nonparametric statistics** and require no assumptions about the shape of the sampling distributions (for tables, see Appendix Tables 6, 7, and 8). The most popular of these tests are the Wald-Wolfowitz runs test, the Mann-Whitney (or Wilcoxon) test, the Kolmogorov-Smirnov test, and the Wilcoxon matched-pairs signed-rank test (see Siegel, 1956; Loether and McTavish, 1980). All of these are simply hypothesis tests used to test for the significance of relationships between ordinal variables that take the form of ranks. They are typically used in conjunction with a rank-order measure of association such as Kendall's tau or Spearman's r_s (see Chapter 6).

Hypothesis testing can be used with regression, correlation, and path analysis to establish the significance of correlation coefficients, standardized and unstandardized regression coefficients, partial correlation and regression coefficients, and path coefficients. The usual procedure is to test the null hypothesis that a coefficient is zero in the population. If that hypothesis is rejected, then the substantive hypothesis that the coefficient is greater than zero (or less than zero in the case of negative coefficients) is supported with a known probability of error (see Loether and McTavish, 1980, pp. 597–605).

The typical form of such results is, "There is a significant correlation between X and Y ($r = +.20$, $p = .02$)." As with other results of hypothesis testing, it is important to always look beyond the significance level to the actual size of the coefficients. "Statistical significance" only expresses a degree of confidence that a coefficient is not exactly equal to zero, and as we have seen before, with a large enough sample, even the most trivial coefficient can emerge as statistically significant.

It is also important to keep in mind, however, that if a sample is small, a substantial and important relationship between two variables or a difference between two groups may not be *detected* as significant simply because the standard error is so large. With a small sample and a large standard error, for example, the difference between two groups must be very large in order for us to reject the null hypothesis with a high degree of confidence. Thus, with large samples we are in danger of attributing importance to findings that are statistically significant but trivial. With small samples, we run the opposite risk of failing to acknowledge the existence of important findings simply because they do not achieve the required level of statistical significance.

On the Uses and Limitations of Hypothesis Testing

Hypothesis testing serves several useful functions. If we find that there are no significant differences between groups, for example, this can be important in two ways. It can indicate that we have reached a dead end for a particular research direction and need to redirect our thinking towards other hypotheses. It can also be an important finding in itself. If researchers found no significant differences in the IQ scores of blacks and whites in the U.S., it would challenge important cultural assumptions about racial differences.

When a finding does achieve a level of statistical significance, it provides support for the substantive hypothesis; but while significance tests are a necessary condition for establishing the importance of a finding, they are not a sufficient condition. In other words, hypothesis tests can only confirm that something is going on in a set of data; they cannot tell us how much importance we should attribute to it. To demonstrate that an educational program results in "significant improvement" in student performance does not mean that it justifies whatever costs go with it. The only way to answer that question is to find out how *much* of an improvement is involved, information that confidence intervals, but not hypothesis tests, are designed to provide.

The most unfortunate aspect of hypothesis testing is that it formulates research questions in terms of an either-or decision-making model, in spite of the fact that the statistical techniques rely on probabilities that range continuously from 0.0 to 1.0 (see, for example, Rozeboom, 1971). The setting of a significance level can never be anything but an arbitrary decision. When researchers decide to reject a null hypothesis only when the probability of error is .05 or less, they are necessarily deciding to not reject when the probability of error is anything greater than .05, including .06 or .07, both of which are still quite small probabilities.

The results of traditional hypothesis tests, then, typically fall into two categories: null hypotheses that are rejected and those that are not rejected. By contrast, confidence intervals offer a very different approach in which a finding is stated in terms of an estimate to which a probability is attached. It is then left to the reader to decide what to make of the estimate and how to evaluate the probability of error.

Increasingly, researchers are choosing to present the results of hypothesis tests in a way that allows greater flexibility and incorporates some of the advantages of confidence intervals. Rather than setting and rigidly adhering to an arbitrary significance level, this method simply reports the probability of error that would be associated with rejecting a null hypothesis. Instead of saying, for example, "the groups are not significantly different at the .05 level," we could report the exact significance levels—as in "the groups are significantly different at the .09 level"—and leave it to the reader to decide whether this probability of error is small enough to justify rejecting the null hypothesis.

This approach shows some appreciation for the fact that although it has been conventional to rely on fixed significance levels of .05, .01, and .001, there is nothing scientifically compelling about these arbitrary cutoff points. It does, of course, make the evaluation of hypotheses more complicated and less automatic, but this is probably as it should be given the nature of probabilities and the necessity of relying on human judgment in even the most objective scientific research.

Summary

1. In statistical inference, the purpose of hypothesis testing is to decide whether or not sample results can be used to make statements about the populations from which they are drawn.

2. To test a hypothesis, we assume a specific value for the mean of the sampling distribution (the null hypothesis) and then test to see how inconsistent the sample result is with that assumption. If the probability of getting the sample result given the null hypothesis is small enough, then we reject the null hypothesis in favor of the substantive hypothesis, with a known probability of error (the significance level).

3. The logic of hypothesis testing is the same regardless of which characteristics of populations are involved (proportions, means, differences between means or proportions, relationships between variables, etc.).

4. When a null hypothesis is rejected at a given level of significance, this means only that we are confident that the null hypothesis is false. A significant difference between two means is not necessarily large or important.

5. The chi-square statistic is used to test null hypotheses about relationships between variables presented in the form of cross-tabulations. Chi-square tests the null hypothesis of no relationship between the variables

by comparing the observed frequency in each cell of the sample table with the frequency we would expect to find if the variables were independent. The more inconsistent the observed sample frequencies are with the expected frequencies under the null hypothesis, the more likely we are to reject the null hypothesis.

6. Goodness-of-fit tests use the chi-square statistic to test a null hypothesis about the shape of a frequency distribution for a single variable. If the sample distribution is significantly different from the population distribution we assume under the null hypothesis, then we reject the null hypothesis with a known probability of error. Having done this, we can then use the index of dissimilarity to estimate the actual difference between the two distributions. We do this by calculating the percentage of cases in either distribution that would have to be placed in different categories in order for the two distributions to be identical.

Key Terms

chi-square 342
   expected frequencies 343
   observed frequencies 343
degrees of freedom 348
goodness-of-fit test 350
hypothesis 325
   null 326
   substantive 326

hypothesis test 325
index of dissimilarity 352
nonparametric statistics 354
one-tailed hypothesis test 334
significance level 328

Problems

1. Define and give an example of each of the key terms listed at the end of the chapter.
2. In a criminal trial, we assume the defendant is innocent. We then gather evidence to see if we can reject the assumption of innocence in favor of a conclusion of guilt. If we looked at criminal trials as hypothesis tests, what would be the null hypothesis? What would be the substantive hypothesis? How would you interpret the phrase, "guilty beyond a shadow of a doubt"?
3. Nationally, about 12 percent of the population live at or below the official poverty level. We decide to conduct a survey in our community to decide if the percentage of our population living in poverty is lower than the national figure. In this case, what is the null hypothesis? What is the substantive hypothesis?

4. We use the significance level to express the probability of error associated with rejecting a null hypothesis. What can we do about estimating the probability of error associated with *not* rejecting the null hypothesis?

5. Some farmers calculate that under current market conditions, the only way they can make a profit is by producing a mean yield of more than 900 bushels per acre. To estimate their final income when the harvesting season is over, they sample the crops from 100 acres and find a mean yield of 941 bushels per acre with a standard deviation of 150 bushels.
   a. What is the null hypothesis?
   b. What is the substantive hypothesis?
   c. Estimate the standard error for sample means.
   d. At what significance level could you reject the null hypothesis? Would you? Why or why not?
   e. State your conclusion in a sentence or two.
   f. Suppose the mean yield per acre in the sample had been 875 bushels? At what level of significance could you reject the null hypothesis?

6. From problem 3: We select a sample of 300 residents and find that the proportion who are living at or below the poverty level is .10.
   a. Estimate the standard error for sample proportions.
   b. At what significance level could you reject the null hypothesis? Would you? Why or why not?
   c. State your conclusion in a sentence or two.
   d. Suppose the proportion at or below the poverty level had been .12. At what level of significance could you reject the null hypothesis?

7. We want to find out if men are more likely than women to support capital punishment. In 1984, the GENSOC survey found that 80 percent of men approved ($N = 569$) and 71 percent of women approved ($N = 807$).
   a. What is the null hypothesis?
   b. What is the substantive hypothesis?
   c. Estimate the standard error for the difference between sample proportions.
   d. At what level of significance could you reject the null hypothesis? Would you? Why or why not?
   e. State your conclusion in a sentence or two.
   f. Suppose women in the sample were *more* likely than men to support capital punishment. At what level of significance could you reject the null hypothesis?

8. We want to find out if Democrats and Independents are more likely than Republicans to believe that welfare programs make people not want to work. In the 1984 GENSOC survey, 43 percent of Democrats and

Independents agreed ($N = 1,030$) compared with 55 percent of Republicans ($N = 366$).
   a. What is the null hypothesis?
   b. What is the substantive hypothesis?
   c. Estimate the standard error for the difference between sample proportions.
   d. At what level of significance could you reject the null hypothesis? Would you? Why or why not?
   e. State your conclusion in a sentence or two.

9. We want to find out if people who identify themselves as lower or working class ("Low Class") tend to watch more television than do people who identify themselves as middle or upper class ("High Class"). In the 1984 GENSOC survey, people in the "Low Class" category ($N = 587$) watched an average of 3.1 hours a day with a standard deviation of 2.2 hours. Those in the "High Class" category ($N = 569$) watched an average of 2.8 hours with a standard deviation of 2.1 hours.
   a. What is the null hypothesis?
   b. What is the substantive hypothesis?
   c. Estimate the standard error for the difference between the two means.
   d. At what level of significance could you reject the null hypothesis? Would you? Why or why not?
   e. State your conclusion in a sentence or two.

10. We want to find out if people who earn less than $15,000 a year ("Low Income") have more children than do those who earn $15,000 a year or more ("High Income"). In the 1984 GENSOC survey, the mean number of children born to "Low Income" adults ($N = 565$) was 1.66 with a standard deviation of 1.68 children. The mean number of children born to "High Income" adults ($N = 314$) was 1.91 children with a standard deviation of 1.7 children.
    a. What is the null hypothesis?
    b. What is the substantive hypothesis?
    c. Estimate the standard error for the difference between two means.
    d. At what significance level could you reject the null hypothesis? Would you? Why or why not?
    e. State your conclusion in a sentence or two.

11. Table 13-8 shows the relationship between marital status and answers to the question, "Taken all together, how would you say things are these days—would you say that you are very happy, pretty happy, or not too happy?" The data are drawn from the 1984 GENSOC survey. Use chi-square to test for the existence of a relationship between these two variables in the population.

Table 13-8 Reported Level of Happiness by Marital Status, United States, 1984

Reported Happiness	Marital Status			
	Married	Widowed/ Divorced	Single	All
Very	347	97	58	502
Less Than Very	467	256	220	943
TOTAL*	814	353	278	1,445

SOURCE: Computed from 1984 General Social Survey raw data.
* There are 28 cases of nonresponse for the total sample of 1,473.

a. What is the null hypothesis?
b. What is the substantive hypothesis?
c. At what significance level could you reject the null hypothesis? Would you? Why or why not?
d. State your conclusion in a sentence or two.

12. We do a study in which we examine the religious composition of boards of directors of major corporations. Suppose we find that of 1,000 randomly selected board members, 650 are Protestant, 260 are Catholic, 10 are Jewish, and 80 express no religious preference. We know from national studies that 63.5 percent of the adult population are Protestant, 26.9 percent are Catholic, 2.4 percent are Jewish, and 7.2 percent express no religious preference. We want to know if the composition of boards of directors constitutes a representative group in terms of religious preference.
    a. What is the null hypothesis?
    b. What is the substantive hypothesis?
    c. Use a goodness-of-fit test to test the null hypothesis. At what level of significance could you reject the null hypothesis? Would you? Why or why not?
    d. State your conclusion in a sentence or two.
    e. Use the index of dissimilarity to compare the two distributions. What does this tell you?

13. Suppose the sample results in problems 5–12 were based on quota samples. How would you apply the inference techniques described in this chapter to data from such samples? Explain your answer.

Epilogue

In this book I have tried to introduce you to statistical techniques that are fundamental to an appreciation of the uses of quantitative data in scientific research and the problems faced by its consumers. Inevitably, there are techniques that have been omitted, in most cases because they are beyond the scope of an introductory course in statistics. It would be useful, however, for you to have at least passing familiarity with some of these techniques and the situations in which they are used.

Analysis of variance, for example, is a standard set of techniques used to analyze relationships between interval- or ratio-scale dependent variables and nominal- or ordinal-scale independent variables. Its purpose is to explain variance in the dependent variable using several different independent variables at once, to identify the relative contribution of each independent variable, and to make statistical inferences about the results (see Blalock, 1979; Edwards, 1960; Hays, 1981).

Recently there has been considerable progress in the development of sophisticated techniques that take analysis of variance a step further. With *multiple-classification analysis*, for example, we can analyze relationships between nominal- or ordinal-scale independent variables and interval- or ratio-scale dependent variables, and compute statistics that are comparable to multiple correlation coefficients and partial regression coefficients. We can also make statistical inferences about the results (see Andrews, Morgan, and Sonquist, 1973).

A more recent development is log-linear analysis which makes it possible to describe and analyze multiple relationships that involve only nominal or ordinal variables (Goodman, 1984). This general approach is rapidly emerging as one of the most versatile and powerful set of statistical techniques in the social scientist's inventory (see Duncan and Duncan, 1978; Knoke and Burke, 1980).

Some Parting Thoughts on Statistics

You have by now been exposed to the major ideas that lie behind the use of statistical techniques in the social and behavioral sciences. In the end, however, what you stand to gain by understanding all of this will depend on the perspective from which you approach it.

As individuals, we all have ideas about ourselves, others, and the social and physical environments we live and act in, and these, coupled with our own experience, often form the basis for generalizations about how the world works. As participants in a culture, we cannot help but adopt many views of reality whose main support is our shared store of cultural beliefs, values, and attitudes supplemented by personal experience. While these sources of knowledge are unquestionably real and valuable, they often constitute a poor and biased sample of the complex populations we try to understand and describe.

The value of systematic data gathering is that it enables us to extend our experience beyond the confines of our own lives or the legacy of cultural stereotypes. In spite of their many limitations (and there are many), statistical techniques help us test many of our ideas about the world in a broader and more rigorously representative framework. The rigor of the techniques and their application, however, should not be allowed to hide the fact that ultimately the results are generated and interpreted by human beings whose experiences and values give scientific research its meaning. Without a human context of value and meaning, numbers are nothing more than numbers; they have no life of their own.

There is an ongoing debate in the social and behavioral sciences about the scientific status of research. A substantial portion of this book has focused on the problems associated with studying human behavior and social environments, for error creeps in at so many points in the research process that it often makes it difficult to draw conclusions with the kind of confidence we would like to have. This prompts some to attack the use of scientific techniques by behavioral and social scientists as premature at best.

Science, however, does not pretend to be a set of unassailable findings (which it is not), but rather a set of methods and standards for getting certain kinds of answers to certain kinds of questions. Our job is to formulate testable, logically related sets of propositions about how the various aspects of the world work, propositions that we can test with data that yield replicable, unambiguous results. Because the world we are trying to understand is so complex, however, and because the causes of human behavior are usually hidden from direct observation, testing such propositions is at best a very difficult job.

To see what I mean more concretely, consider a chemist who is trying to conduct a simple experiment. A window is open and the furnace in the basement is turning on and off at irregular intervals, causing the temperature in the room to fluctuate unpredictably. The chemicals contain impurities that are difficult to identify, not to mention remove. The chemist still approaches the research from a scientific perspective, but under conditions that make it difficult to draw definitive conclusions from the experiment.

Social and behavioral scientists are in somewhat the same position, which helps to explain why it is necessary to rely so heavily on probabilities in reaching conclusions. As anthropologist Marvin Harris (1980) wrote,

> After Hume, science could no longer be considered a distinctive way of knowing because it alone among ways of knowing can achieve certainty. Rather, it is a distinctive way of knowing because it claims to be able to distinguish between different degrees of uncertainty. In judging scientific theories one does not inquire which theory leads to accurate predictions in all instances, but rather which theories lead to accurate predictions in more instances. Failure to achieve complete predictability does not invalidate a scientific theory; it merely constitutes an invitation to do better. (p. 11)

While I hope that you will be increasingly less in awe of statistics, I also hope that you will resist the temptation to be hypercritical or cynical. As a human enterprise, empirical research is bound to be flawed in one way or another because it is impossible to study human behavior and the human condition and identify, much less control, all potential sources of error. This means that we are always in a position to criticize or dismiss a piece of research as imperfect. Somewhere between the cynicism of those who believe that all statements supported by statistics are suspect and the blind positivism of those who believe that only those statements supported by statistics are acceptable, there lies a more balanced and intelligent approach to the problem of gathering and using data to better understand ourselves and the world. It is toward that middle ground that I hope this book has helped to take us. In short, I hope that I have helped to make the gullible less gullible, the cynical less cynical, and those who want to learn from quantitative information more capable of doing so.

Appendix Tables

Table 1 Areas under the Normal Curve (the Distribution of Z)

Z	.00	.01	.02	.03	.04	.05	.06	.07	.08	.09
0.0	0000	0040	0080	0120	0159	0199	0239	0279	0319	0359
0.1	0398	0438	0478	0517	0557	0596	0636	0675	0714	0753
0.2	0793	0832	0871	0910	0948	0987	1026	1064	1103	1141
0.3	1179	1217	1255	1293	1331	1368	1406	1443	1480	1517
0.4	1554	1591	1628	1664	1700	1736	1772	1808	1844	1879
0.5	1915	1950	1985	2019	2054	2088	2123	2157	2190	2224
0.6	2257	2291	2324	2357	2389	2422	2454	2486	2518	2549
0.7	2580	2612	2642	2673	2704	2734	2764	2794	2823	2852
0.8	2881	2910	2939	2967	2995	3023	3051	3078	3106	3133
0.9	3159	3186	3212	3238	3264	3289	3315	3340	3365	3389
1.0	3413	3438	3461	3485	3508	3531	3554	3577	3599	3621
1.1	3643	3665	3686	3718	3729	3749	3770	3790	3810	3830
1.2	3849	3869	3888	3907	3925	3944	3962	3980	3997	4015
1.3	4032	4049	4066	4083	4099	4115	4131	4147	4162	4177
1.4	4192	4207	4222	4236	4251	4265	4279	4292	4306	4319
1.5	4332	4345	4357	4370	4382	4394	4406	4418	4430	4441
1.6	4452	4463	4474	4485	4495	4505	4515	4525	4535	4545
1.7	4554	4564	4573	4582	4591	4599	4608	4616	4625	4633
1.8	4641	4649	4656	4664	4671	4678	4686	4693	4699	4706
1.9	4713	4719	4726	4732	4738	4744	4750	4758	4762	4767
2.0	4773	4778	4783	4788	4793	4798	4803	4808	4812	4817
2.1	4821	4826	4830	4834	4838	4842	4846	4850	4854	4857
2.2	4861	4865	4868	4871	4875	4878	4881	4884	4887	4890
2.3	4893	4896	4898	4901	4904	4906	4909	4911	4913	4916
2.4	4918	4920	4922	4925	4927	4929	4931	4932	4934	4936
2.5	4938	4940	4941	4943	4945	4946	4948	4949	4951	4952
2.6	4953	4955	4956	4957	4959	4960	4961	4962	4963	4964
2.7	4965	4966	4967	4968	4969	4970	4971	4972	4973	4974
2.8	4974	4975	4976	4977	4977	4978	4979	4980	4980	4981
2.9	4981	4982	4983	4984	4984	4984	4985	4985	4986	4986
3.0	4986.5	4987	4987	4988	4988	4988	4989	4989	4989	4990
3.1	4990.0	4991	4991	4991	4992	4992	4992	4992	4993	4993
3.2	4993.129									
3.3	4995.166									
3.4	4996.631									
3.5	4997.674									
3.6	4998.409									
3.7	4998.922									
3.8	4999.277									
3.9	4999.519									
4.0	4999.683									
4.5	4999.966									
5.0	4999.997133									

SOURCE: Rugg, 1917, Appendix Table III, pp. 389–90.

Table 2 Student's t-Distribution

	Level of Significance for One-tailed Test					
	.10	.05	.025	.01	.005	.0005
	Level of Significance for Two-tailed Test					
df	.20	.10	.05	.02	.01	.001
1	3.078	6.314	12.706	31.821	63.657	636.619
2	1.886	2.920	4.303	6.965	9.925	31.598
3	1.638	2.353	3.182	4.541	5.841	12.941
4	1.533	2.132	2.776	3.747	4.604	8.610
5	1.476	2.015	2.571	3.365	4.032	6.859
6	1.440	1.943	2.447	3.143	3.707	5.959
7	1.415	1.895	2.365	2.998	3.499	5.405
8	1.397	1.860	2.306	2.896	3.355	5.041
9	1.383	1.833	2.262	2.821	3.250	4.781
10	1.372	1.812	2.228	2.764	3.169	4.587
11	1.363	1.796	2.201	2.718	3.106	4.437
12	1.356	1.782	2.179	2.681	3.055	4.318
13	1.350	1.771	2.160	2.650	3.012	4.221
14	1.345	1.761	2.145	2.624	2.977	4.140
15	1.341	1.753	2.131	2.602	2.947	4.073
16	1.337	1.746	2.120	2.583	2.921	4.015
17	1.333	1.740	2.110	2.567	2.898	3.965
18	1.330	1.734	2.101	2.552	2.878	3.922
19	1.328	1.729	2.093	2.539	2.861	3.883
20	1.325	1.725	2.086	2.528	2.845	3.850
21	1.323	1.721	2.080	2.518	2.831	3.819
22	1.321	1.717	2.074	2.508	2.819	3.792
23	1.319	1.714	2.069	2.500	2.807	3.767
24	1.318	1.711	2.064	2.492	2.797	3.745
25	1.316	1.708	2.060	2.485	2.787	3.725
26	1.315	1.706	2.056	2.479	2.779	3.707
27	1.314	1.703	2.052	2.473	2.771	3.690
28	1.313	1.701	2.048	2.467	2.763	3.674
29	1.311	1.699	2.045	2.462	2.756	3.659
30	1.310	1.697	2.042	2.457	2.750	3.646
40	1.303	1.684	2.021	2.423	2.704	3.551
60	1.296	1.671	2.000	2.390	2.660	3.460
120	1.289	1.658	1.980	2.358	2.617	3.373
∞	1.282	1.645	1.960	2.326	2.576	3.291

SOURCE: Fisher and Yates, 1974, Table III.

Table 3 The Distribution of Chi-Square

Level of Significance

df	.99	.98	.95	.90	.80	.70	.50	.30	.20	.10	.05	.02	.01	.001
1	.0³157	.0³628	.00393	.0158	.0642	.148	.455	1.074	1.642	2.706	3.841	5.412	6.635	10.827
2	.0201	.0404	.103	.211	.446	.713	1.386	2.408	3.219	4.605	5.991	7.824	9.210	13.815
3	.115	.185	.352	.584	1.005	1.424	2.366	3.665	4.642	6.251	7.815	9.837	11.341	16.268
4	.297	.429	.711	1.064	1.649	2.195	3.357	4.878	5.989	7.779	9.488	11.668	13.277	18.465
5	.554	.752	1.145	1.610	2.343	3.000	4.351	6.064	7.289	9.236	11.070	13.388	15.086	20.517
6	.872	1.134	1.635	2.204	3.070	3.828	5.348	7.231	8.558	10.645	12.592	15.033	16.812	22.457
7	1.239	1.564	2.167	2.833	3.822	4.671	6.346	8.383	9.803	12.017	14.067	16.622	18.475	24.322
8	1.646	2.032	2.733	3.490	4.594	5.527	7.344	9.524	11.030	13.362	15.507	18.168	20.090	26.125
9	2.088	2.532	3.325	4.168	5.380	6.393	8.343	10.656	12.242	14.684	16.919	19.679	21.666	27.877
10	2.558	3.059	3.940	4.865	6.179	7.267	9.342	11.781	13.442	15.987	18.307	21.161	23.209	29.588
11	3.053	3.609	4.575	5.578	6.989	8.148	10.341	12.899	14.631	17.275	19.675	22.618	24.725	31.264
12	3.571	4.178	5.226	6.304	7.807	9.034	11.340	14.011	15.812	18.549	21.026	24.054	26.217	32.909
13	4.107	4.765	5.892	7.042	8.634	9.926	12.340	15.119	16.985	19.812	22.362	25.472	27.688	34.528
14	4.660	5.368	6.571	7.790	9.467	10.821	13.339	16.222	18.151	21.064	23.685	26.873	29.141	36.123
15	5.229	5.985	7.261	8.547	10.307	11.721	14.339	17.322	19.311	22.307	24.996	28.259	30.578	37.697
16	5.812	6.614	7.962	9.312	11.152	12.624	15.338	18.418	20.465	23.542	26.296	29.633	32.000	39.252
17	6.408	7.255	8.672	10.085	12.002	13.531	16.338	19.511	21.615	24.769	27.587	30.995	33.409	40.790
18	7.015	7.906	9.390	10.865	12.857	14.440	17.338	20.601	22.760	25.989	28.869	32.346	34.805	42.312
19	7.633	8.567	10.117	11.651	13.716	15.352	18.338	21.689	23.900	27.204	30.144	33.687	36.191	43.820
20	8.260	9.237	10.851	12.443	14.578	16.266	19.337	22.775	25.038	28.412	31.410	35.020	37.566	45.315
21	8.897	9.915	11.591	13.240	15.445	17.182	20.337	23.858	26.171	29.615	32.671	36.343	38.932	46.797
22	9.542	10.600	12.338	14.041	16.314	18.101	21.337	24.939	27.301	30.813	33.924	37.659	40.289	48.268
23	10.196	11.293	13.091	14.848	17.187	19.021	22.337	26.018	28.429	32.007	35.172	38.968	41.638	49.728
24	10.856	11.992	13.848	15.659	18.062	19.943	23.337	27.096	29.553	33.196	36.415	40.270	42.980	51.179
25	11.524	12.697	14.611	16.473	18.940	20.867	24.337	28.172	30.675	34.382	37.652	41.566	44.314	52.620
26	12.198	13.409	15.379	17.292	19.820	21.792	25.336	29.246	31.795	35.563	38.885	42.856	45.642	54.052
27	12.879	14.125	16.151	18.114	20.703	22.719	26.336	30.319	32.912	36.741	40.113	44.140	46.963	55.476
28	13.565	14.847	16.928	18.939	21.588	23.647	27.336	31.391	34.027	37.916	41.337	45.419	48.278	56.893
29	14.256	15.574	17.708	19.768	22.475	24.577	28.336	32.461	35.139	39.087	42.557	46.693	49.588	58.302
30	14.953	16.306	18.493	20.599	23.364	25.508	29.336	33.530	36.250	40.256	43.773	47.962	50.892	59.703

SOURCE: Fisher and Yates, 1974, Table IV.

Table 4 Some Random Numbers

10	09	73	25	33	76	52	01	35	86	34	67	35	48	76	80	95	90	91	17	39	29	27	49	45
37	54	20	48	05	64	89	47	42	96	24	80	52	40	37	20	63	61	04	02	00	82	29	16	65
08	42	26	89	53	19	64	50	93	03	23	20	90	25	60	15	95	33	47	64	35	08	03	36	06
99	01	90	25	29	09	37	67	07	15	38	31	13	11	65	88	67	67	43	97	04	43	62	76	59
12	80	79	99	70	80	15	73	61	47	64	03	23	66	53	98	95	11	68	77	12	17	17	68	33
66	06	57	47	17	34	07	27	68	50	36	69	73	61	70	65	81	33	98	85	11	19	92	91	70
31	06	01	08	05	45	57	18	24	06	35	30	34	26	14	86	79	90	74	39	23	40	30	97	32
85	26	97	76	02	02	05	16	56	92	68	66	57	48	18	73	05	38	52	47	18	62	38	85	79
63	57	33	21	35	05	32	54	70	48	90	55	35	75	48	28	46	82	87	09	83	49	12	56	24
73	79	64	57	53	03	52	96	47	78	35	80	83	42	82	60	93	52	03	44	35	27	38	84	35
98	52	01	77	67	14	90	56	86	07	22	10	94	05	58	60	97	09	34	33	50	50	07	39	98
11	80	50	54	31	39	80	82	77	32	50	72	56	82	48	29	40	52	42	01	52	77	56	78	51
83	45	29	96	34	06	28	89	80	83	13	74	67	00	78	18	47	54	06	10	68	71	17	78	17
88	68	54	02	00	86	50	75	84	01	36	76	66	79	51	90	36	47	64	93	29	60	91	10	62
99	59	46	73	48	87	51	76	49	69	91	82	60	89	28	93	78	56	13	68	23	47	83	41	13
65	48	11	76	74	17	46	85	09	50	58	04	77	69	74	73	03	95	71	86	40	21	81	65	44
80	12	43	56	35	17	72	70	80	15	45	31	82	23	74	21	11	57	82	53	14	38	55	37	63
74	35	09	98	17	77	40	27	72	14	43	23	60	02	10	45	52	16	42	37	96	28	60	26	55
69	91	62	68	03	66	25	22	91	48	36	93	68	72	03	76	62	11	39	90	94	40	05	64	18
09	89	32	05	05	14	22	56	85	14	46	42	75	67	88	96	29	77	88	22	54	38	21	45	98
91	49	91	45	23	68	47	92	76	86	46	16	28	35	54	94	75	08	99	23	37	08	92	00	48
80	33	69	45	98	26	94	03	68	58	70	29	73	41	35	53	14	03	33	40	42	05	08	23	41
44	10	48	19	49	85	15	74	79	54	32	97	92	65	75	57	60	04	08	81	22	22	20	64	13
12	55	07	37	42	11	10	00	20	40	12	86	07	46	97	96	64	48	94	39	28	70	72	58	15
63	60	64	93	29	16	50	53	44	84	40	21	95	25	63	43	65	17	70	82	07	20	73	17	90
61	19	69	04	46	26	45	74	77	74	51	92	43	37	29	65	39	45	95	93	42	58	26	05	27
15	47	44	52	66	95	27	07	99	53	59	36	78	38	48	82	39	61	01	18	33	21	15	94	66
94	55	72	85	73	67	89	75	43	87	54	62	24	44	31	91	19	04	25	92	92	92	74	59	73
42	48	11	62	13	97	34	40	87	21	16	86	84	87	67	03	07	11	20	59	25	70	14	66	70
23	52	37	83	17	73	20	88	98	37	68	93	59	14	16	26	25	22	96	63	05	52	28	25	62
04	49	35	24	94	75	24	63	38	24	45	86	25	10	25	61	96	27	93	35	65	33	71	24	72
00	54	99	76	54	64	05	18	81	59	96	11	96	38	96	54	69	28	23	91	23	28	72	95	29
35	96	31	53	07	26	89	80	93	54	33	35	13	54	62	77	97	45	00	24	90	10	33	93	33
59	80	80	83	91	45	42	72	68	42	83	60	94	97	00	13	02	12	48	92	78	56	52	01	06
46	05	88	52	36	01	39	09	22	86	77	28	14	40	77	93	91	08	36	47	70	61	74	29	41
32	17	90	05	97	87	37	92	52	41	05	56	70	70	07	86	74	31	71	57	85	39	41	18	38
69	23	46	14	06	20	11	74	52	04	15	95	66	00	00	18	74	39	24	23	97	11	89	63	38
19	56	54	14	30	01	75	87	53	79	40	41	92	15	85	66	67	43	68	06	84	96	28	52	07
45	15	51	49	38	19	47	60	72	46	43	66	79	45	43	59	04	79	00	33	20	82	66	95	41
94	86	43	19	94	36	16	81	08	51	34	88	88	15	53	01	54	03	54	56	05	01	45	11	76
98	08	62	48	26	45	24	02	84	04	44	99	90	88	96	39	09	47	34	07	35	44	13	18	80
33	18	51	62	32	41	94	15	09	49	89	43	54	85	81	88	69	54	19	94	37	54	87	30	43
80	95	10	04	06	96	38	27	07	74	20	15	12	33	87	25	01	62	52	98	94	62	46	11	71
79	75	24	91	40	71	96	12	82	96	69	86	10	25	91	74	85	22	05	39	00	38	75	95	79
18	63	33	25	37	98	14	50	65	71	31	01	02	46	74	05	45	56	14	27	77	93	89	19	36
74	02	94	39	02	77	55	73	22	70	97	79	01	71	19	52	52	75	80	21	80	81	45	17	48
54	17	84	56	11	80	99	33	71	43	05	33	51	29	69	56	12	71	92	55	36	04	09	03	24
11	66	44	98	83	52	07	98	48	27	59	38	17	15	39	09	97	33	34	40	88	46	12	33	56
48	32	47	79	28	31	24	96	47	10	02	29	53	68	70	32	30	75	75	46	15	02	00	99	94
69	07	49	41	38	87	63	79	19	76	35	58	40	44	01	10	51	82	16	15	01	84	87	69	38
09	18	82	00	97	32	82	53	95	27	04	22	08	63	04	83	38	98	73	74	64	27	85	80	44
90	04	58	54	97	51	98	15	06	54	94	93	88	19	97	91	87	07	61	50	68	47	66	46	59
73	18	95	02	07	47	67	72	62	69	62	29	06	44	64	27	12	46	70	18	41	36	18	27	60
75	76	87	64	90	20	97	18	17	49	90	42	91	22	72	95	37	50	58	71	93	82	34	31	78
54	01	64	40	56	66	28	13	10	03	00	68	22	73	98	20	71	45	32	95	07	70	61	78	13

SOURCE: Dixon and Massey, 1969, pp. 446–47.

Table 4 (continued)

```
08 35 86 99 10    78 54 24 27 85    13 66 15 88 73    04 61 89 75 53    31 22 30 84 20
28 30 60 32 64    81 33 31 05 91    40 51 00 78 93    32 60 46 04 75    94 11 90 18 40
53 84 08 62 33    81 59 41 36 28    51 21 59 02 90    28 46 66 87 95    77 76 22 07 91
91 75 75 37 41    61 61 36 22 69    50 26 39 02 12    55 78 17 65 14    83 48 34 70 55
89 41 59 26 94    00 39 75 83 91    12 60 71 76 46    48 94 97 23 06    94 54 13 74 08

77 51 30 38 20    86 83 42 99 01    68 41 48 27 74    51 90 81 39 80    72 89 35 55 07
19 50 23 71 74    69 97 92 02 88    55 21 02 97 73    74 28 77 52 51    65 34 46 74 15
21 81 85 93 13    93 27 88 17 57    05 68 67 31 56    07 08 28 50 46    31 85 33 84 52
51 47 46 64 99    68 10 72 36 21    94 04 99 13 45    42 83 60 91 91    08 00 74 54 49
99 55 96 83 31    62 53 52 41 70    69 77 71 28 30    74 81 97 81 42    43 86 07 28 34

33 71 34 80 07    93 58 47 28 69    51 92 66 47 21    58 30 32 98 22    93 17 49 39 72
85 27 48 68 93    11 30 32 92 70    28 83 43 41 37    73 51 59 04 00    71 14 84 36 43
84 13 38 96 40    44 03 55 21 66    73 85 27 00 91    61 22 26 05 61    62 32 71 84 23
56 73 21 62 34    17 39 59 61 31    10 12 39 16 22    85 49 65 75 60    81 60 41 88 80
65 13 85 68 06    87 64 88 52 61    34 31 36 58 61    45 87 52 10 69    85 64 44 72 77

38 00 10 21 76    81 71 91 17 11    71 60 29 29 37    74 21 96 40 49    65 58 44 96 98
37 40 29 63 97    01 30 47 75 86    56 27 11 00 86    47 32 46 26 05    40 03 03 74 38
97 12 54 03 48    87 08 33 14 17    21 81 53 92 50    75 23 76 20 47    15 50 12 95 78
21 82 64 11 34    47 14 33 40 72    64 63 88 59 02    49 13 90 64 41    03 85 65 45 52
73 13 54 27 42    95 71 90 90 35    85 79 47 42 96    08 78 98 81 56    64 69 11 92 02

07 63 87 79 29    03 06 11 80 72    96 20 74 41 56    23 82 19 95 38    04 71 36 69 94
60 52 88 34 41    07 95 41 98 14    59 17 52 06 95    05 53 35 21 39    61 21 20 64 55
83 59 63 56 55    06 95 89 29 83    05 12 80 97 19    77 43 35 37 83    92 30 15 04 98
10 85 06 27 46    99 59 91 05 07    13 49 90 63 19    53 07 57 18 39    06 41 01 93 62
39 82 09 89 52    43 62 26 31 47    64 42 18 08 14    43 80 00 93 51    31 02 47 31 67

59 58 00 64 78    75 56 97 88 00    88 83 55 44 86    23 76 80 61 56    04 11 10 84 08
38 50 80 73 41    23 79 34 87 63    90 82 29 70 22    17 71 90 42 07    95 95 44 99 53
30 69 27 06 68    94 68 81 61 27    56 19 68 00 91    82 06 76 34 00    05 46 26 92 00
65 44 39 56 59    18 28 82 74 37    49 63 22 40 41    08 33 76 56 76    96 29 99 08 36
27 26 75 02 64    13 19 27 22 94    07 47 74 46 06    17 98 54 89 11    97 34 13 03 58

91 30 70 69 91    19 07 22 42 10    36 69 95 37 28    28 82 53 57 93    28 97 66 62 52
68 43 49 46 88    84 47 31 36 22    62 12 69 84 08    12 84 38 25 90    09 81 59 31 46
48 90 81 58 77    54 74 52 45 91    35 70 00 47 54    83 82 45 26 92    54 13 05 51 60
06 91 34 51 97    42 67 27 86 01    11 88 30 95 28    63 01 19 89 01    14 97 44 03 44
10 45 51 60 19    14 21 03 37 12    91 34 23 78 21    88 32 58 08 51    43 66 77 08 83

12 88 39 73 43    65 02 76 11 84    04 28 50 13 92    17 97 41 50 77    90 71 22 67 69
21 77 83 09 76    38 80 73 69 61    31 64 94 20 96    63 28 10 20 23    08 81 64 74 49
19 52 35 95 15    65 12 25 96 59    86 28 36 82 58    69 57 21 37 98    16 43 59 15 29
67 24 55 26 70    35 58 31 65 63    79 24 68 66 86    76 46 33 42 22    26 65 59 08 02
60 58 44 73 77    07 50 03 79 92    45 13 42 65 29    26 76 08 36 37    41 32 64 43 44

53 85 34 13 77    36 06 69 48 50    58 83 87 38 59    49 36 47 33 31    96 24 04 36 42
24 63 73 87 36    74 38 48 93 42    52 62 30 79 92    12 36 91 86 01    03 74 28 38 73
83 08 01 24 51    38 99 22 28 15    07 75 95 17 77    97 37 72 75 85    51 97 23 78 67
16 44 42 43 34    36 15 19 90 73    27 49 37 09 39    85 13 03 25 52    54 84 65 47 59
60 79 01 81 57    57 17 86 57 62    11 16 17 85 76    45 81 95 29 79    65 13 00 48 60
```

Table 5 The Distribution of $F(.05)$

Degrees of Freedom of the Numerator

df_1 / df_2	1	2	3	4	5	6	7	8	9
1	161.4	199.5	215.7	224.6	230.2	234.0	236.8	238.9	240.5
2	18.51	19.00	19.16	19.25	19.30	19.33	19.35	19.37	19.38
3	10.13	9.55	9.28	9.12	9.01	8.94	8.89	8.85	8.81
4	7.71	6.94	6.59	6.39	6.26	6.16	6.09	6.04	6.00
5	6.61	5.79	5.41	5.19	5.05	4.95	4.88	4.82	4.77
6	5.99	5.14	4.76	4.53	4.39	4.28	4.21	4.15	4.10
7	5.59	4.74	4.35	4.12	3.97	3.87	3.79	3.73	3.68
8	5.32	4.46	4.07	3.84	3.69	3.58	3.50	3.44	3.39
9	5.12	4.26	3.86	3.63	3.48	3.37	3.29	3.23	3.18
10	4.96	4.10	3.71	3.48	3.33	3.22	3.14	3.07	3.02
11	4.84	3.98	3.59	3.36	3.20	3.09	3.01	2.95	2.90
12	4.75	3.89	3.49	3.26	3.11	3.00	2.91	2.85	2.80
13	4.67	3.81	3.41	3.18	3.03	2.92	2.83	2.77	2.71
14	4.60	3.74	3.34	3.11	2.96	2.85	2.76	2.70	2.65
15	4.54	3.68	3.29	3.06	2.90	2.79	2.71	2.64	2.59
16	4.49	3.63	3.24	3.01	2.85	2.74	2.66	2.59	2.54
17	4.45	3.59	3.20	2.96	2.81	2.70	2.61	2.55	2.49
18	4.41	3.55	3.16	2.93	2.77	2.66	2.58	2.51	2.46
19	4.38	3.52	3.13	2.90	2.74	2.63	2.54	2.48	2.42
20	4.35	3.49	3.10	2.87	2.71	2.60	2.51	2.45	2.39
21	4.32	3.47	3.07	2.84	2.68	2.57	2.49	2.42	2.37
22	4.30	3.44	3.05	2.82	2.66	2.55	2.46	2.40	2.34
23	4.28	3.42	3.03	2.80	2.64	2.53	2.44	2.37	2.32
24	4.26	3.40	3.01	2.78	2.62	2.51	2.42	2.36	2.30
25	4.24	3.39	2.99	2.76	2.60	2.49	2.40	2.34	2.28
26	4.23	3.37	2.98	2.74	2.59	2.47	2.39	2.32	2.27
27	4.21	3.35	2.96	2.73	2.57	2.46	2.37	2.31	2.25
28	4.20	3.34	2.95	2.71	2.56	2.45	2.36	2.29	2.24
29	4.18	3.33	2.93	2.70	2.55	2.43	2.35	2.28	2.22
30	4.17	3.32	2.92	2.69	2.53	2.42	2.33	2.27	2.21
40	4.08	3.23	2.84	2.61	2.45	2.34	2.25	2.18	2.12
60	4.00	3.15	2.76	2.53	2.37	2.25	2.17	2.10	2.04
120	3.92	3.07	2.68	2.45	2.29	2.17	2.09	2.02	1.96
∞	3.84	3.00	2.60	2.37	2.21	2.10	2.01	1.94	1.88

Degrees of Freedom of the Denominator

SOURCE: Pearson and Hartley, 1958, pp. 171–73.

10	12	15	20	24	30	40	60	120	∞
241.9	243.9	245.9	248.0	249.1	250.1	251.1	252.2	253.3	254.3
19.40	19.41	19.43	19.45	19.45	19.46	19.47	19.48	19.49	19.50
8.79	8.74	8.70	8.66	8.64	8.62	8.59	8.57	8.55	8.53
5.96	5.91	5.86	5.80	5.77	5.75	5.72	5.69	5.66	5.63
4.74	4.68	4.62	4.56	4.53	4.50	4.46	4.43	4.40	4.36
4.06	4.00	3.94	3.87	3.84	3.81	3.77	3.74	3.70	3.67
3.64	3.57	3.51	3.44	3.41	3.38	3.34	3.30	3.27	3.23
3.35	3.28	3.22	3.15	3.12	3.08	3.04	3.01	2.97	2.93
3.14	3.07	3.01	2.94	2.90	2.86	2.83	2.79	2.75	2.71
2.98	2.91	2.85	2.77	2.74	2.70	2.66	2.62	2.58	2.54
2.85	2.79	2.72	2.65	2.61	2.57	2.53	2.49	2.45	2.40
2.75	2.69	2.62	2.54	2.51	2.47	2.43	2.38	2.34	2.30
2.67	2.60	2.53	2.46	2.42	2.38	2.34	2.30	2.25	2.21
2.60	2.53	2.46	2.39	2.35	2.31	2.27	2.22	2.18	2.13
2.54	2.48	2.40	2.33	2.29	2.25	2.20	2.16	2.11	2.07
2.49	2.42	2.35	2.28	2.24	2.19	2.15	2.11	2.06	2.01
2.45	2.38	2.31	2.23	2.19	2.15	2.10	2.06	2.01	1.96
2.41	2.34	2.27	2.19	2.15	2.11	2.06	2.02	1.97	1.92
2.38	2.31	2.23	2.16	2.11	2.07	2.03	1.98	1.93	1.88
2.35	2.28	2.20	2.12	2.08	2.04	1.99	1.95	1.90	1.84
2.32	2.25	2.18	2.10	2.05	2.01	1.96	1.92	1.87	1.81
2.30	2.23	2.15	2.07	2.03	1.98	1.94	1.89	1.84	1.78
2.27	2.20	2.13	2.05	2.01	1.96	1.91	1.86	1.81	1.76
2.25	2.18	2.11	2.03	1.98	1.94	1.89	1.84	1.79	1.73
2.24	2.16	2.09	2.01	1.96	1.92	1.87	1.82	1.77	1.71
2.22	2.15	2.07	1.99	1.95	1.90	1.85	1.80	1.75	1.69
2.20	2.13	2.06	1.97	1.93	1.88	1.84	1.79	1.73	1.67
2.19	2.12	2.04	1.96	1.91	1.87	1.82	1.77	1.71	1.65
2.18	2.10	2.03	1.94	1.90	1.85	1.81	1.75	1.70	1.64
2.16	2.09	2.01	1.93	1.89	1.84	1.79	1.74	1.68	1.62
2.08	2.00	1.92	1.84	1.79	1.74	1.69	1.64	1.58	1.51
1.99	1.92	1.84	1.75	1.70	1.65	1.59	1.53	1.47	1.39
1.91	1.83	1.75	1.66	1.61	1.55	1.50	1.43	1.35	1.25
1.83	1.75	1.67	1.57	1.52	1.46	1.39	1.32	1.22	1.00

Table 5 (continued)   The Distribution of $F(.025)$

Degrees of Freedom of the Numerator

df_1 / df_2	1	2	3	4	5	6	7	8	9
1	647.8	799.5	864.2	899.6	921.8	937.1	948.2	956.7	963.3
2	38.51	39.00	39.17	39.25	39.30	39.33	39.36	39.37	39.39
3	17.44	16.04	15.44	15.10	14.88	14.73	14.62	14.54	14.47
4	12.22	10.65	9.98	9.60	9.36	9.20	9.07	8.98	8.90
5	10.01	8.43	7.76	7.39	7.15	6.98	6.85	6.76	6.68
6	8.81	7.26	6.60	6.23	5.99	5.82	5.70	5.60	5.52
7	8.07	6.54	5.89	5.52	5.29	5.12	4.99	4.90	4.82
8	7.57	6.06	5.42	5.05	4.82	4.65	4.53	4.43	4.36
9	7.21	5.71	5.08	4.72	4.48	4.32	4.20	4.10	4.03
10	6.94	5.46	4.83	4.47	4.24	4.07	3.95	3.85	3.78
11	6.72	5.26	4.63	4.28	4.04	3.88	3.76	3.66	3.59
12	6.55	5.10	4.47	4.12	3.89	3.73	3.61	3.51	3.44
13	6.41	4.97	4.35	4.00	3.77	3.60	3.48	3.39	3.31
14	6.30	4.86	4.24	3.89	3.66	3.50	3.38	3.29	3.21
15	6.20	4.77	4.15	3.80	3.58	3.41	3.29	3.20	3.12
16	6.12	4.69	4.08	3.73	3.50	3.34	3.22	3.12	3.05
17	6.04	4.62	4.01	3.66	3.44	3.28	3.16	3.06	2.98
18	5.98	4.56	3.95	3.61	3.38	3.22	3.10	3.01	2.93
19	5.92	4.51	3.90	3.56	3.33	3.17	3.05	2.96	2.88
20	5.87	4.46	3.86	3.51	3.29	3.13	3.01	2.91	2.84
21	5.83	4.42	3.82	3.48	3.25	3.09	2.97	2.87	2.80
22	5.79	4.38	3.78	3.44	3.22	3.05	2.93	2.84	2.76
23	5.75	4.35	3.75	3.41	3.18	3.02	2.90	2.81	2.73
24	5.72	4.32	3.72	3.38	3.15	2.99	2.87	2.78	2.70
25	5.69	4.29	3.69	3.35	3.13	2.97	2.85	2.75	2.68
26	5.66	4.27	3.67	3.33	3.10	2.94	2.82	2.73	2.65
27	5.63	4.24	3.65	3.31	3.08	2.92	2.80	2.71	2.63
28	5.61	4.22	3.63	3.29	3.06	2.90	2.78	2.69	2.61
29	5.59	4.20	3.61	3.27	3.04	2.88	2.76	2.67	2.59
30	5.57	4.18	3.59	3.25	3.03	2.87	2.75	2.65	2.57
40	5.42	4.05	3.46	3.13	2.90	2.74	2.62	2.53	2.45
60	5.29	3.93	3.34	3.01	2.79	2.63	2.51	2.41	2.33
120	5.15	3.80	3.23	2.89	2.67	2.52	2.39	2.30	2.22
∞	5.02	3.69	3.12	2.79	2.57	2.41	2.29	2.19	2.11

Degrees of Freedom of the Denominator

10	12	15	20	24	30	40	60	120	∞
968.6	976.7	984.9	993.1	997.2	1000	1006	1010	1014	1018
39.40	39.41	39.43	39.45	39.46	39.46	39.47	39.48	39.49	39.50
14.42	14.34	14.25	14.17	14.12	14.08	14.04	13.99	13.95	13.90
8.84	8.75	8.66	8.56	8.51	8.46	8.41	8.36	8.31	8.26
6.62	6.52	6.43	6.33	6.28	6.23	6.18	6.12	6.07	6.02
5.46	5.37	5.27	5.17	5.12	5.07	5.01	4.96	4.90	4.85
4.76	4.67	4.57	4.47	4.42	4.36	4.31	4.25	4.20	4.14
4.30	4.20	4.10	4.00	3.95	3.89	3.84	3.78	3.73	3.67
3.96	3.87	3.77	3.67	3.61	3.56	3.51	3.45	3.39	3.33
3.72	3.62	3.52	3.42	3.37	3.31	3.26	3.20	3.14	3.08
3.53	3.43	3.33	3.23	3.17	3.12	3.06	3.00	2.94	2.88
3.37	3.28	3.18	3.07	3.02	2.96	2.91	2.85	2.79	2.72
3.25	3.15	3.05	2.95	2.89	2.84	2.78	2.72	2.66	2.60
3.15	3.05	2.95	2.84	2.79	2.73	2.67	2.61	2.55	2.49
3.06	2.96	2.86	2.76	2.70	2.64	2.59	2.52	2.46	2.40
2.99	2.89	2.79	2.68	2.63	2.57	2.51	2.45	2.38	2.32
2.92	2.82	2.72	2.62	2.56	2.50	2.44	2.38	2.32	2.25
2.87	2.77	2.67	2.56	2.50	2.44	2.38	2.32	2.26	2.19
2.82	2.72	2.62	2.51	2.45	2.39	2.33	2.27	2.20	2.13
2.77	2.68	2.57	2.46	2.41	2.35	2.29	2.22	2.16	2.09
2.73	2.64	2.53	2.42	2.37	2.31	2.25	2.18	2.11	2.04
2.70	2.60	2.50	2.39	2.33	2.27	2.21	2.14	2.08	2.00
2.67	2.57	2.47	2.36	2.30	2.24	2.18	2.11	2.04	1.97
2.64	2.54	2.44	2.33	2.27	2.21	2.15	2.08	2.01	1.94
2.61	2.51	2.41	2.30	2.24	2.18	2.12	2.05	1.98	1.91
2.59	2.49	2.39	2.28	2.22	2.16	2.09	2.03	1.95	1.88
2.57	2.47	2.36	2.25	2.19	2.13	2.07	2.00	1.93	1.85
2.55	2.45	2.34	2.23	2.17	2.11	2.05	1.98	1.91	1.83
2.53	2.43	2.32	2.21	2.15	2.09	2.03	1.96	1.89	1.81
2.51	2.41	2.31	2.20	2.14	2.07	2.01	1.94	1.87	1.79
2.39	2.29	2.18	2.07	2.01	1.94	1.88	1.80	1.72	1.64
2.27	2.17	2.06	1.94	1.88	1.82	1.74	1.67	1.58	1.48
2.16	2.05	1.94	1.82	1.76	1.69	1.61	1.53	1.43	1.31
2.05	1.94	1.83	1.71	1.64	1.57	1.48	1.39	1.27	1.00

Table 5 (continued)   The Distribution of $F(.01)$

Degrees of Freedom of the Denominator / Degrees of Freedom of the Numerator

df_2 \ df_1	1	2	3	4	5	6	7	8	9
1	4052	4999.5	5403	5625	5764	5859	5928	5981	6022
2	98.50	99.00	99.17	99.25	99.30	99.33	99.36	99.37	99.39
3	34.12	30.82	29.46	28.71	28.24	27.91	27.67	27.49	27.35
4	21.20	18.00	16.69	15.98	15.52	15.21	14.98	14.80	14.66
5	16.26	13.27	12.06	11.39	10.97	10.67	10.46	10.29	10.16
6	13.75	10.92	9.78	9.15	8.75	8.47	8.26	8.10	7.98
7	12.25	9.55	8.45	7.85	7.46	7.19	6.99	6.84	6.72
8	11.26	8.65	7.59	7.01	6.63	6.37	6.18	6.03	5.91
9	10.56	8.02	6.99	6.42	6.06	5.80	5.61	5.47	5.35
10	10.04	7.56	6.55	5.99	5.64	5.39	5.20	5.06	4.94
11	9.65	7.21	6.22	5.67	5.32	5.07	4.89	4.74	4.63
12	9.33	6.93	5.95	5.41	5.06	4.82	4.64	4.50	4.39
13	9.07	6.70	5.74	5.21	4.86	4.62	4.44	4.30	4.19
14	8.86	6.51	5.56	5.04	4.69	4.46	4.28	4.14	4.03
15	8.68	6.36	5.42	4.89	4.56	4.32	4.14	4.00	3.89
16	8.53	6.23	5.29	4.77	4.44	4.20	4.03	3.89	3.78
17	8.40	6.11	5.18	4.67	4.34	4.10	3.93	3.79	3.68
18	8.29	6.01	5.09	4.58	4.25	4.01	3.84	3.71	3.60
19	8.18	5.93	5.01	4.50	4.17	3.94	3.77	3.63	3.52
20	8.10	5.85	4.94	4.43	4.10	3.87	3.70	3.56	3.46
21	8.02	5.78	4.87	4.37	4.04	3.81	3.64	3.51	3.40
22	7.95	5.72	4.82	4.31	3.99	3.76	3.59	3.45	3.35
23	7.88	5.66	4.76	4.26	3.94	3.71	3.54	3.41	3.30
24	7.82	5.61	4.72	4.22	3.90	3.67	3.50	3.36	3.26
25	7.77	5.57	4.68	4.18	3.85	3.63	3.46	3.32	3.22
26	7.72	5.53	4.64	4.14	3.82	3.59	3.42	3.29	3.18
27	7.68	5.49	4.60	4.11	3.78	3.56	3.39	3.26	3.15
28	7.64	5.45	4.57	4.07	3.75	3.53	3.36	3.23	3.12
29	7.60	5.42	4.54	4.04	3.73	3.50	3.33	3.20	3.09
30	7.56	5.39	4.51	4.02	3.70	3.47	3.30	3.17	3.07
40	7.31	5.18	4.31	3.83	3.51	3.29	3.12	2.99	2.89
60	7.08	4.98	4.13	3.65	3.34	3.12	2.95	2.82	2.72
120	6.85	4.79	3.95	3.48	3.17	2.96	2.79	2.66	2.56
∞	6.63	4.61	3.78	3.32	3.02	2.80	2.64	2.51	2.41

10	12	15	20	24	30	40	60	120	∞
6056	6106	6157	6209	6235	6261	6287	6313	6339	6366
99.40	99.42	99.43	99.45	99.46	99.47	99.47	99.48	99.49	99.50
27.23	27.05	26.87	26.69	26.60	26.50	26.41	26.32	26.22	26.13
14.55	14.37	14.20	14.02	13.93	13.84	13.75	13.65	13.56	13.46
10.05	9.89	9.72	9.55	9.47	9.38	9.29	9.20	9.11	9.02
7.87	7.72	7.56	7.40	7.31	7.23	7.14	7.06	6.97	6.88
6.62	6.47	6.31	6.16	6.07	5.99	5.91	5.82	5.74	5.65
5.81	5.67	5.52	5.36	5.28	5.20	5.12	5.03	4.95	4.86
5.26	5.11	4.96	4.81	4.73	4.65	4.57	4.48	4.40	4.31
4.85	4.71	4.56	4.41	4.33	4.25	4.17	4.08	4.00	3.91
4.54	4.40	4.25	4.10	4.02	3.94	3.86	3.78	3.69	3.60
4.30	4.16	4.01	3.86	3.78	3.70	3.62	3.54	3.45	3.36
4.10	3.96	3.82	3.66	3.59	3.51	3.48	3.34	3.25	3.17
3.94	3.80	3.66	3.51	3.43	3.35	3.27	3.18	3.09	3.00
3.80	3.67	3.52	3.37	3.29	3.21	3.13	3.05	2.96	2.87
3.69	3.55	3.41	3.26	3.18	3.10	3.02	2.93	2.84	2.75
3.59	3.46	3.31	3.16	3.08	3.00	2.92	2.83	2.75	2.65
3.51	3.37	3.23	3.08	3.00	2.92	2.84	2.75	2.66	2.57
3.43	3.30	3.15	3.00	2.92	2.84	2.76	2.67	2.58	2.49
3.37	3.23	3.09	2.94	2.86	2.78	2.69	2.61	2.52	2.42
3.31	3.17	3.03	2.88	2.80	2.72	2.64	2.55	2.46	2.36
3.26	3.12	2.98	2.83	2.75	2.67	2.58	2.50	2.40	2.31
3.21	3.07	2.93	2.78	2.70	2.62	2.54	2.45	2.35	2.26
3.17	3.03	2.89	2.74	2.66	2.58	2.49	2.40	2.31	2.21
3.13	2.99	2.85	2.70	2.62	2.54	2.45	2.36	2.27	2.17
3.09	2.96	2.81	2.66	2.58	2.50	2.42	2.33	2.23	2.13
3.06	2.93	2.78	2.63	2.55	2.47	2.38	2.29	2.20	2.10
3.03	2.90	2.75	2.60	2.52	2.44	2.35	2.26	2.17	2.06
3.00	2.87	2.73	2.57	2.49	2.41	2.33	2.23	2.14	2.03
2.98	2.84	2.70	2.55	2.47	2.39	2.30	2.21	2.11	2.01
2.80	2.66	2.52	2.37	2.29	2.20	2.11	2.02	1.92	1.80
2.63	2.50	2.35	2.20	2.12	2.03	1.94	1.84	1.73	1.60
2.47	2.34	2.19	2.03	1.95	1.86	1.76	1.66	1.53	1.38
2.32	2.18	2.04	1.88	1.79	1.70	1.59	1.47	1.32	1.00

Table 6 Critical Values for the Wilcoxon Signed-Rank Test

One-sided	Two-sided	n = 5	n = 6	n = 7	n = 8	n = 9	n = 10	n = 11	n = 12	n = 13	n = 14	n = 15	n = 16
α = .05	α = .10	1	2	4	6	8	11	14	17	21	26	30	36
α = .025	α = .05		1	2	4	6	8	11	14	17	21	25	30
α = .01	α = .02			0	2	3	5	7	10	13	16	20	24
α = .005	α = .02				0	2	3	5	7	10	13	16	19

One-sided	Two-sided	n = 17	n = 18	n = 19	n = 20	n = 21	n = 22	n = 23	n = 24	n = 25	n = 26	n = 27	n = 28
α = .05	α = .10	41	47	54	60	68	75	83	92	101	110	120	130
α = .025	α = .05	35	40	46	52	59	66	73	81	90	98	107	117
α = .01	α = .02	28	33	38	43	49	56	62	69	77	85	93	102
α = .005	α = .01	23	28	32	37	43	49	55	61	68	76	84	92

One-sided	Two-sided	n = 29	n = 30	n = 31	n = 32	n = 33	n = 34	n = 35	n = 36	n = 37	n = 38	n = 39
α = .05	α = .10	141	152	163	175	188	201	214	228	242	256	271
α = .025	α = .05	127	137	148	159	171	183	195	208	222	235	250
α = .01	α = .02	111	120	130	141	151	162	174	186	198	211	224
α = .005	α = .01	100	109	118	128	138	149	160	171	183	195	208

One-sided	Two-sided	n = 40	n = 41	n = 42	n = 43	n = 44	n = 45	n = 46	n = 47	n = 48	n = 49	n = 50
α = .05	α = .10	287	303	319	336	353	371	389	408	427	446	466
α = .25	α = .05	264	279	295	311	327	344	361	379	397	415	434
α = .01	α = .02	238	252	267	281	297	313	329	345	362	380	398
α = .005	α = .01	221	234	248	262	277	292	307	323	339	356	373

SOURCE: Wilcoxon and Wilcox, 1964, p. 28.

Table 7 Critical Values for the Mann-Whitney Test

Critical values for a one-tailed test at the .005 level of significance or a two-tailed test at the .01 level (boldfaced type).

Critical values for a one-tailed test at the .01 level of significance or a two-tailed test at the .02 level (roman type).

n_2 \ n_1	1	2	3	4	5	6	7	8	9	10	11	12	13	14	15	16	17	18	19	20
1	—[b]	—	—	—	—	—	—	—	—	—	—	—	—	—	—	—	—	—	—	—
2	—	—	—	—	—	—	—	—	—	—	—	—	**0**	**0**	**0**	**0**	**0**	**0**	**1**	**1**
													—	—	—	—	—	—	0	0
3	—	—	—	—	—	—	**0**	**0**	**1**	**1**	**1**	**2**	**2**	**2**	**3**	**3**	**4**	**4**	**4**	**5**
							—	—	0	0	0	1	1	1	2	2	2	2	3	3
4	—	—	—	—	**0**	**1**	**1**	**2**	**3**	**3**	**4**	**5**	**5**	**6**	**7**	**7**	**8**	**9**	**9**	**10**
					—	0	0	1	1	2	2	3	3	4	5	5	6	6	7	8
5	—	—	—	**0**	**1**	**2**	**3**	**4**	**5**	**6**	**7**	**8**	**9**	**10**	**11**	**12**	**13**	**14**	**15**	**16**
				—	0	1	1	2	3	4	5	6	7	7	8	9	10	11	12	13
6	—	—	—	**1**	**2**	**3**	**4**	**6**	**7**	**8**	**9**	**11**	**12**	**13**	**15**	**16**	**18**	**19**	**20**	**22**
				0	1	2	3	4	5	6	7	9	10	11	12	13	15	16	17	18
7	—	—	**0**	**1**	**3**	**4**	**6**	**7**	**9**	**11**	**12**	**14**	**16**	**17**	**19**	**21**	**23**	**24**	**26**	**28**
			—	0	1	3	4	6	7	9	10	12	13	15	16	18	19	21	22	24
8	—	—	**0**	**2**	**4**	**6**	**7**	**9**	**11**	**13**	**15**	**17**	**20**	**22**	**24**	**26**	**28**	**30**	**32**	**34**
			—	1	2	4	6	7	9	11	13	15	17	18	20	22	24	26	28	30
9	—	—	**1**	**3**	**5**	**7**	**9**	**11**	**14**	**16**	**18**	**21**	**23**	**26**	**28**	**31**	**33**	**36**	**38**	**40**
			0	1	3	5	7	9	11	13	16	18	20	22	24	27	29	31	33	36
10	—	—	**1**	**3**	**6**	**8**	**11**	**13**	**16**	**19**	**22**	**24**	**27**	**30**	**33**	**36**	**38**	**41**	**44**	**47**
			0	2	4	6	9	11	13	16	18	21	24	26	29	31	34	37	39	42
11	—	—	**1**	**4**	**7**	**9**	**12**	**15**	**18**	**22**	**25**	**28**	**31**	**34**	**37**	**41**	**44**	**47**	**50**	**53**
			0	2	5	7	10	13	16	18	21	24	27	30	33	36	39	42	45	48
12	—	—	**2**	**5**	**8**	**11**	**14**	**17**	**21**	**24**	**28**	**31**	**35**	**38**	**42**	**46**	**49**	**53**	**56**	**60**
			1	3	6	9	12	15	18	21	24	27	31	34	37	41	44	47	51	54
13	—	**0**	**2**	**5**	**9**	**12**	**16**	**20**	**23**	**27**	**31**	**35**	**39**	**43**	**47**	**51**	**55**	**59**	**63**	**67**
		—	1	3	7	10	13	17	20	24	27	31	34	38	42	45	49	53	56	60
14	—	**0**	**2**	**6**	**10**	**13**	**17**	**22**	**26**	**30**	**34**	**38**	**43**	**47**	**51**	**56**	**60**	**65**	**69**	**73**
		—	1	4	7	11	15	18	22	26	30	34	38	42	46	50	54	58	63	67
15	—	**0**	**3**	**7**	**11**	**15**	**19**	**24**	**28**	**33**	**37**	**42**	**47**	**51**	**56**	**61**	**66**	**70**	**75**	**80**
		—	2	5	8	12	16	20	24	29	33	37	42	46	51	55	60	64	69	73
16	—	**0**	**3**	**7**	**12**	**16**	**21**	**26**	**31**	**36**	**41**	**46**	**51**	**56**	**61**	**66**	**71**	**76**	**82**	**87**
		—	2	5	9	13	18	22	27	31	36	41	45	50	55	60	65	70	74	79
17	—	**0**	**4**	**8**	**13**	**18**	**23**	**28**	**33**	**38**	**44**	**49**	**55**	**60**	**66**	**71**	**77**	**82**	**88**	**93**
		—	2	6	10	15	19	24	29	34	39	44	49	54	60	65	70	75	81	86
18	—	**0**	**4**	**9**	**14**	**19**	**24**	**30**	**36**	**41**	**47**	**53**	**59**	**65**	**70**	**76**	**82**	**88**	**94**	**100**
		—	2	6	11	16	21	26	31	37	42	47	53	58	64	70	75	81	87	92
19	—	**1**	**4**	**9**	**15**	**20**	**26**	**32**	**38**	**44**	**50**	**56**	**63**	**69**	**75**	**82**	**88**	**94**	**101**	**107**
		0	3	7	12	17	22	28	33	39	45	51	56	63	69	74	81	87	93	99
20	—	**1**	**5**	**10**	**16**	**22**	**28**	**34**	**40**	**47**	**53**	**60**	**67**	**73**	**80**	**87**	**93**	**100**	**107**	**114**
		0	3	8	13	18	24	30	36	42	48	54	60	67	73	79	86	92	99	105

SOURCE: Kirk, 1978, pp. 423–24.

Table 7 (continued)   Critical Values for the Mann-Whitney Test

Critical values for a one-tailed test at the .025 level of significance or a two-tailed test at the .05 level (boldfaced type).

Critical values for a one-tailed test at the .05 level of significance or a two-tailed test at the .10 level (roman type).

n_2 \ n_1	1	2	3	4	5	6	7	8	9	10	11	12	13	14	15	16	17	18	19	20
1	—	—	—	—	—	—	—	—	—	—	—	—	—	—	—	—	—	—	0	0
2	—	—	—	—	0	0	0	1	1	1	1	2	2	2	3	3	3	4	4	4
	—	—	—	—	—	—	—	0	0	0	0	1	1	1	1	1	2	2	2	2
3	—	—	0	0	1	2	2	3	3	4	5	5	6	7	7	8	9	9	10	11
	—	—	—	—	0	1	1	2	2	3	3	4	4	5	5	6	6	7	7	8
4	—	—	0	1	2	3	4	5	6	7	8	9	10	11	12	14	15	16	17	18
	—	—	—	0	1	2	3	4	4	5	6	7	8	9	10	11	11	12	13	13
5	—	0	1	2	4	5	6	8	9	11	12	13	15	16	18	19	20	22	23	25
	—	—	0	1	2	3	5	6	7	8	9	11	12	13	14	15	17	18	19	20
6	—	0	2	3	5	7	8	10	12	14	16	17	19	21	23	25	26	28	30	32
	—	—	1	2	3	5	6	8	10	11	13	14	16	17	19	21	22	24	25	27
7	—	0	2	4	6	8	11	13	15	17	19	21	24	26	28	30	33	35	37	39
	—	—	1	3	5	6	8	10	12	14	16	18	20	22	24	26	28	30	32	34
8	—	1	3	5	8	10	13	15	18	20	23	26	28	31	33	36	39	41	44	47
	—	0	2	4	6	8	10	13	15	17	19	22	24	26	29	31	34	36	38	41
9	—	1	3	6	9	12	15	18	21	24	27	30	33	36	39	42	45	48	51	54
	—	0	2	4	7	10	12	15	17	20	23	26	28	31	34	37	39	42	45	48
10	—	1	4	7	11	14	17	20	24	27	31	34	37	41	44	48	51	55	58	62
	—	0	3	5	8	11	14	17	20	23	26	29	33	36	39	42	45	48	52	55
11	—	1	5	8	12	16	19	23	27	31	34	38	42	46	50	54	57	61	65	69
	—	0	3	6	9	13	16	19	23	26	30	33	37	40	44	47	51	55	58	62
12	—	2	5	9	13	17	21	26	30	34	38	42	47	51	55	60	64	68	72	77
	—	1	4	7	11	14	18	22	26	29	33	37	41	45	49	53	57	61	65	69
13	—	2	6	10	15	19	24	28	33	37	42	47	51	56	61	65	70	75	80	84
	—	1	4	8	12	16	20	24	28	33	37	41	45	50	54	59	63	67	72	76
14	—	2	7	11	16	21	26	31	36	41	46	51	56	61	66	71	77	82	87	92
	—	1	5	9	13	17	22	26	31	36	40	45	50	55	59	64	67	74	78	83
15	—	3	7	12	18	23	28	33	39	44	50	55	61	66	72	77	83	88	94	100
	—	1	5	10	14	19	24	29	34	39	44	49	54	59	64	70	75	80	85	90
16	—	3	8	14	19	25	30	36	42	48	54	60	65	71	77	83	89	95	101	107
	—	1	6	11	15	21	26	31	37	42	47	53	59	64	70	75	81	86	92	98
17	—	3	9	15	20	26	33	39	45	51	57	64	70	77	83	89	96	102	109	115
	—	2	6	11	17	22	28	34	39	45	51	57	63	67	75	81	87	93	99	105
18	—	4	9	16	22	28	35	41	48	55	61	68	75	82	88	95	102	109	116	123
	—	2	7	12	18	24	30	36	42	48	55	61	67	74	80	86	93	99	106	112
19	0	4	10	17	23	30	37	44	51	58	65	72	80	87	94	101	109	116	123	130
	—	2	7	13	19	25	32	38	45	52	58	65	72	78	85	92	99	106	113	119
20	0	4	11	18	25	32	39	47	54	62	69	77	84	92	100	107	115	123	130	138
	—	2	8	13	20	27	34	41	48	55	62	69	76	83	90	98	105	112	119	127

Table 8 Critical Values for the Wald-Wolfowitz Runs Test (.05 Level of Significance)

n_1 \ n_2	5	6	7	8	9	10	11	12	13	14	15
2	*	*	*	*	*	*	*	2/6	2/6	2/6	2/6
3	*	2/8	2/8	2/8	2/8	2/8	2/8	2/8	2/8	2/8	3/8
4	2/9	2/9	2/10	3/10	3/10	3/10	3/10	3/10	3/10	3/10	3/10
5	2/10	3/10	3/11	3/11	3/12	3/12	4/12	4/12	4/12	4/12	4/12
6	3/10	3/11	3/12	3/12	4/13	4/13	4/13	4/13	5/14	5/14	5/14
7	3/11	3/12	3/13	4/13	4/14	5/14	5/14	5/14	5/15	5/15	6/15
8	3/11	3/12	4/13	4/14	5/14	5/15	5/15	6/16	6/16	6/16	6/16
9	3/12	4/13	4/14	5/14	5/15	5/16	6/16	6/16	6/17	7/17	7/18
10	3/12	4/13	5/14	5/15	5/16	6/16	6/17	7/17	7/18	7/18	7/18
11	4/12	4/13	5/14	5/15	6/16	6/17	7/17	7/18	7/19	8/19	8/19
12	4/12	4/13	5/14	6/16	6/16	7/17	7/18	7/19	8/19	8/20	8/20
13	4/12	5/14	5/15	6/16	6/17	7/18	7/19	8/19	8/20	9/20	9/21
14	4/12	5/14	5/15	6/16	7/17	7/18	8/19	8/20	9/20	9/21	9/22
15	4/12	5/14	6/15	6/16	7/18	7/18	8/19	8/20	9/21	9/22	10/22

SOURCE: Weiss and Hassett, 1982, p. 594.

Glossary

For each of the following, the number in parentheses refers to the page on which the term is first discussed. Where appropriate, the mathematical symbol is given to the right of the definition.

    Symbol

Analysis of variance   A technique for analyzing relationships between interval- or ratio-scale dependent variables and nominal- or ordinal-scale independent variables. (361)

Antecedent variable   In a relationship between two variables, a variable that causes the independent variable. (208)

Asymmetrical measure of association   A measure of association for which it is necessary to specify which variable is independent and which is dependent. (137)

Bar graph   A graphic display in which frequencies, proportions, or percentages of cases falling in each category of a discrete variable are represented by the height of a vertical bar. (49)

Beta weight   (see *standardized regression coefficient*) β

Bias   The tendency for some kinds of error to occur more often than others. (20)

Biased estimator   A sample statistic such as a standard deviation which, in the long run, tends to underestimate or overestimate the corresponding population characteristic. (296)

Symbol

Case Any unit (such as a person or group) that we observe for the purpose of gathering information. (31)
Categorical variable (see *nominal variable*)
Cell In a cross-tabulation, a category representing a combination of two or more characteristics. (36)
Census A gathering of observations of all members of a population. (10)
Central-limit theorem A mathematical law which states that regardless of the shape of the population distribution, the larger samples are, the closer the sampling distribution for means, proportions, and percentages will be to a normal curve. (280)
Chi-square A statistic used to test for the significance of relationships between variables and differences between frequency distributions. (342) χ^2
Cluster sample A complex sample in which observations are concentrated in small geographic areas such as city blocks in order to make the most efficient use of time and money. (261)
Code A value, usually a number, assigned to the categories of a variable for purposes of processing, summarizing, and analyzing data. (8–9)
Coefficient of alienation A statistical measure of the degree to which a regression line does *not* describe a set of data (the opposite of a correlation). It measures the correlation between Y and all variables *except* the independent variable(s). (181)
Coefficient of determination In regression analysis, the proportion of the variation in Y that is statistically accounted for by its relationship to the independent variable(s). (177) r^2 R^2
Cohort A group of people who share a common experience (such as birth or graduation from college) at roughly the same time. (83)
Cohort effect A difference between individuals or groups that results from membership in different cohorts. (83)
Complex sample A sample design that is drawn in stages, from the largest unit down to the smallest. (260)
Component variable A variable which conceptually forms part of another variable. For example, education, income, and occupation are generally viewed as components of the variable "social class." (207)
Compound event An event consisting of two or more events which may occur singly or in any combination. (116) $(A \text{ or } B)$
Concordant pair In relationships between variables, a pair of cases that differ on the dependent variable in the same direction as their difference on the independent variable. (138)

Symbol

Conditional probability The probability that one event will occur given that another event has occurred. (107) $p(A|B)$

Conditionally perfect association A relationship between variables that is perfect within some but not all categories of the independent variable. (141)

Confidence interval A range of scores with a confidence level (in the form of a percentage) attached that is used to estimate a population characteristic such as a mean or proportion. (298)

Constant A characteristic that is the same for all observations. (14)

Continuous variable A variable that can take on any value within its range. (22)

Correlation coefficient (Pearson's product-moment) A symmetrical measure of association for interval- or ratio-scale variables, used especially in regression analysis. The square of the correlation coefficient is the coefficient of determination. (171) r

Correlation matrix A display showing how each variable in a set of variables is correlated with each of the other variables in the set. (232)

Covariance A statistical measure of the degree to which variation in the value of Y is associated with variation in the value of X. (168)

Cross-tabulation A display of data in which the cases falling in each category of one variable are subdivided by the categories of one or more additional variables. (39)

Curvilinear relationship A relationship between two variables in which the pattern of change in one variable associated with changes in the other takes the shape of a curved line. (129)

***d*, Somers'** (see *Somers' d*)

Data (plural of *datum*) Information gathered by observation. (10)

Degrees of freedom In the calculation of chi-square for cross-tabulations, a measure of table size calculated as the number of rows minus one *times* the number of columns minus one; in goodness-of-fit tests, the number of categories minus one. (307) *df*

Dependent variable In a relationship involving two or more variables, the variable that is considered to be affected by one or more other variables. (125)

Symbol

Descriptive statistics   Statistical techniques used to portray a population or sample in terms of one or more variables. (10)
Deviation from the mean   The difference between a score in a distribution and the mean of that distribution. (88)
Dichotomy   A variable that has only two categories. (26)
Discordant pair   In relationships between variables, a pair of cases that differ on the dependent variable in the opposite direction as their difference on the independent variable. (138)
Discrete variable   A variable that can take on only a limited number of values within its range. (22)
Distorter variable   A variable that, when uncontrolled, makes a relationship between two variables appear to be in the wrong direction. (214)

Efficiency, sampling   The long-run average accuracy of a sample design relative to its size. (255)
Error   (see *bias* and *random error*; see also *standard error* and *standard error of estimate*)
Event   In probability, a possible outcome for a trial. (102)
Exogenous variable   In path analysis, background variables that are statistically related to but unaffected by other variables in the model. (240)
Expected frequency   In the calculation of chi-square the number of sample cases we would expect to find in a category or cell under the assumption contained in the null hypothesis. (343) f_e
Expected value   Using probabilities, the number of cases in the long run we would expect to have a given characteristic. (118)
Explained variance   In the regression of Y on X, a variance based on the squared differences between the points on a least-squares regression line and the mean of Y. (177)
Extraneous variable   For a relationship between two variables, an extraneous variable is one which causes both the independent and dependent variables and is used to demonstrate spuriousness. (200)

Failure   In probability, any outcome in which the event for which the probability is being calculated does not occur. (103)
Finding   A statistical result. (7)
Frame   In sampling, either a list of all members of the population or a precise definition of the population that unambiguously identifies all members of the population. (250)

	Symbol

Frequency distribution   A display showing the number of cases that fall in each of the categories of a variable. (30)

Frequency polygon   A line graph formed by connecting the midpoints of the tops of the bars in a histogram. (56)

Gamma, Goodman and Kruskal's   (see *Goodman and Kruskal's gamma*)

GENSOC survey   The General Social Survey conducted annually by the National Opinion Research Center at the University of Chicago. (8)

Goodman and Kruskal's gamma   A symmetrical measure of association with a PRE interpretation used to describe linear cross-tabulated relationships involving ordinal, ratio, or interval variables expressed as frequencies. (149) γ

Goodman and Kruskal's tau   An asymmetrical measure of association with a PRE interpretation used with cross-tabulated variables regardless of their scale of measurement or the number of cells in the table. (146) τ_a τ_b

Goodness-of-fit test   The use of chi-square to test the null hypothesis that the observed differences between a sample distribution and an assumed population are due only to chance. (350)

Graphics   Pictorial representations of statistical information. (48)

Grouped data   A distribution in which some of the categories that comprise a variable have been combined to make a smaller number of larger categories. The process is called *grouping*. (33)

Grouping   (see *grouped data*)

GSS   (see *GENSOC survey*)

Histogram   A graphic display in which frequencies, proportions, or percentages of cases falling in each category of a continuous variable are represented by the height of a vertical bar. (55)

Hypothesis   A proposition that can be tested through scientific observation. (325)

Independence   A condition in which the occurrence of one event has no effect on the probability that a second event will occur. (112)

Independent variable   A variable that is considered to cause changes in another variable. (125) X

	Symbol
Index of dissimilarity A statistical measure which indicates the percentage of cases in either of two distributions that would have to be redistributed in order for the two distributions to be identical. (352)	Δ

Inductive statistics (see *inferential statistics*)
Inferential statistics The set of statistical techniques for making statements about populations on the basis of sample information, and attaching a probability of error to the result. (10)
Interaction effect A condition in which the strength or direction of a relationship between X and Y differs from one category of Z to another. (215)
Interval variable The second highest scale of measurement, allowing comparisons of scores in terms of similarity, order, and the magnitude of the differences (intervals) between scores. (24)
Intervening variable In a relationship between two variables, a third variable that is caused by the independent variable and which, in turn, has a causal effect on the dependent variable. (200)

Joint event In probability, the simultaneous occurrence of two or more events as the outcome of a trial. (102)	(A,B)
Joint probability The probability that two or more events will occur simultaneously. (110)	$p(A,B)$
Kendall's tau A symmetric rank-order measure of association used to indicate the degree to which two sets of ranks agree or disagree with each other. (156)	τ

Kurtosis The degree to which the shape of a bell-shaped curve is flatter or more peaked than a normal curve. (286)

Law of large numbers The mathematical law according to which as sample size increases, the proportion of cases with a particular characteristic in samples will more closely resemble the proportion found in the population. (118)
Least-squares regression line The straight line function that best fits a set of points that constitute a relationship between interval- or ratio-scale variables. (166)
Leptokurtic A bell-shaped curve that is more peaked than a normal curve. (286)
Line graph A graphic display in which frequencies, proportions, or percentages of cases falling in each category are represented by the height of a continuous line. (52)

Symbol

Linear relationship   A relationship between two variables in which as the scores for the first variable increase or decrease, scores on the second variable generally change consistently in only one direction and at a relatively constant rate. (128)

Log-linear analysis   A set of techniques for analyzing multiple relationships among nominal- or ordinal-scale variables. (361)

Longitudinal effect   A trend in which the characteristics of a specific set of cases change over time. (83)

Marginal   In a cross-tabulation, the number of cases that fall in a particular category of one of the variables. (36)

Mathematically perfect association   In a cross-tabulation, a relationship between variables in which one of the diagonals is empty. In regression analysis, a relationship in which all of the data points lie on the least-squares regression line. (141)

Mean   A measure of central tendency found by adding together a group of scores and dividing by the number of scores. (67)

Population: μ
Sample: $\bar{X}$

Mean independence   A condition in which the mean for one variable (Y) is identical within all categories of a second variable (X). (184)

Measure of association   A statistic used to measure the strength of relationships between variables. (133)

Measurement instrument   A set of procedures for making observations and classifying the results. (14)

Median   A measure of central tendency that consists of the score corresponding to the middle case in an ordered distribution of scores. (75)

Md

Mode   The most frequent score in a distribution. (81)

Mo

Multiple classification analysis   A set of techniques for analyzing multiple relationships between interval- or ratio-scale dependent variables and nominal- or ordinal-scale independent variables. (361)

Multiple correlation coefficient   A symmetrical measure of association which indicates the strength of association between an interval- or ratio-scale dependent variable and two or more interval- or ratio-scale independent variables. (228)

R

Multiple regression analysis   A form of regression analysis in which two or more independent variables are used simultaneously to predict a dependent variable. (227)

Multistage sample   (see *complex sample*)

Mutually exclusive events   Two or more events that cannot occur simultaneously. (117)

Symbol

Negative cross-product The total number of discordant pairs in a cross-tabulation. (140)

Negative skewness For distributions, the characteristic of having a small minority of cases at the lower end of the distribution with most of the cases at the upper end and in the middle (also known as *skewed to the left*). (94)

Nominal variable The lowest scale of measurement, which allows comparisons of scores based only on similarity. (22)

Nonparametric statistics Statistical inference procedures that require no assumptions about the shape of sampling or population distributions. (354)

Normal curve A bell-shaped curve dictated by a mathematical function which describes important distributions, especially the sampling distributions used in statistical inference. (280)

Null hypothesis A testable assumption about a population, whose rejection with a known probability of error is used to support the substantive hypothesis. (326) H_0

Observed frequency In the calculation of chi-square, the number of sample cases in a category or cell. (343) f_o

Ogive A line graph that displays a cumulative frequency or percentage distribution showing for each score the percentage of cases whose score is equal to or less than that score. (59)

Open-ended confidence interval A confidence interval consisting of a range of scores above (or below) a given value. (316)

Operationalization The process of creating a measurement instrument. (14)

Order of relationship The number of variables being controlled. (193)

Ordinal variable The second lowest scale of measurement, which allows comparisons of scores on the basis of similarity and order. (23)

Oversampling The practice of selecting a disproportionately large number of cases from a subgroup of a population in order to ensure a large enough number of cases for analysis and inference purposes. (258)

Panel study A research design in which data are gathered from the same set of cases at different times. (83)

Partial correlation A measure of the strength of an association between two variables after controlling for one or more additional variables. (228)

Partial regression coefficient In multiple regression analysis, a measure of the impact that each independent variable has on the dependent variable over and above the effects of the remaining independent variables. (228)

Partial slope (see *partial regression coefficient*)

Path analysis The use of causal models consisting of a series of multiple regressions that describe relationships not only between independent and dependent variables but also among the independent variables. (238)

Path coefficient In path analysis, a standardized partial regression coefficient which indicates the independent effect of one variable on another in a causal model. (240) β

Pearson's r (see *correlation coefficient*)

Percentage The number of observations with a given characteristic divided by the total number of observations, and then multiplied by 100. (37)

Percentile The score below which a specified percentage of cases in an ordered distribution lie. (80)

Periodicity In systematic sampling, a problematic characteristic of lists in which important characteristics recur at regular intervals thus resulting in a biased sample. (254)

Phi (Pearson's) A symmetrical measure of association used with 2×2 cross-tabulations in the form of frequencies. Phi is mathematically equivalent to the correlation coefficient used in regression analysis. (143) ϕ

Platykurtic A bell-shaped curve that is flatter than a normal curve. (286)

Point estimate A single number that is used to estimate a population characteristic such as a mean or proportion. (296)

Population Any precisely defined collection of people, things, events, or observations of their characteristics. (10)

Positive cross-product The total number of concordant pairs in a cross-tabulation. (140)

Positive skewness For distributions, the characteristic of having a small minority of cases at the upper end of the distribution with most of the cases at the lower end and in the middle (also known as *skewed to the right*). (93)

PRE interpretation An interpretation of a measure of association in terms of a proportional reduction in error. (142)

Probabilities proportional to size (PPS) In complex, multistage sample designs, the practice of selecting units such as city blocks with probabilities that vary directly according to the size of the unit. (263)

Symbol

Probability   In the long run, the proportion of trials that are expected to have a particular event as an outcome. Also defined as the number of potentially successful outcomes divided by the total number of possible outcomes. (103)

Probability density function   A mathematical function in the form of a graph that shows probabilities as proportions of the area contained beneath a curve (such as the normal curve). (271)

Probability distribution   The distribution of probabilities for all events that can occur on a given trial. (107)

Product-moment correlation coefficient   (see *correlation coefficient*)

Proportion   The number of cases in a category divided by the total number of cases in a distribution. (37)

Proportional reduction in error   (see *PRE interpretation*)

Q, **Yule's**   (see *Yule's Q*)

Qualitative variable   A variable that can be used to make comparisons between observations only in terms of similarity or difference. (22)

Quantitative variable   A variable that can be used to compare observations in terms of similarity and order, and, in some cases, the interval between scores, or the ratio of one score to another. (22)

Quota sample   A sample design in which interviewers nonrandomly select respondents who are representative on certain specified characteristics such as race, gender, and age. (264)

Random error   A pattern of error in which all possible kinds of error are equally likely. (20)

Range   The difference between the highest and lowest score in a distribution. (92)

Rank-order correlation   The general term for the degree to which two sets of ordered ranks agree or disagree with each other. (154)

Ratio   One number divided by another, used as a measure of the relative magnitude of the numerator relative to the denominator. (45)

Ratio variable   A variable that can be used to compare observations in terms of similarity, order, the relative magnitude of intervals between scores, and the relative magnitude of the scores themselves. (25)

	Symbol

Regression analysis A set of techniques used to describe relationships involving interval or ratio variables using the best fitting straight line (linear regression) or curve (curvilinear regression). (164)

Regression coefficient (slope) In a regression analysis, the number of units of change in Y associated with each unit change in X. A measure of the steepness of the least-squares regression line. (167) b

Regression constant In a regression analysis, the value of Y when the value of all independent variables is zero. (167) a

Reliability A characteristic of measurement instruments that refers to the degree to which they produce the same result over repeated measurements when the case being observed does not change. (16)

Replacement, sampling with A sample design in which members of the population are eligible to be selected more than once. (111)

Replacement, sampling without A sample design in which members of the population are eligible to be selected only once. (111)

Residual In regression analysis, the difference between the actual value of Y for a given value of X and the value of Y predicted by the least-squares regression line. (180)

Response rate The percentage of cases selected in a sample from which observations are actually obtained. (256)

Sample A subgroup selected from a population. (10)

Sampling distribution A theoretical distribution of sample statistics such as means, proportions, percentages, or variances for all possible samples of a given size that could be drawn from a population. (274)

Sampling weights Correction factors that are applied to compensate for oversampling and other sources of unequal selection probabilities in the sampling process. (259)

Scatterplot A graphic display that uses a system of coordinates to represent combinations of scores on interval- and ratio-scale variables as ordered pairs. (59)

Significance level The probability of error associated with rejection of the null hypothesis. (328)

Simple event In probability theory, the occurrence of a single event as the outcome of a trial. (102)

Simple random sample A sample design in which (1) all members of the population are equally likely to be selected;

	Symbol

(2) all combinations of members of the population are equally likely; and (3) all selections are made independently of one another. (252)

Skewness The degree to which scores in a distribution tend to be concentrated at either the high or low end of the variable's range. (93)

Skip interval In a systematic sample design, the number of cases on the population list that separate cases selected into the sample. (253)

Slope (see *regression coefficient*)

Somers' d An asymmetric measure of association used with cross-tabulated ordinal-scale variables in the form of frequencies, which takes into account pairs that are tied on the dependent variable but which differ on the independent variable. (153) — d

Spearman's r_s A symmetrical rank-order measure of association used to indicate the degree to which two sets of ranks agree or disagree with each other. (154) — r_s

Specification A statistical procedure in which we see if the strength and direction of a relationship between X and Y differs from one category of Z to another. (214)

Spuriousness A condition in which a statistical relationship between variables is incorrectly interpreted as causal. (194)

Standard deviation A measure of variation in distributions of interval or ratio variables, with several mathematical properties that make it useful both for describing distributions of variables and for using sampling distributions in statistical inference. The square root of the variance. (88) — Population: σ Sample: s

Standard error The standard deviation in a sampling distribution. It measures the degree to which the characteristics of samples (such as their mean), for samples of a given size, cluster about the corresponding population characteristic. (276) — $\sigma_{\bar{X}}, \sigma_{P_s},$ etc.

Standard error of estimate In regression analysis, a measure of the degree to which two variables are *not* related, calculated as the coefficient of alienation times the standard deviation of Y. (181) — $s_{est.y}$

Standardized regression coefficient In regression analysis, the number of standard deviations of change in Y that are associated with each change of one standard deviation in X. It is used especially in multiple regression to measure the impact of one independent variable on Y over and above the effects of all the other independent variables. (230) — β

Statistical inference The set of techniques designed to use sample results in order to make statements about populations, with known probabilities of error. (268)

Statistics The set of techniques used to organize, summarize, and interpret quantitative information, as well as the results of applying those techniques. (3)

Stem and leaf display A graphic display for showing distributions of interval- or ratio-scale variables in terms of actual scores. (56)

Stochastic independence A condition in which the distribution of one variable is identical within all categories of a second variable. (127)

Stratification In sampling, the drawing of samples within subgroups of the population in order to force the resulting sample to contain a representative percentage of cases from those subgroups. (255)

Stratum (singular of *strata*) In stratified sampling designs, a subgroup of a population within which sampling is carried out. (255)

Strength of association The degree to which conditional probability distributions of a dependent variable differ from one category to another of an independent variable. (133)

Student's t A statistic used in place of Z when estimating sample means using samples of less than 100 cases. (307) t

Substantive hypothesis A proposition that is supported through rejection of the null hypothesis. (326) H_1

Success In probability, any outcome in which the event for which the probability is being calculated occurs. (103)

Suppressor variable A variable which, when not controlled, makes a relationship between X and Y appear weaker than it in fact is. (211)

Symmetrical distribution A distribution in which one-half of the distribution is a mirror image of the other. (93)

Symmetrical measure of association A measure of association for which it is not necessary to specify which variable is independent and which is dependent. (137)

Systematic sample In sampling, a design in which members of a population are selected from a list, by going through the list and selecting members whose names occur at specified intervals. (253)

t, Student's (see *Student's t*)

Table Any arrangement of numbers in columns and rows. (31)

Tail In a distribution with a relatively small percentage of

                                                                    Symbol

very high and/or very low scores, the upper and/or lower ends of the distribution are called tails. Most often used to describe the upper and lower portions of a normal distribution. (285)
Tau, Goodman and Kruskal's   (see *Goodman and Kruskal's tau*)
Tau, Kendall's   (see *Kendall's tau*)
Trial   In probability, a single observation for which one or more events can occur. (102)

Unbiased estimator   A sample statistic such as a mean that, in the long run, has an average value that equals the population characteristic. (296)
Unconditional probability   The probability that an event will occur regardless of whether any other event has occurred. (106)
Unexplained variance   In regression analysis, a variance calculated as the squared difference between the actual value of Y for a given value of X and the value of Y predicted by the least-squares regression line. (177)

Validity   An aspect of measurement instruments that refers to whether they measure what they are intended to measure. (17)
Validity, construct   An aspect of measurement instruments that refers to their appropriateness for the cases that are being observed. (17)
Validity, predictive   The degree to which observations on one variable allow us to predict observations on another. (19)
Variable   A characteristic that differs from one observation to another. (14)
Variance   A measure of variation in distributions of interval and ratio variables which rests on the idea of squared deviations of scores from the mean of the distribution. It has several mathematical properties that make it useful not only for describing distributions, but also for analyzing relationships between variables and using sample data to make inferences about populations. (88)

Population: σ^2
Sample: s^2

Venn diagram   A graphic technique for displaying probabilities with circles that may or may not overlap. (106)

Weight   (see *sampling weights*)

***Y*-intercept**   The point at which a regression line crosses the Y-axis (the value of Y when X has a value of zero). (167)

a

	Symbol
Yule's Q A symmetrical measure of association with a PRE interpretation used to measure the strength of relationships in 2×2 cross-tabulations. (138)	Q
Z-score In a distribution of interval or ratio variables (or in a sampling distribution), the distance between a score and the mean of the distribution, measured in units of standard deviations (or standard errors). (287)	Z

Zero-order relationship   A relationship between two variables with no control variables. (193)

References

Ackoff, R.L. (1953). *The design of social research.* Chicago: University of Chicago Press.

Andrews, F., Morgan, J., and Sonquist, J. (1973). *Multiple classification analysis* (rev. ed.). Ann Arbor, MI: Institute for Social Research.

Babbie, E.R. (1982). *The practice of social research* (3rd ed.). Belmont, CA: Wadsworth.

Baldus, D. (1987). Findings reported by K.B. Nobel in the *New York Times* (March 23).

Bishop, Y.M.M., Fienberg, S.E., and Holland, P.W. (1975). *Discrete multivariate analysis: Theory and practice.* Cambridge, MA: MIT Press.

Blalock, H.M. (1964). *Causal inferences in nonexperimental research.* Chapel Hill, NC: University of North Carolina Press.

Blalock, H.M. (1979). *Social statistics* (rev. ed.). New York: McGraw-Hill.

Blalock, H.M. (1985). *Causal models in experimental and panel designs.* Chicago: IL: Aldine.

Blalock, H.M., and Blalock, A.B. (eds.) (1968). *Methodology in social research.* New York: McGraw-Hill.

Boocock, S.S. (1972). *An introduction to the sociology of education.* Boston: Houghton Mifflin. (See also 1980, 2nd ed.)

Bordua, D.J., and Somers, R.H. (1962). Comments on "Alienation and integration of student intellectuals." *American Sociological Review* 27 (June).

Bowles, S., and Nelson, V. (1974). The "inheritance of IQ" and the intergenerational reproduction of economic inequality. *Review of Economics and Statistics* 56, no. 1 (February).

Braudel, F. (1982). *The wheels of commerce: Civilization and capitalism, 15th–18th century,* vol. 2. New York: Harper and Row.

Brownmiller, S. (1975). *Against our will: Men, women, and rape.* New York: Simon and Schuster.

Bureau of Labor Statistics, U.S. Department of Commerce (1986). Reported in the *New York Times* (November 30).

Cohen, J., and Cohen, P. (1983). *Applied multiple regression/correlation analysis for the behavioral sciences.* Hillsdale, NJ: Lawrence Erlbaum Associates.

Cole, S. (1980). *The sociological method* (3rd ed.). Boston: Houghton Mifflin.

Coleman, J.S. (1964). *Introduction to mathematical sociology*. New York: Free Press.

Coleman, J.S. (1981). *Longitudinal data analysis*. New York: Basic Books.

Crowne, D.P., and Marlowe, D. (1964). *The approval motive*. New York: Wiley.

Crowne, D.P., and Marlowe, D. (1980). *The approval motive*. Westport, CT: Greenwood Press.

Davis, J.A. (1971). *Elementary survey analysis*. Englewood Cliffs, NJ: Prentice-Hall.

Davis, J.A. (1984). *General social surveys, 1972–1984: Cumulative codebook*. Chicago: National Opinion Research Center.

Davis, J.A. (1986). *General social surveys, 1972–1986: Cumulative codebook*. Chicago: National Opinion Research Center.

Davis, J.A., and Jacobs, A.M. (1968). Tabular presentation. In D.L. Sills (ed.), *International encyclopedia of the social sciences*, vol. 15, pp. 497–509. New York: Macmillan Company and Free Press.

Dixon, W.J., and Massey, J., Jr. (1969). *Introduction to statistical analysis* (3rd ed.). New York: McGraw-Hill.

Duncan, B., and Duncan, O.D. (1978). *Sex typing and social roles: A research report*, Appendix A. New York: Academic Press.

Duncan, O.D. (1966). Path analysis: Sociological examples. *American Journal of Sociology* 72 (July), pp. 1–16.

Duncan, O.D. (1969). Inheritance of poverty or inheritance of race? In D.P. Moynihan (ed.), *On understanding poverty*, pp. 85–110. New York: Basic Books.

Duncan, O.D. (1984). *Notes on social measurement: Historical and critical*. New York: Russell Sage.

Durkheim, E. (1951). *Suicide*. New York: Free Press. (Originally published in 1897.)

Edwards, A.L. (1960). *Experimental design in psychological research*. New York: Holt, Rinehart, and Winston.

Edwards, A.L. (1979). *Multiple regression and the analysis of variance and covariance*. San Francisco: W.H. Freeman.

Edwards, A.L. (1984). *An introduction to linear regression and correlation* (2nd ed.). New York: W.H. Freeman.

Edwards, A.L. (1985). *Multiple regression and the analysis of variance and covariance* (2nd ed.). New York: W.H. Freeman.

Farley, R., Bianchi, S., and Colasanto, D. (1979). Barriers to the racial integration of neighborhoods: The Detroit case. *Annals of the American Academy of Political and Social Science* 441 (January), pp. 97–113.

Fienberg, S.E. (1977). *The analysis of cross-classified data*. Cambridge, MA: MIT Press.

Fisher, R.A., and Yates, F. (1974). *Statistical tables for biological, agricultural, and medical research* (6th ed.). London: Longman.

Fox, J. (1984). *Linear statistical models and related methods*. New York: Wiley.

Garfinkel, H. (1971). "Passing" as a woman: A study of sex change. In A.S. Skolnick and J.H. Skolnick (eds.), *Family in transition*. Boston: Little Brown.

Goodman, L.A. (1984). *The analysis of cross-classified data having ordered categories.* Cambridge, MA: Harvard University Press.

Goodman, L.A., and Kruskal, W.H. (1954). Measures of association for cross-classifications, Part I. *Journal of the American Statistical Association* 49, pp. 732–64.

Goodman, L.A., and Kruskal, W.H. (1959). Measures of association for cross-classifications, Part II. *Journal of the American Statistical Association* 54, pp. 123–63.

Goodman, L.A., and Kruskal, W.H. (1963). Measures of association for cross-classifications, Part III. *Journal of the American Statistical Association* 58, pp. 310–64.

Hajda, J. (1961). Alienation and integration of student intellectuals. *American Sociological Review* 26 (October), pp. 758–77.

Hajda, J. (1962). Reply to Bordua and Somers. *American Sociological Review* 27 (June), p. 416.

Harris, M. (1980). *Cultural materialism.* New York: Vintage.

Hays, W.L. (1981). *Statistics for psychologists* (3rd ed.). New York: Holt, Rinehart, and Winston.

Heise, D.R. (1974). Some issues in sociological measurement. In H.L. Costner (ed.), *Sociological methodology, 1973–1974.* San Francisco: Jossey-Bass.

Huff, D., and Geis, I. (1954). *How to lie with statistics.* New York: Norton.

Hyman, H.H. (1955). *Survey design and analysis.* Glencoe, IL: Free Press.

Hyman, H.H. (1972). *Secondary analysis of sample surveys.* New York: Wiley.

Institute for Social Research (1968). *American national election study.* Ann Arbor, MI: University of Michigan.

Katzer, J., Cook, K.H., and Crouch, W.W. (1978). *Evaluating information: A guide to users of social science research.* Reading, MA: Addison-Wesley.

Kemeny, J.G., et al. (1974). *Introduction to finite mathematics* (3rd ed.). Englewood Cliffs, NJ: Prentice-Hall.

Kendall, M.G. (1962). *Rank correlation methods* (4th ed.). New York: Hafner.

Kirk, R. (1978). *Introductory statistics.* Monterey, CA: Wadsworth.

Kirschenbaum, H., et al. (1971). *Wad-ja-get?* New York: A & W.

Kish, L. (1965). *Survey sampling.* New York: Wiley.

Knoke, D., and Burke, P.J. (1980). *Log-linear models.* Beverly Hills, CA: Sage.

Labovitz, S. (1968). Criteria for selecting a significance level: A note on the sacredness of .05. *American Sociologist* 3, pp. 220–22.

Land, K.C. (1969). Principles of path analysis. In E.F. Borgatta (ed.), *Sociological methodology, 1969.* San Francisco: Jossey-Bass.

Lazarsfeld, P.F. (1955). Interpretation of statistical relations as a research operation. In P.F. Lazarsfeld and M. Rosenberg (eds.), *The language of social research.* New York: Free Press.

Lazarsfeld, P.F., Berelson, B., and Gaudet, H. (1960). *The people's choice: How the voter makes up his mind in a presidential campaign* (2nd ed). New York: Columbia University Press.

Lazarsfeld, P.F., and Fiske, M. (1938). The "panel" as a new tool for measuring opinion. *Public Opinion Quarterly* 2, pp. 596–612.

Lazarsfeld, P.F., and Rosenberg, M. (eds.) (1955). *The language of social research*. New York: Free Press.

Lenski, G., and Lenski, J. (1978). *Human societies* (3rd ed.). New York: McGraw-Hill.

Lenski, G., and Lenski, J. (1982). *Human societies* (4th ed.). New York: McGraw-Hill.

Levenson, B. (1968). Panel studies. In D.L. Sills (ed.), *The international encyclopedia of the social sciences*, vol. 11, pp. 371–79. New York: Macmillan Company and Free Press.

Lieberman, B. (ed.) (1971). *Contemporary problems in statistics*. New York: Oxford.

Lieberson, S. (1985). *Making it count: The improvement of social research and theory*. Berkeley: University of California Press.

Loether, H.J., and McTavish, D.G. (1980). *Descriptive and inferential statistics: An introduction*. Boston: Allyn and Bacon.

Massey, D.S. (1981). Hispanic residential segregation: A comparison of Mexicans, Cubans, and Puerto Ricans. *Sociology and Social Research* 65.

Matras, J. (1984). *Social inequality, stratification and mobility*. Englewood Cliffs, NJ: Prentice-Hall.

Money, J., and Ehrhardt, A. (1972). *Man and woman, boy and girl*. Baltimore, MD: Johns Hopkins University Press.

Mosteller, F.R., et al. (1949). *The pre-election polls of 1948*. Social Science Research Council.

Mosteller, F.R., Rourke, R.E.K., and Thomas, G.B. (1961). *Probability with statistical applications*. Reading, MA: Addison-Wesley.

Mosteller, F.R., et al. (1976). *Statistics by example*. Reading, MA: Addison-Wesley.

National Center for Health Statistics (1975). *Vital statistics of the United States, 1973*. Washington, D.C.: U.S. Government Printing Office.

Pearson, E.S., and Hartley, H.O. (1958). *Biometrika tables for statisticians*, vol. 1. Forestburgh, NY: Lubrecht and Cramer.

Pindyck, R.S., and Rubinfeld, D.L. (1981). *Econometric models and economic forecasts*. New York: McGraw-Hill.

Rand Corporation (1955). *A million random digits*. Glencoe, IL: Free Press.

Reichmann, W.J. (1961). *Use and abuse of statistics*. Baltimore, MD: Penguin Books.

Riley, M.W., and Foner, A. (1972). *Aging and society*, vol. 3. New York: Russell Sage Foundation.

Robinson, W.S. (1950). Ecological correlation and the behavior of individuals. *American Sociological Review* 15, pp. 351–57.

Rosenberg, M. (1968). *The logic of survey analysis*. New York: Basic Books.

Rozeboom, W.W. (1971). The fallacy of the null-hypothesis significance test. In B. Lieberman (ed.), *Contemporary problems in statistics*. New York: Oxford University Press.

Rugg, H.O. (1917). *Statistical methods applied to education*. Boston: Houghton Mifflin.

Russell, D.E.H. (1984). *Sexual exploitation: Rape, child sexual abuse, and workplace harassment*. Beverly Hills, CA: Sage.

Ryder, N.B. (1965). The cohort as a concept in the study of social change. *American Sociological Review* 30, pp. 843–61.

Rytina, J.H., Form, W.H., and Pease, J. (1970). Income and stratification ideology: Beliefs about the American opportunity structure. *American Journal of Sociology* 75 (January), pp. 703–16.

Schuman, H., and Converse, J.M. (1971). The effects of black and white interviewers on black responses in 1968. *Public Opinion Quarterly* 35, pp. 44–68.

Schuman, H., and Presser, S. (1981). *Questions and answers in attitude surveys.* New York: Academic Press.

Selvin, H. (1957). A critique of tests of significance in survey research. *American Sociological Review* 22, no. 5 (October), pp. 519–27.

Siegel, S. (1956). *Nonparametric statistics.* New York: McGraw-Hill.

Somers, R.H. (1962). A new asymmetric measure of association for ordinal variables. *American Sociological Review* 27, pp. 799–811.

Sorensen, A., et al. (1974). *Indexes of racial residential segregation for 109 cities in the United States, 1940 to 1970.* Madison, WI: University of Wisconsin Institute for Research on Poverty.

Stern, P.C. (1979). *Evaluating social science research.* New York: Oxford University Press.

Stouffer, S.A. (1962). *Social research to test ideas.* Glencoe, IL: Free Press.

Summers, G.F., and Hammonds, A.D. (1966). Effect of racial characteristics of investigator on self-enumerated responses to a Negro prejudice scale. *Social Forces* 44, pp. 515–18.

Taeuber, K.E. (1965). Residential segregation. *Scientific American* 213, no. 2, pp. 12–19.

Tanur, J.M., et al. (1978). *Statistics: A guide to the unknown.* San Francisco: Holden-Day.

Tufte, E.R. (ed.) (1970). *The quantitative analysis of social problems.* Reading, MA: Addison-Wesley.

Tufte, E.R. (1974). *Data analysis for politics and policy.* Englewood Cliffs, NJ: Prentice-Hall.

Tufte, E.R. (1982). *The visual display of quantitative information.* Cheshire, CT: Graphics Press.

Tukey, J.W. (1977). *Exploratory data analysis.* Reading, MA: Addison-Wesley.

U.S. Bureau of the Census (1974). *Current population reports: Estimates of the population of New York counties and metropolitan areas: July 1, 1972 and 1973.* Series P-25, No. 527. Washington, D.C.: U.S. Government Printing Office.

U.S. Bureau of the Census (1981). *Statistical abstract of the United States: 1982.* Washington, D.C.: U.S. Government Printing Office.

U.S. Bureau of the Census (1983). *Statistical abstract of the United States: 1984.* Washington, D.C.: U.S. Government Printing Office.

U.S. Bureau of the Census (1986). *Current population reports: Evaluation of population estimation procedures for counties: 1980,* Series P-25, No. 984. Washington, D.C.: U.S. Government Printing Office.

U.S. Department of Health, Education, and Welfare (1961). *U.S. health examination survey: 1959–1961.* Washington, D.C.: U.S. Government Printing Office.

U.S. Department of Justice (1976). *Criminal victimization in the United States: A comparison of 1973 and 1974 findings.* Washington, D.C.: U.S. Government Printing Office.

Wallach, M.A. (1972). The psychology of talent and graduate education. Paper presented at the Invitational Conference on Cognitive Styles and Creativity in Higher Education, Montreal (November).

Weiss, N., and Hassett, M. (1982). *Introductory statistics.* Reading, MA: Addison-Wesley.

Wilcoxon, F., and Wilcox, R.A. (1964). *Some rapid approximate statistical procedures.* Pearl River, NY: Lederle Laboratories, American Cyanamid Company.

Yule, G.U. (1912). On the methods of measuring association between two attributes. *Journal of the Royal Statistical Society* 75, pp. 579–652.

Zeisel, H. (1985). *Say it with figures* (rev. ed.). New York: Harper and Row.

Answers to Selected Problems

In this section you will find answers to most of the problems found at the end of each chapter. The only ones that have been omitted are those that (1) lack a clear-cut correct answer; (2) have so many possible answers that listing them all is impractical; or (3) ask you to provide your own examples.

Chapter 2

3a. nominal
3b. ordinal
3c. interval
3d. ordinal
3e. interval
3f. ratio
3g. ordinal
3h. nominal
3i. ratio
3j. ordinal
3k. ratio
3l. ratio
3m. ordinal
3n. ordinal
3o. ordinal
3p. nominal

3q. nominal
3r. ratio
4. Continuous: c, i, k, l. All others are discrete.
6. Because the scale of measurement determines which statistical techniques are appropriate.
7. Because it can be treated as an interval-scale variable even when it otherwise seems to have only nominal or ordinal properties.

Chapter 3

2a. Percentage down; more than high school (78% vs. 52%)
2b. Percentage down; high school only (30% vs. 22%)
2c. Percentage across; someone who approves (20% vs. 11%)
2d. Percentage down; most approve (66%)
2e. Percentage across; high school only (53%)
2f. Percentage down; more than high school, high school only, less than high school (78%, 70%, 52%)
2g. Percentage down; less than high school (52%)
2h. Percentage down; more than high school (78%)
2i. Yes. No figures for nonresponse and missing cases. Depending on how many such cases there are, the results could be considerably biased.
2j. Two, educational attainment and answers to the question on hitting.
3. To compare two categories, percentaging must be done within those categories.
4. $(5553 - 4154)/4154 = .337$; or $5553/4154 = 1.337$. Each shows a 33.7 percent increase in the overall crime rate.
4a. No, because you do not have rates for the intervening years.
5. Ratios can be misleading if we do not pay attention to the actual magnitude of the numerator and the denominator of the ratio. A large ratio can represent a relatively small absolute difference, just as a small ratio can represent a relatively large absolute difference.

Chapter 4

2a. $21/14 = 1.5$
2b. With 14 cases, halfway between the 7th and 8th cases: $(1 + 2)/2 = 1.5$
2c. 2
2d. 0 to 4
2e. 1.25
2f. 1.12
2g. It cuts off the lower 12 scores out of 14. It is the $(12/14)100 = 86$th percentile.

ANSWERS TO SELECTED PROBLEMS / 403

2h. It is not skewed. The mean and median are the same.
3a. Means: 1972: 4,193/1,455 = 2.88; 1986: 3,570/1,370 = 2.61
3b. Medians: 1972 = 728th case, or 3; 1986 = halfway between 685th and 686th cases, or 2
3c. Modes: 1972 = 2; 1986 = 2
3d. Ranges: same for both years: 0 to 7
3e. Variances: 1972 = 1.45; 1986 = 0.92
3f. Standard deviations: 1972 = 1.20; 1986 = .96
3g. Skewness: 1972: because the median is slightly larger than the mean, slightly skewed to the left (negatively); 1986 just the opposite
4. Since the mean summarizes television watching for the population, it does not tell us how many hours were watched by any individual.
5. No. The median and mean are not comparable. The mean income could increase simply because a few individuals experienced a dramatic increase in income, leaving everyone else (including the "typical person") behind.
6. Yes. The variance is a ratio-scale measure of variation in a distribution.
6a. No. The standard deviation is an ordinal scale.
7a. The mean can be used with interval or ratio scales only.
7b. The median can be used with interval, ratio, or ordinal scales only.
7c. The mode can be used with any scale.
7d. The range can be used with interval, ratio, or ordinal scales only.
7e. Percentiles can be used with interval, ratio, or ordinal scales only.
7f. The variance can be used with interval or ratio scales only.
7g. The standard deviation can be used with interval or ratio scales only.
8. It has no effect since the middle score in the distribution remains unchanged.
9. With an even number of cases: halfway between the middle pair of scores: $N/2$ and $(N/2) + 1$. With an odd number of cases, the middle case: $(N + 1)/2$.
10. Yes.
11. The scale of measurement, the shape of the distribution, and the kinds of statements you want to make about it.
12. Because data for missing years can distort the true trend line.
13. Because the mean is defined as the point at which the average distance between itself and all of the scores is zero (the point at which the balance beam balances).
14. The mean is greater than the median when the distribution is positively skewed; the mean is smaller than the median when the distribution is negatively skewed.
15. Distribution (a): unimodal; mean = median; not skewed; symmetrical.
    Distribution (b): unimodal; mean = median; not skewed; symmetrical.

Distribution (c): no mode (all scores equally likely); mean = median; not skewed; symmetrical (rectangular).

Distribution (d): bimodal; mean greater than median; positively skewed; asymmetrical.

Distribution (e): unimodal; median is greater than mean; negatively skewed.

Chapter 5

2. Five. Having less than a high school education (470/1529 = .307); a high school education (804/1529 = .526); college or more (255/1529 = .167; approve of hitting (1006/1529) = .658); disapprove of hitting (523/1529 = .342).

3. Six (one for each cell). $p(A,C) = .160$; $p(A,D) = .368$; $p(A,E) = .130$; $p(B,C) = .148$; $p(B,D) = .158$; $p(B,E) = .037$.

4.
First Draw	Second Draw	Probability
less than high school	less than high school	.094
less than high school	high school	.162
less than high school	college or more	.051
high school	high school	.276
high school	less than high school	.162
high school	college or more	.088
college or more	college or more	.028
college or more	less than high school	.051
college or more	high school	.088

9. p(less than high school = 470/1529 = .307; p(less than high school given approval of punching) = 244/1006 = .243. Since the conditional and unconditional probabilities are not equal, the two events are not independent of each other. An alternate way to answer the same problem is: p(approves of punching) = 1006/1529 = .658; p(approves of punching given less than high school = 244/470 = .519. Again, the conditional and unconditional probabilities are not equal, and the events are therefore not independent of each other.

p(college or more) = 255/1529 = .167; p(college or more given disapproval of punching) = 56/523 = .107. Or, p(disapproval) = 523/1529 = .342; p(disapproval given college or more) = 56/255 = .220.

The events "approves of punching" and "disapproves of punching" are mutually exclusive and therefore not independent. The probability that one will occur given that the other has occurred is always zero.

ANSWERS TO SELECTED PROBLEMS / 405

10. p(disapproves) $= 523/1529 = .342$. Expected number who disapprove is $p(n) = .342(100) = 34$.
11. Has less than a high school education *and* has college or more.
12. If event A is independent of event B, then event B is also independent of event A.

Chapter 6

2. No. A statistical relationship does not necessarily imply a cause-and-effect relationship.
3. Percentaging is done within categories of the independent variable in order to see if differences on the independent variable correspond to differences on the dependent variable.
5. Measures of association should have a value of ± 1 when the relationship is perfect; a value of zero when the variables are independent of each other; and a meaningful interpretation in terms of the degree of dependence between the variables.
6. The variables are dependent on each other. First, if you percentage down the columns, you can see that the distributions are different (women are more likely than men to support gun control laws). Second, the unconditional probability of favoring gun control laws is not the same as the conditional probability of favoring such laws given that one is a man or given that one is a woman. For example, p(favor) $= 1034/1430 = .72$. p(favor | man) $= 371/589 = .63$.
7. A negative relationship between being a man and favoring gun control; a negative relationship between being a woman and opposing gun control; a positive relationship between being a man and opposing gun control; a positive relationship between being a woman and favoring gun control.
8a. Yule's $Q = -.37$
   (1) Used only with 2 × 2 tables.
   (2) symmetrical
   (3) In guessing how men and women differ on gun control, we will do 37 percent better than chance if we always guess that women will favor gun control and men will oppose it (a PRE interpretation).
   (4) Q tends to give relatively large values compared with other measures; has a value of ± 1.0 even when relationships are only conditionally perfect.
8b. phi $= -.17$
   (1) Used only with 2 × 2 tables expressed in frequencies.

(2) symmetrical
(3) no PRE interpretation
(4) Has a value of ±1.0 only when relationship is mathematically perfect.

8c. $phi^2 = .03$
   (1) Used only with 2 × 2 tables expressed in frequencies.
   (2) symmetrical
   (3) Represents the proportion of the variation in the dependent variable that is statistically explained by its relationship to the independent variable.
   (4) Has a value of ±1.0 only when relationship is mathematically perfect.

8d. Goodman and Kruskal's tau = .03
   (1) Can be used only with tables expressed in frequencies.
   (2) asymmetrical
   (3) Has a PRE interpretation: proportional reduction of error in predicting the dependent variable from knowing the independent variable.
   (4) Advantage: can be used with any size table.

9. Collapsing a variable is justified when it does not distort the direction or strength of the relationship; otherwise it should not be done.

10. No. The percentage distributions in the columns are not the same. Also, the conditional and unconditional probabilities are not equal. For example, $p(yes) = 864/1408 = .61$; $p(yes|HS\ grad) = 268/466 = .58$.

11. The more education people have the more likely they are to express tolerance for homosexual college teachers. Or, there is a positive relationship between education and expressions of tolerance. The relationship is roughly linear.

12a. Goodman and Kruskal's gamma = +.50
   (1) Can be used only with linear relationships involving variables of at least an ordinal scale (unless one of the variables is a dichotomy) with cells expressed in terms of frequencies.
   (2) symmetrical
   (3) In guessing how people with different educations differ on tolerance, we will do 50 percent better than chance if we always guess that the person with a higher education expresses tolerance and the person with a lower education does not (a PRE interpretation).
   (4) Gamma tends to give relatively large values compared with other measures; can have a value of ±1.0 even when relationships are only conditionally perfect.

12b. Somers' $d = .25$
  (1) Can be used only with linear relationships involving variables of at least an ordinal scale (unless one of the variables is a dichotomy) with cells expressed in terms of frequencies.
  (2) asymmetrical
  (3) no PRE interpretation
  (4) Advantage is that it takes tied pairs into account. Disadvantage is lack of a PRE interpretation.
13. Yes. You can use tau with any cross-tabulation so long as it is expressed in frequencies.
14. To use phi, we would have to collapse the education variable into a dichotomy by combining either the two highest or two lowest educational categories into a single category. Since the relationship is roughly linear, this would be acceptable.
15. Spearman's $r_s = -.40$
Kendall's tau $= -.40$
The judges tend to disagree more than they agree, and the tendency is moderately strong.
16. Because one of the variables is a three-category nominal scale, the only measure of association that is appropriate is Goodman and Kruskal's tau. All of the other measures discussed in this chapter must be used either with higher scales of measurement or with dichotomies.
17. Yule's Q, phi, and phi^2 if the variables are in the form of dichotomies. Goodman and Kruskal's tau, gamma, and Somers' d if the variables are cross-tabulated.
18. Yule's Q, phi, phi^2 if the variables are in the form of dichotomies; Goodman and Kruskal's tau, gamma, Somers' d if the variables are cross-tabulated; Spearman's r_s, Kendall's tau if the variables are in the form of ranks.
19. phi, phi^2, Goodman and Kruskal's tau, Kendall's tau, and Spearman's r_s.
20. Yule's Q, phi^2, Goodman and Kruskal's tau, and gamma.

Chapter 7

2a. Each equation represents the least-squares line for the relationship between the two variables.
2b. $\hat{Y}$ is the dependent variable, the predicted value for the number of years of schooling the child has; 8.43 and 9.45 are the regression constants, the points at which the least-squares lines cross the Y axis (which

happens when X equals 0); .42 and .35 are the regression coefficients (slopes) which show how many units of change in $\hat{Y}$ are associated with each unit change in X_1 or X_2; X_1 and X_2 represent the independent variables, mother's education and father's education.

2c. Neither is the better predictor. Since the correlations are the same for both (.48), both predict child's education equally well.

2d. Mother's education has the greater impact because its slope (.42) is greater than that for father's education (.35).

2e. The coefficient of determination is the same for both equations: $(.48)^2 = .23$. It represents the proportion of the variation in Y that is statistically accounted for by its linear relationship with X.

2f. The coefficient of alienation is the same for both equations: $\sqrt{1 - (.48)^2} = .88$. The closer the coefficient of alienation is to 1.0, the weaker is the relationship between the two variables.

2g. The standard error of estimate is simply the coefficient of alienation multiplied times the standard deviation of Y, or $1.1(.88) = .97$. Since the coefficient of alienation is the same for both equations, so is the standard error of estimate.

2h. See answer to 2e.

2i. $\hat{Y} = 8.43 + .42(12)$
    $\hat{Y} = 13.47$

2j. $\hat{Y} = 9.45 + .35(12)$
    $\hat{Y} = 13.65$

3a. $b = -.23$

3b. $a = 5.72$

3c. $r = -.77$

3d. Coefficient of determination = $(-.77)^2 = .59$

3e. Coefficient of alienation = $\sqrt{1 - (-.77)^2} = .64$

3f. Unexplained variance = $1 - (-.77)^2 = .41$ (or 1 − coefficient of determination)

3g. Variance of $Y = 3.03$; standard deviation of $Y = 1.74$. Standard error of estimate = (standard deviation of Y)(coefficient of alienation), or $1.74(.64) = 1.11$

3h. $\hat{Y} = 5.72 - .23X$

3i. 2.27

3j. The prediction is for the mean for those with 15 years of schooling. Since the correlation is not perfect, not all people with a score of 15 on X will have the same score for Y. In other words, there will be variation about the least-squares line, representing unexplained variance.

4. $a = 55$.

5. Yes, if the relationship is weak enough. Even though the line must pass through the mean of Y and the mean of X, none of the cases must necessarily have either of these scores.
6a. 0
6b. the standard deviation for Y
6c. 0
6d. 1.00
6e. 0
6f. the mean of Y
7. the mean of Y
8. the slope or regression coefficient
9. With stochastic independence, the distribution of Y is the same for every value of X; with mean independence, the mean of Y is the same for every value of X even though the distribution of Y may differ from one value of X to another. When $r = 0.00$, we know there is mean independence. There may or may not be stochastic independence.
10. the correlation coefficient or the coefficient of determination
11. In addition to the obvious danger of incorrectly describing the direction of the relationship, the main danger lies with underestimating the strength of the relationship.

Chapter 8

2. To ensure that the relationship is not spurious; to understand causal mechanisms; to see how a relationship changes under different conditions.
5a. Income by education

	Education		
Income	Less Than College (−)	Some College or More (+)	All
Low (−)	335	185	520
High (+)	159	253	412
TOTAL	494	438	932

5b. Occupational prestige by education

	Education		
Prestige	Less Than College (−)	Some College or More (+)	All
Low (−)	170	41	211
High (+)	324	397	721
TOTAL	494	438	932

5c. Income by occupation

	Occupational Prestige		
Income	Low (−)	High (+)	All
Low (−)	171	349	520
High (+)	40	372	412
TOTAL	211	721	932

5d. An intervening variable.
    Education ⟶ Occupation ⟶ Income
5f. The same. Z is an intervening variable.
5g. The partials would have been the same as the zero-order association.
5h. The results would have been the same.
5i. The partials would have been larger than the zero-order association, not smaller.
5j. The partials would have been in the opposite direction (negative).
5k. The results would have been the same.
5l. The partial would have been smaller than the zero-order association.
5m. The association within the two control categories might have been of differing strengths or in different directions.
6. Add the corresponding cells in Table 8-9 together and the result will be Table 8-6.

7a. Position on segregation by region

	Region		
Position on Segregation	South (+)	Nonsouth (−)	All
For (+)	65	66	131
Against (−)	167	456	623
TOTAL	232	522	754

7b. Position on segregation by occupation

	Occupation		
Position on Segregation	White-Collar (+)	Blue-Collar (−)	All
For (+)	48	83	131
Against (−)	374	249	623
TOTAL	422	332	754

7c. Occupation by region

	Region		
Occupation	South (+)	Nonsouth (−)	All
White-Collar (+)	125	297	422
Blue-Collar (−)	107	225	332
TOTAL	232	522	754

8. We tend to run out of cases to represent the large number of categories formed by the cells in the table.

9a. Control for health.

9b. An extraneous variable.
9c. Testing for spuriousness.
9d. Partials that are zero or close to zero would indicate spuriousness.
10a. An interaction variable.
10b. Study time should make less of a difference among those who study at the last minute and right up to the moment of the exam.
11a. Component variable.
11b. The partial should be lower (closer to zero) than the zero-order association.
12a. Antecedent variable.
12b. When you control for Z, the partial association between X and Y should be roughly as strong as the zero-order association. When you control for X, however, the partial association between Z and Y should be considerably weaker than the zero-order association.
13. Z is a suppressor variable.
14. Z is a distorter variable.

Chapter 9

2a. $a = 20.36$
2b. The equation is the best fitting function for the relationship between Y and the four independent variables acting at once.
2c. $\hat{Y}$ represents the predicted value of the dependent variable; a represents the regression constant, the value of $\hat{Y}$ when all four independent variables are zero; .12, .12, .24, and .31 are the unstandardized partial slopes for each independent variable, representing the amount of change in $\hat{Y}$ associated with each change in each independent variable over and above those changes associated with changes in the other independent variables; each X represents an independent variable.
2d. The four exams explain 96 percent of the variation in semester grades. R^2 is less than 1.00 because some of the variation in semester grades is due to other variables (the final exam and the short written assignments).
2e. for X_1: $(.12)(17/13) = .16$
for X_2: $(.12)(14/13) = .13$
for X_3: $(.24)(20/13) = .37$
for X_4: $(.31)(11/13) = .26$
Each beta weight is a standardized regression coefficient. It tells us how many standard deviations of change in $\hat{Y}$ are associated with each change of one standard deviation in each independent variable over and above the effects of the other independent variables.
2f. X_3 has the greatest independent impact, judging from the beta weight in 2e.

2g. X_2 has the smallest independent impact, judging from the beta weight in 2e.
2h. $\hat{Y} = 20.36 + .12(65) + .12(74) + .24(88) + .31(92) = 86.68$
3a. $a = 7.60$
3b. That the best predictor of education using parents' education as independent variables is to take 7.60 years and add to it .26 years for every year of education the mother had plus .21 years for every year of education the father had.
3c. $\hat{Y}$ is the predicted value of the dependent variable; a is the regression constant (the value of $\hat{Y}$ when X_1 and X_2 equal zero); .26 and .21 are the regression coefficients, representing the amount of change in Y associated with each change of one unit of X; X_1 and X_2 are the independent variables used to predict Y.
3d. The value of R^2 is .28, telling us that parents' education explains just over a quarter of the variation in education.
3e. for X_1: $(.26)(4.24/3.43) = .32$
for X_2: $(.21)(3.18/3.43) = .19$
Each beta weight is a standardized regression coefficient. It tells us how many standard deviations of change in Y are associated with each change of one standard deviation in each independent variable over and above the effects of the other independent variables.
3f. Mother's education (X_1), judging from the beta weights in 3e.
3g. $\hat{Y} = 7.60 + .26(16) + .21(11) = 14.07$
4a. occupation
4b occupation
4c. Number of siblings has *no* direct effect on son's income.
4d. $(-.13)(-.04) + (-.13)(-.21)(.53) + (.27)(.53) = .16$
4e. $(.20)(.09) + (.20)(.53)(.31) + (.02)(.31) + (-.21)(-.04)(.31) + (-.21)(-.21)(.09) + (-.21)(-.21)(.53)(.31) = .07$
4f. $(.27)(.09) + (.27)(.53)(.31) + (.13)(.31) + (-.13)(-.04)(.31) + (-.13)(-.21)(.09) + (-.13)(-.21)(.53)(.31) = .12$
4g. $.09 + (.53)(.31) = .25$
4h. Seventeen percent of the variance is explained $(1 - \sqrt{.91})$. The coefficient of alienation is .91.
4i. The curved arrow indicates that the two variables are correlated, but no causal connection is being specified.

Chapter 10

2. The size of the sample, by affecting the size of the standard error. Unless the sample constitutes a substantial portion of the population (at least 10 percent), the size of the population is irrelevant.

3. By making a complete list of all of the elements in the population or by defining the population in such a way that it is possible to tell unambiguously what is and is not included.
4. (1) The sampling frame must be complete, giving all members of the population a chance to be included in the sample; (2) the probability of selection for all members of the population must be known; (3) selections must be made independently of one another.
5. In any random process (such as selecting a sample), all events must have equal probabilities of occurring on any given trial. It is often confused with "haphazard" sampling in which probabilities are unknown and most likely unequal.
6. With replacement, members of the population that are selected are then returned to the population and are therefore eligible to be selected again. Without replacement, members of the population can be selected only once. In general, sampling without replacement can only hurt a sample if the sample constitutes a fairly large proportion of the population.
7. Unlike random sampling, systematic sampling does not allow all combinations of members of the population to be selected. It is most likely to be used to save time when a list of the population is available. It causes problems with lists that have periodicity (are not well-shuffled). The solution is to shuffle the list.
8. Stratification tends to lower both random error and bias. Low response rates tend to increase the chances for bias, depending on the reason for the nonresponse. Stratification can never hurt the accuracy of a sample since it forces it to be representative for a particular variable.
9. Oversampling is most likely to be used when we want to study a small subgroup of a population. Because oversampling leads to an overrepresentation of the subgroup, however, in order to make statements based on the entire sample it is necessary to use sampling weights.
10. Complex samples are most likely to be used when selecting samples from populations that have large, complex frames, such as large geographic areas or large complex organizations. Cluster sampling is most likely to be used in multistage samples whose next-to-last stage of sampling involves widely separated units such as city blocks. Cluster sampling saves travel time and money but tends to result in samples that are more homogeneous than the populations they come from.
11. Because they do not conform to the rules of scientific sampling, on which statistical inference techniques are based, there is no mathematical basis for relying on them for inference purposes.
14. To the extent that the telephone directories constitute an incomplete frame of the population (lots of households do not have telephones), the

Chapter 11

2. Probability density functions use the proportion of the area beneath a curve to represent probabilities rather than the height of the curve at any given point.

3. The population mean is the result of adding up the weight of every 12 to 16-year-old adolescent in the U.S. and dividing by the number of adolescents. The sample mean is found in the same way as the population mean except that it is based on a sample of those adolescents. The sampling distribution of sample means is the theoretical distribution of the infinite number of possible sample means we could have drawn from the population.

4. It is necessary to estimate the standard error when the population standard deviation is unknown. With means, the sample standard deviation is used to estimate the unknown population standard deviation.

5. .30

6. $\hat{\sigma}_{\bar{x}} = 10.8/\sqrt{900 - 1} = .36$

7. The larger the sample size, the more the sampling distribution approximates a normal curve regardless of the shape of the population. It is important because without knowing the shape of the sampling distribution, we cannot make statistical inferences.

8a. .50

8b. .50

8c. $Z = (60 - 80)/20 = -1.0$ standard deviation; proportion is .3413.

8d. $Z = (90 - 80)/20 = +.5$ standard deviation; proportion is .1915.

8e. $Z = (75 - 80)/20 = -.25$ standard deviation; proportion is .0987. Must add this to .5000 (for area above 80) to get .5987.

8f. $Z = (50 - 80)/20 = -1.5$ standard deviations; proportion is .4332. Must *subtract* this from .5000 to get .0668.

8g. $Z = 20/20 = 1$ standard deviation; proportion is .3413. Must double this to get .6826.

8h. Subtract the answer for 8g from 1.000 to get .3714.

8i. $Z = 30/20 = 1.5$ standard deviations; proportion is .4332. Must double this to get .8664.

8j. $Z = 40/20 = 2.0$ standard deviations; proportion is .4773. Must subtract this from .5000 to get .0227 and then double this to get .0454.

8k. As a probability density function, the normal distribution cannot be used to calculate probabilities for exact scores, only ranges of scores.

8l. See 8k.

8m. Must be done in two pieces. First, $Z = (70 - 80)/20 = -.5$ standard deviation; proportion is .1915. Then, $Z = (90 - 80)/20 = +.5$ standard deviation; proportion is .1915. Add the two together to get .3830.

8n. Must be done in two pieces. First, $Z = (60 - 80)/20 = -1.0$ standard deviation; proportion is .3413. Then, $Z = (85 - 80)/20 = +.25$ standard deviation; proportion is .0987. Add the two together to get .4400.

9. Since a normal curve is symmetrical, the mean and the median will have the same value, 35.

10a. $\hat{\sigma}_{\bar{x}} = 10/\sqrt{401 - 1} = .50$

10b. $Z = +1.0$ standard error; the probability is .3413.

10c. $Z = -1.0$ standard error; the probability is .3413.

10d. $Z = \pm 1.5$ standard errors; the probability is $.4332 + .4332 = .8664$.

10e. $Z = \pm 2.5$ standard errors; the probability is $.4938 + .4938 = .9876$.

10f. $Z = +2.25$ standard errors; the probability is .4878 which must then be subtracted from .5000 to get .0122.

10g. $Z = -2.75$ standard errors; the probability is .4970 which must then be subtracted from .5000 to get .0030.

10h. $Z = \pm 1.96$ standard errors; the probability is .4750 which must then be subtracted from .5000 to get .0250, and then doubled to get .0500.

10i. $Z = +.5/.5 = 1.0$ standard error; the probability is .3413 which must then be subtracted from .5000 to get .1587.

10j. $Z = -1/.5 = -2.0$ standard errors; the probability is .4773 which must then be subtracted from .5000 to get .0227.

10k. $Z = \pm 1.5/.5 = \pm 3.0$ standard errors; the probability is $.49865 + .49865 = .9973$.

10l. Subtract the answer for 10k from 1.0000 to get .0027.

10m. $Z = \pm 1.0$ standard error; the probability is $.3413 + .3413 = .6826$.

10n. .5000

10o. .5000

10p. The normal distribution cannot be used to calculate probabilities for exact scores, only ranges of scores.

10q. The normal distribution cannot be used to calculate probabilities for exact scores, only ranges of scores.

Chapter 12

2a. $\hat{\sigma}_{\bar{x}} = 12/\sqrt{1,600 - 1} = .3$ hour

2b. For a 95-percent confidence interval, add and subtract 1.96 standard errors to the sample mean: $42.5 \pm 1.96(.3)$, or 41.9 to 43.1.

2c. With a confidence level of 95 percent, we estimate that the mean number of hours worked each week is somewhere between 41.9 and 43.1.
3a. $\hat{\sigma}_X = 2.0/\sqrt{401 - 1} = .10$ hour
3b. For a 99-percent confidence interval, add and subtract 2.575 standard errors to the sample mean: $3.7 \pm 2.575(.1)$, or 3.4 to 4.0.
3c. With a confidence level of 99 percent, we estimate that the mean number of hours spent listening to the radio is somewhere between 3.4 and 4.0.
3d. With only 41 cases, we must use the t-distribution. Now the 99-percent confidence interval is $3.7 \pm 2.704(.1)$, or 3.4 to 4.0, slightly wider but within rounding error of what we got before.
4a. Using $P = Q = .5$: $\hat{\sigma}_{P_s} = \sqrt{(.5)(.5)/200} = .035$
4b. For a 98-percent confidence interval, add and subtract 2.33 standard errors to the sample proportion: $.40 \pm 2.33(.035)$, or .32 to .48.
4c. With a confidence level of 98 percent, we estimate that the proportion of chips that are defective is somewhere between .32 and .48, or, in percentage terms, between 32 and 48 percent.
5a. Using $P = Q = .5$: $\hat{\sigma}_{P_s} = \sqrt{(.5)(.5)/1,470} = .013$
5b. For 95-percent confidence interval, add and subtract 1.96 standard errors to sample proportion: $.56 \pm 1.96(.013)$, or .53 to .59.
5c. With a confidence level of 95 percent, we estimate that the proportion of adults who approve of birth control methods being made available to teenagers whose parents do not approve is somewhere between .53 and .59, or, in percentage terms, between 53 and 59 percent.
6a. Using $P_1 = Q_1 = P_2 = Q_2 = .5$:

$$\hat{\sigma}_{P_{s_1} - P_{s_2}} = \sqrt{\frac{(.5)(.5)}{569} + \frac{(.5)(.5)}{807}} = .027$$

6b. For a 99-percent confidence interval, add and subtract 2.575 standard errors to the difference between the sample proportions: $(.80 - .71) \pm 2.575(.027)$, or $.09 \pm .07$, or from .02 to .16.
6c. With a confidence level of 99 percent, we estimate that men are more likely than women to approve of capital punishment by a margin that falls somewhere between .02 and .16, or, in percentage terms, men are between 2 and 16 percent more likely than women to approve of capital punishment.
7a. Using $P_1 = Q_1 = P_2 = Q_2 = .5$:

$$\hat{\sigma}_{P_{s_1} - P_{s_2}} = \sqrt{\frac{(.5)(.5)}{1,030} + \frac{(.5)(.5)}{366}} = .03$$

7b. For a 95-percent confidence interval, add and subtract 1.96 standard errors to the difference between the sample proportions: $(.55 - .43) \pm 1.96(.03)$, or $.12 \pm .06$, or from .06 to .18.

7c. With a confidence level of 95 percent, we estimate that Republicans are more likely than Democrats and Independents to agree that welfare makes people not want to work, by a margin that falls somewhere between .06 and .18; or, in percentage terms, Republicans are between 6 and 18 percent more likely than Democrats and Independents to hold this belief.

8a. Using the sample *variance* to estimate the population variance, the estimated standard error is:

$$\hat{\sigma}_{\bar{x}_1 - \bar{x}_2} = \sqrt{\frac{4.84}{587 - 1} + \frac{4.41}{569 - 1}} = .13 \text{ hour}$$

8b. For a 98-percent confidence interval, add and subtract 2.33 standard errors to the difference between the sample means:
$(3.1 - 2.8) \pm 2.33(.13)$, or $.3 \pm .3$, or from 0.0 to 0.6 hour.

8c. With a confidence level of 98 percent, we estimate that people who identify themselves as lower or working class tend to watch between zero and .6 more hours of television each day than do those who identify themselves as middle or upper class. Notice that this estimate includes zero (no class difference).

9a. Using the sample *variance* to estimate the population variance, the estimated standard error is:

$$\hat{\sigma}_{\bar{x}_1 - \bar{x}_2} = \sqrt{\frac{2.82}{565 - 1} + \frac{2.89}{314 - 1}} = .12 \text{ child}$$

9b. For a 95-percent confidence interval, add and subtract 1.96 standard errors to the difference between the sample means: $(1.91 - 1.66) \pm 1.96(.12)$, or $.25 \pm .24$, or from 0.01 to 0.49 child.

9c. With a confidence level of 95 percent, we estimate that people who earn $15,000 or more have an average of between .01 and .49 more children than do those who earn less than $15,000.

10. Quadrupling the sample size cuts the standard error in half, which, in turn, would cut the confidence interval in half. It would have the same effect on *any* confidence interval regardless of the confidence level involved.

ANSWERS TO SELECTED PROBLEMS / 419

11. You could not apply these inference techniques to a quote sample because the techniques assume random sampling or its equivalent, a condition that is not met by quota sampling techniques.

Chapter 13

2. H_0: the defendant is innocent.
   H_1: the defendant is guilty.
3. H_0: the percentage living in poverty in our community = the percentage living in poverty in the population.
   H_1: the percentage living in poverty in our community is less than the percentage living in poverty in the population.
4. In most cases, nothing. We cannot estimate this probability.
5a. H_0: the population mean = 900 bushels per acre.
5b. H_1: the population mean is greater than 900 bushels per acre.
5c. Using the sample standard deviation for the population standard deviation:

$$\hat{\sigma}_{\bar{x}} = s/\sqrt{(N-1)} = 150/\sqrt{99} = 15 \text{ bushels}$$

5d. The sample mean is (941 − 900) = 41 bushels above the assumed population mean. This is equivalent to 41/15 = 2.73 standard errors above the null hypothesis mean of 900. The probability of getting a result that is 2.73 standard errors or more above the population mean by chance is (.5000 − .4968) = .0032. We could reject H_0 at the .0032 level of significance. Since this is a very low probability of error, I would reject H_0.
5e. We are very confident that the total yield will exceed 900 bushels per acre ($p = .0032$). Or, the yield is significantly greater than 900 bushels per acre ($p = .0032$).
5f. With a sample mean *less* than the null hypothesis mean, it would make no sense to test H_0 unless we changed H_1.
6a. Since the sample proportion is a long way from .5, I would use $P = .2$ and $Q = .8$ for the estimated standard error (to be conservative, you could also use $P = Q = .5$; it is a matter of judgment).

$$\hat{\sigma}_{P_s} = \sqrt{(.2)(.8)/300} = .02$$

6b. The sample proportion is (.12 − .10) = .02 below the assumed population proportion. This is equivalent to .02/.02 = 1.00 standard error below the population proportion. The probability of getting a result that is 1.00 standard error or more below the population proportion by chance is (.5000 − .3413) = .1587. We could reject H_0 at the .1587 level of significance. Since this is a fairly high probability of error, I would not reject H_0.

6c. The data do not justify the conclusion that the proportion living in poverty in the community is lower than the national figure. Put another way, the proportion living below poverty in the community is not significantly below the national level. This does *not* mean that it is the same or higher than the national figure.

6d. Since the sample proportion exactly equals the proportion assumed by the null hypothesis, it would be impossible to use the sample result as a basis for rejecting the null hypothesis.

7a. H_0: the difference between the proportion of men who support capital punishment and the proportion of women who support it is zero.

7b. H_1: the proportion of men who support capital punishment is greater than the proportion of women who support it (the difference is greater than zero).

7c. Using $P_1 = Q_1 = P_2 = Q_2 = .5$:

$$\hat{\sigma}_{P_{s_1} - P_{s_2}} = \sqrt{\frac{(.5)(.5)}{569} + \frac{(.5)(.5)}{807}} = .03$$

7d. The difference between the sample proportions is (.80 − .71) = .09 above the null hypothesis difference of zero. This is equivalent to .09/.03 = 3.00 standard errors above the mean of the sampling distribution. The probability of getting a result that is 3.00 standard errors or more above the mean of the sampling distribution by chance is (.5000 − .49865) = .001. We could reject H_0 at the .001 level of significance. Since this is a very small probability of error, I would reject H_0.

7e. We are very certain that men are more likely than women to support capital punishment ($p = .001$); or, men are significantly more likely than women to support capital punishment ($p = .001$).

7f. With a sample proportion less than the null hypothesis proportion, it makes no sense to test H_0 unless we change H_1.

8a. H_0: there is no difference (a difference of zero) between the proportion of Republicans who agree and the proportion of Democrats and Independents who agree.

8b. H_1: the difference between the proportion of Republicans who agree and the proportion of Democrats and Independents who agree is greater than zero (Republicans are more likely to agree).

8c. Using $P_1 = Q_1 = P_2 = Q_2 = .5$:

$$\hat{\sigma}_{P_{s_1} - P_{s_2}} = \sqrt{\frac{(.5)(.5)}{1,030} + \frac{(.5)(.5)}{366}} = .03$$

8d. The difference between the sample proportions is $(.55 - .43) = .12$ above the null hypothesis difference of zero. This is equivalent to $.12/.03 = 4.00$ standard errors above the mean of the sampling distribution. The probability of getting a result that is 4.00 standard errors or more above the mean of the sampling distribution by chance is $(.5000 - .4999683) = .00003$. We could reject H_0 at the .00003 level of significance. Since this is a very small probability of error, I would reject H_0.

8e. We are very sure that Republicans are more likely than Democrats and Independents to agree that welfare makes people not want to work. Or, Republicans are significantly more likely than Democrats and Independents to agree with this ($p = .00003$).

9a. H_0: the difference between the mean for lower- and working-class people and the mean for middle- and upper-class people is zero (there is no difference).

9b. H_1: the difference between the mean for lower- and working-class people and the mean for middle- and upper-class people is greater than zero (the lower-/working-class mean is greater than the middle-/upper-class mean).

9c. Using the sample *variance* to estimate the population variance, the estimated standard error is:

$$\hat{\sigma}_{\bar{X}_1 - \bar{X}_2} = \sqrt{\frac{4.84}{587 - 1} + \frac{4.41}{569 - 1}} = .13 \text{ hour}$$

9d. The difference between the sample means is $(3.1 - 2.8) = 0.3$ hours above the null hypothesis difference of zero. This is equivalent to $0.3/.13 = 2.31$ standard errors above the mean of the sampling distribution. The probability of getting a result that is 2.31 standard errors or more above the mean of the sampling distribution by chance is $(.5000 - .4896) = .0104$. We could reject H_0 at the .0104 level of significance. Since this is a very small probability of error, I would reject H_0.

9e. We are very sure that people who identify themselves as lower or working class report watching more television each day than do those who identify themselves as middle or upper class ($p = .0104$). Or, people who identify themselves as lower or working class report watching significantly more television than do those who identify themselves as middle or upper class ($p = .0104$).

10a. H_0: that the difference between the average number of children born to higher income people and the average born to lower income people is zero (there is no difference).

10b. H_1: that those earning lower incomes have more children than do those earning higher incomes.

10c. Using the sample *variance* to estimate the population variance, the estimated standard error is:

$$\hat{\sigma}_{\bar{x}_1 - \bar{x}_2} = \sqrt{\frac{2.82}{565 - 1} + \frac{2.89}{314 - 1}} = .12 \text{ child}$$

10d. Since the sample difference is in the *opposite* direction to that predicted by the substantive hypothesis (the mean for higher income people is larger than the mean for lower income people), there is no point in testing the null hypothesis. (We might then want to change the substantive hypothesis and then test the null.)

10e. We cannot conclude from these data that lower income people have more children than higher income people.

11a. H_0: marital status and reported happiness are independent of each other.

11b. H_1: there is a relationship between marital status and reported happiness.

11c. $\chi^2 = \frac{(347 - 283)^2}{283} + \frac{(97 - 123)^2}{123} + \frac{(58 - 97)^2}{97} + \frac{(467 - 531)^2}{531} + \frac{(256 - 230)^2}{230} + \frac{(220 - 181)^2}{181}$

$\chi^2 = 14.47 + 5.50 + 15.68 + 7.71 + 2.94 + 8.40 = 54.70$

With $(3 - 2)(2 - 1) = 2$ degrees of freedom, a value of chi-square of 54.70 would occur by chance with a probability far smaller than .001 (since 54.70 is much greater than 13.815). Therefore, we can reject H_0 at the $p < .001$ level of significance.

11d. We are very certain that there is a relationship between marital status and reported happiness ($p < .001$). Or, there is a significant relationship between marital status and reported happiness.

ANSWERS TO SELECTED PROBLEMS / 423

12a. H_0: the religious compositions of the corporate board membership and the adult population are the same.
12b. H_1: the religious compositions of the corporate board membership and the adult population are different.

12c. $$x^2 = \frac{(650-635)^2}{635} + \frac{(260-269)^2}{269} + \frac{(10-24)^2}{24} + \frac{(80-72)^2}{72}$$

$x^2 = .35 + 1.30 + 8.17 + .89 = 9.71$

With $(4-1) = 3$ degrees of freedom, a value of chi-square of 9.71 would occur by chance with a probability less than .05 (since 9.71 is greater than 7.815 but less than 9.837). Therefore, we can reject H_0 at the $p < .05$ level of significance.

12d. We are certain that there is a relationship between religious affiliation and membership on corporate boards. ($p < .05$). Or, there is a significant relationship between religious affiliation and board membership ($p < .05$).

13. You could not apply these inference techniques to a quota sample because the techniques assume random sampling or its equivalent, a condition that is not met by quota sampling techniques.

Copyrights and Acknowledgments and Illustration Credits

Tables

3-5, 3-6, 3-7, and **3-8**
Lenski, G., and Lenski, J. (1978). *Human Societies* (3rd ed.), Table 4.5, p. 101. New York: McGraw-Hill. Reprinted by permission of the publisher.

5-4
Bowles, S., and Nelson, V. (1974). "The 'Inheritance of IQ' and the Intergenerational Reproduction of Economic Inequality." *Review of Economics and Statistics* 56, no. 1 (February). North Holland Publishing Company. Reprinted by permission of the publisher and the authors.

11-8 and **Appendix 1**
Rugg, Harold O. *Statistical Methods Applied to Education*, Appendix Table III, pp. 389–90. Copyright © 1917, renewed 1945 by Houghton Mifflin Company. Used by permission.

12-2 and **Appendix 2**
Fisher, R.A., and Yates, F. (1974). *Statistical Tables for Biological, Agricultural, and Medical Research* (6th ed.), Table III. Published by Longman Group Ltd., London (previously published by Oliver & Boyd, Edinburgh). Reprinted by permission of the literary executor of the late Sir Ronald A. Fisher, F.R.S., Dr. Frank Yates, F.R.S., and Longman Group, Ltd., London.

13-5 and **Appendix 3**
Fisher, R.A., and Yates, F. (1974). *Statistical Tables for Biological, Agricultural, and Medical Research* (6th ed.), Table IV. Published by Longman Group Ltd., London (previously published by Oliver & Boyd, Edinburgh). Reprinted by permission of the literary executor of the late Sir Ronald A. Fisher, F.R.S., Dr. Frank Yates, F.R.S., and Longman Group Ltd., London.

Appendix 4
Reprinted from p. 2 of *A Million Random Digits with 100,000 Normal Deviates* by the RAND Corporation (New York: The Free Press, 1955). Copyright © 1955 and 1980 by the RAND Corporation. Used by permission. Adapted in Dixon, W.J., and Massey, J., Jr. (1969). *Introduction to Statistical Analysis* (3rd ed.), Table A-1, pp. 446–47. New York: McGraw-Hill.

Appendix 5
Pearson, E.S., and Hartley, H.O. (1958). *Biometrika Tables for Statisticians*, vol. 1, Table 18, pp. 171–73. Reprinted by permission of the Biometrika Trustees.

Appendix 6

Wilcoxon, F., and Wilcox, R.A. (1964). *Some Rapid Approximate Statistical Procedures*, p. 28. Pearl River, NY: Lederle Laboratories, American Cyanamid Company. Reproduced with the permission of the American Cyanamid Company.

Appendix 7

From *Introductory Statistics* by Roger E. Kirk, pp. 423–24. Copyright © 1978 by Wadsworth Publishing Company, Inc. Reprinted by permission of Brooks/Cole Publishing Company, Pacific Grove, California 93950.

Appendix 8

Adapted in Weiss, N., and Hassett, M. (1982). *Introductory Statistics*, p. 594. Reading, MA: Addison-Wesley. From Owen, D.B. (1962). *Handbook of Statistical Tables*. Reading, MA: Addison-Wesley. Reprinted with permission.

Figures

2-1

From "Barriers to the Racial Integration of Neighborhoods: The Detroit Case," by R. Farley, S. Bianchi, and D. Colasanto, *Annals of the American Academy of Political and Social Sciences*, 441 (January), pp. 97–113.

3-2 and 3-3

Boocock, S. *Sociology of Education* (2nd ed.). Boston: Houghton Mifflin, 1980, p. 73.

3-4

Braudel, F. *The Wheels of Commerce*. New York: Harper and Row, 1979, p. 472.

9-2 and 9-3

Duncan, Otis D., "Inheritance of Poverty or Inheritance of Race?" in Daniel P. Moynihan (ed.), *On Understanding Poverty*. New York: Basic Books, 1969, p. 90.

11-7

Blalock, Hubert M., Jr. *Social Statistics*, Second Edition. New York: McGraw-Hill, 1972, pp. 181–84.

Index

A

Accuracy, of estimate, 319
Alienation, coefficient of, 181, 240
Analysis of variance, 361
Antecedent variables, 208–9
Association, measures of. *See* Measures of association
Asymmetrical measures of association, 137, 146, 153
Average. *See* Central tendency, measures of

B

Bar graphs, 49
Beta weights, 230. *See also* Regression coefficient, partial
Bias. *See* Error
Biased estimators, 296

C

Cases, 31. *See also* Sample, size of
  importance of number of, 37, 118, 218–19, 309–10
  missing, 32
Categorical variables, 22
Causal relationships. *See also* Simple recursive path models; Spuriousness
  time ordering and, 194
Cells, 35
Census, 10
Central-limit theorem, 279–84, 315
Central tendency, measures of, 66–82
  choosing, 82
  means
    calculation, 67, 69, 70–72
    interpretation, 68–70, 80
    and skewness, 93–96
  medians, 75–80
    calculation, 75–76, 77–78
    interpretation, 76, 80
    and skewness, 93–96
  modes, 81
  percentiles, 80–81
Chebyshev's Theorem, 90–91
Chi-square, 341–52
  degrees of freedom, 348
  expected frequencies, 343
  and goodness-of-fit tests, 349–52
  interpretation of, 348–49
  observed frequencies, 343
  and relationships between variables, 341–49
  table of probabilities, 346–47, 367
Cluster samples, 261–63
Codes, 9
Coefficient, correlation. *See* Correlation coefficient
Coefficient, path, 240
Coefficient of alienation, 181, 240
Coefficient of determination, 177
  in linear regression, 177
  in multiple regression, 228
Cohorts, 83
  cohort effects, 83
Complex samples, 260–63
Component variables, 206–8
Compound events, 116–17
Concordant pairs, 138–40, 150–52
  and spuriousness, 199
Conditional probability, 107, 108–9
Conditionally perfect association, 141
Confidence, trade-off with precision and sample size, 319–20
Confidence intervals, 297–320
  as an alternative to hypothesis testing, 339–40
  for differences between means, 317–19
  for differences between proportions, 314–16
  interpretation, 320
  for means, 297–309
  open-ended, 316–17
  for proportions, 310–13
    "instant," 312–13

Constants, 14
   regression. *See* Regression constant
Continuous variables, 22
Controls, idea of, 192–93
   and small samples, 218–19
Correlation coefficient, 171–75
   linear (r), 171–75,
      interpretation of, 182, 184
   matrix, 232–33
   multiple (R), 228
   partial, 200, 228
   rank order, 154–58
Covariance, 167–68, 171
Cross-products, 140, 143, 150–52
   negative, defined, 140
   positive, defined, 140
Cross-tabulations, 39
Curvilinear relationships, 129–31, 185–87

D

d, Somers'. *See* Somers' d
Data (datum), 10
Degrees of freedom
   chi-square, 348
      goodness-of-fit tests, 351
   t-distribution, 307
Density function, probability, 269–72
Dependence
   between events, 114
   between variables, 123–28
Dependent variable, 125
Description, 6
Descriptive statistics. *See* Statistics, descriptive
Determination, coefficient of. *See* Coefficient of determination
Deviation
   from the mean, 88
   standard. *See* Standard deviation
Diagram, Venn, 106, 111, 117
Dichotomy, 26–27
Differences
   absolute vs. relative, 46
   between means, percentages, or proportions. *See* Confidence intervals; Hypothesis testing; Standard error
   significant, 340–41
Direct effects, 203, 240
Direction of relationships. *See* Relationships between variables
Discordant pairs, 138–40, 150–52
   and spuriousness, 199
Discrete variables, 22
Dissimilarity, index of, 352–53
Distorter variables, 213–14
Distributions
   central tendency of. *See* Central tendency; Sampling distributions
   chi-square. *See* Chi-square
   F, 370–75
   frequency, 30
   kurtosis, 286
   normal curve. *See* Normal curve
   percentage, 36–39
   population, 272
   probability, 107–8
   proportional, 36–39
   sample, 272–73
   sampling. *See* Sampling distributions
   shape of, 92–96. *See also* Sampling distributions
      skewness, 93
      symmetry, 93
   Student's t. *See* Student's t-distribution
   tails of, 93, 285

E

Effects
   cohort, 83
   direct, 203, 240
   indirect, 202, 240–41
   interaction, 214–18
   longitudinal, 83
Efficiency, of estimate, 255–56
Error
   measurement, 19–21
      bias, 20
      random, 20
   sampling, 254–55
      bias, 256–57, 296
      random, 255–56. *See also* Statistical inference
      and stratified sampling, 255–56
   standard. *See* Standard error
   type I, 328
   type II, 328
Estimates
   accuracy of, 319
   biased, 296
   interval. *See* Confidence intervals
   point, 296, 298
   precision of, 319–20
   standard error of, 180–81
   unbiased, 296
Estimators. *See* Estimates
Events, 102. *See also* Independence
   compound, 116–17
   failure, 103
   joint, 102, 115–16
   mutually exclusive, 117
   simple, 102
   success, 103
Exogenous variables, 239–40
Expected frequency, 343
Expected values, 118–19
Explained variance. *See* Variance, explained

Explanation, 6
Extraneous variables, 200

F

Failure, 103
F-distribution, 370–75
Finding, 7
Frame, sample, 249–50
Frequency
 distributions, 30
 expected and observed, 343
 polygons, 56

G

Gamma, Goodman and Kruskal's. *See* Goodman and Kruskal's gamma
General Social Survey, 8
GENSOC, 8
Goodman and Kruskal's gamma, 149–53
 calculation, 149–52
 interpretation, 152–53
Goodman and Kruskal's tau, 146–49
 calculation, 146–48
 interpretation, 148–49
Goodness-of-fit test, 349–52
Graphics, 48–60
 bar graph, 49
 histogram, 55–56
 line graph, 52
 ogive, 58–59
 scatterplot, 59–60, 186
 stem and leaf display, 56–58
Grouped data (grouping), 33

H

Heterogeneity. *See* Variance
Histograms, 55–56
Homogeneity. *See* Variance
Hypothesis, 325
 null, 326
 substantive, 326
Hypothesis testing, 295, 325–56
 goodness-of-fit, 349–52
 interpretation of results, 332–34
 limitations of, 355–56
 logic of, 326–29
 about means, 329–34
 one- and two-tailed, 334
 about proportions, 334–36
  about differences between proportions, 337–39
 about relationships between variables, 341–49, 353–55
 significance level of, setting, 330, 332–34

I

Independence
 as condition of scientific sampling, 251
 between events, 111–15
 mean, 184–86
 and significance testing. *See* Hypothesis testing
 stochastic, 127, 184–86
 symmetry of, 113–14
 between variables, 123–28
Independent variables, 125
Index of dissimilarity, 352–53
Indirect effects, 202, 240–41
Inductive statistics, 10
Inference, statistical. *See* Statistical inference
Inferential statistics. *See* Statistical inference
Interaction effects, 214–18
Interval, skip, 253
Interval estimates. *See* Confidence intervals
Interval-scale variables, 24. *See also* Measures of association, Relationships between variables
Intervening variables, 200–206
 and spuriousness, 202, 206
Irregular relationships, 131–33

J

Joint events, 102, 115–16
Joint probability, 110, 115–16

K

Kendall's tau, 156–58
 calculation, 156–58
 interpretation, 158
Kurtosis, 286

L

Law of large numbers, 118
Least-squares regression line, 166. *See also* Regression analysis
Leptokurtic, 286
Levels of measurement. *See* Measurement
Line graphs, 52
Linear regression. *See* Regression analysis
Linear relationships, 128
Listers, 261
Log-linear analysis, 361
Longitudinal effects, 83

M

Mann-Whitney test, 354, 377–78
Marginals, 35

Mathematically perfect association, 141
Matrix, correlation, 232–33
Mean. *See also* Confidence intervals; Hypothesis testing; Standard error
  calculation, 67, 69, 70–72
  interpretation, 68–70, 80
  of sampling distributions, 275
  and skewness, 93–96
Mean independence, 184–86
Measurement, 13–27. *See also* Variables
  error, 19–21
    bias, 20
    random, 20
    instrument, 14
  levels of, 21–25. *See also* Nominal-; Ordinal-; Interval-; Ratio-scale variables
    importance of, 25–26, 134
  operationalization, 14
  properties, summary of, 26
  reliability, 16–17
  validity, 17–19
    construct, 17–18
    predictive, 19
Measures of association, 133–34, 137–39. *See also names of individual measures*
  asymmetrical, 137, 146, 153
  ideal, 137
  interval scales, 171–77, 228
  nominal scales, 137–49
  ordinal scales, 149–58
  PRE interpretation, 142, 148, 152
  rank order, 154–58
  ratio scales, 171–77, 228
  statistical significance of, 353–54
  symmetrical, 137, 138, 143, 149
Medians, 75–80
  calculation, 75–76, 77–78
  interpretation, 76, 80
  and skewness, 93–96
Modes, 81
Multiple classification analysis, 361
Multiple correlation coefficient, 228
Multiple regression analysis, 227–38, 240. *See also* Path analysis; Regression analysis
  equation, 227
Multistage sample, 260–61
Multivariate analysis, idea of, 192, 226. *See also* Multiple regression analysis
Mutually exclusive events, 117

N

Negative cross-product, 140. *See also* Cross-products
Negative skewness, 94
Nominal-scale variables, 22. *See also* Measures of association; Relationships between variables

Nonlinear relationships. *See* Relationships between variables
Nonparametric statistics, 354–55
Normal curve, 285–92
  table of probabilities, 288, 365
  using, 286–92
Null hypothesis, 326

O

Observed frequency, 343
Ogives, 58–59
One-tailed hypothesis tests, 334. *See also* Hypothesis testing
Open-ended confidence intervals, 316–17
Operationalization, 14
Order of relationship, 193
Ordinal-scale variables, 23. *See also* Measures of association; Relationships between variables
Oversampling, 258–60

P

Pairs. *See* Concordant; Discordant; Tied pairs
Panel study, 83. *See also* Trends
Partials
  correlation coefficient, 200, 228
  regression coefficient (slope), 228–30
    standardized, 229–30
    unstandardized, 229–30
  relationships, 200, 228
Path analysis, 238–41
  requirements, 239
  simple recursive, 239
Path coefficient, 240
Pearson product-moment correlation coefficient. *See* Correlation coefficient
Pearson's *r*. *See* Correlation coefficient
Percentage, 37. *See also* Confidence intervals; Hypothesis testing; Standard error
  direction of, 39–42, 124–25
  distributions, 36–39
Percentile, 80–81
Perfect association. *See* Relationships between variables, perfect
Periodicity, 254
Phi coefficient, 143–46
  calculation, 143–44
  interpretation, 144
  compared with Yule's Q, 145–46
Phi-squared coefficient, 144–45
Platykurtic, 286
Point estimate, 296, 298
Population, 10
  distribution, 272
Positive cross-product, 140. *See also* Cross-products
Positive skewness, 93–94
PRE interpretation, 142, 148, 152

Precision, trade-off with sample size and confidence, 319–20
Prediction, 7
 and measures of central tendency, 82
Probabilities proportional to size (PPS), 262–63
Probability, 101–20. *See also* Events
 compound, 116–17
 conditional, 107, 108–9
 defined, 103
 density function, 269–72
 distributions, 107–8
 joint, 110, 115–16
 proportional to size, 262–63
 simple, 105–6
 unconditional, 106, 108
Probability density function, 269–72
Product-moment correlation coefficient. *See* Correlation coefficient
Proportional reduction in error, 142, 148, 152
Proportions, 37. *See also* Confidence intervals; Hypothesis testing; Standard error
 distributions, 36–39

Q

Q, Yule's. *See* Yule's Q
Qualitative variables, 22
Quantitative variables, 22
Quota sample, 264–65

R

r. *See* Correlation coefficient
Random error. *See* Error
Random numbers, 252
 table of, 368–69
Random sample, simple, 252–53
Range, 92
Rank-order correlation, 154–58
Rates, 47
Ratios, 45
Ratio-scale variables, 25. *See also* Measures of association; Relationships between variables
Recursive, simple, 239
Regression analysis, 164–87
 curvilinear, 185
 independence in, 184–86
 least-squares regression line, 166
 linear, 165–85
 multiple, 227–38, 240. *See also* Path analysis
 equations, 227
Regression coefficient, 167
 calculation of, 167–69, 170
 interpretation of, 182, 184
 partial, 228–30
  standardized, 229–30
  unstandardized, 228, 230
 statistical significance of, 234
Regression constant, 167
 calculation of, 169–70
 in multiple regression, 227–28
Relationships between variables, 123, 125. *See also* Measures of association; Regression analysis
 direction of, 128–33, 134–37
 linear, 128
 with nominal variables, 134–37
 nonlinear, 129–33
  curvilinear, 129–31, 185–87
  irregular, 131–33
 independence, 123–28
 order of, 193
 perfect, 141
  conditionally, 141
  mathematically, 141
 statistical significance of, 341–49, 353–55
 strength of, 133–34
Reliability, 16–17
Replacement
 sampling with, 111–12
 sampling without, 113
Representative samples, 249
Residuals, 180
Response rate, 256–58
r_s, Spearman's. *See* Spearman's r_s
Runs test, Wald-Wolfowitz, 354, 379

S

Sample, 10
 cluster, 261–63
 complex, 260–63
 defined, 10
 distributions, 272–73
 efficiency, 255–56
 error. *See* Error, sampling
 frame, 249–50
 multistage, 260–61
 oversampling, 258–59
 PPS, 262–63
 quota, 264–65
 with replacement, 111–12, 252–53
 without replacement, 113, 252–53
 representative, 249
 response rates, 256–58
 rules of, 250–51
 simple random, 252–53
 size of. *See also* Student's *t*-distribution
  choosing, 118, 218–19, 249, 312
  and confidence intervals, 309–10
  and size of population, 249
  trade-off with precision and confidence, 319–20

Sample (*continued*)
  stratified, 255–56
  systematic, 253–54
  weights, 259–60
Sampling distributions, 272–85
  defined, 274
  mean of, 275
  shape of, 276–78, 279–84. *See also* Standard error
  variance of. *See* Standard error
Scales, measurement. *See* Measurement
Scatterplots, 59–60, 186
Scores, standard, 287. *See also* Z-scores
Screeners, 260
Shape, of distributions, 92–96. *See also* Sampling distributions
  skewness, 93
  symmetry, 93
Significance, statistical. *See also* Hypothesis testing
  interpretation of, 332–34, 340
  level, 328
  of measures of association, 353–54
  of regression coefficients, 234
Simple event, 102
  probability of, 105–6
Simple random sample, 252–53
Simple recursive path models, 239
Size of sample. *See* Cases; Sample, size of
Skewness, 93
  negative, 94
  positive, 93
Skip interval, 253
Slope. *See* Regression coefficient
Somers' *d*, 153–54
  calculation, 153–54
  interpretation, 154
Spearman's r_s, 154–56
  calculation, 154–56
  interpretation, 156
Specification, 214–18
Spuriousness, 194–200
  and concordant, discordant pairs, 199
  and extraneous variables, 200
  importance of, 198–99
  and intervening variables, 202, 206
  testing for, 196–98
Squares, sum of, 90
Standard deviation, 86–90
  calculation, 86–90, 91
  interpretation, 90–91
Standard error, 276–79
  for differences between means, 317
  for differences between proportions, 315
  for means, 278–79
  for proportions, 310
Standard error of estimate, 180–81
  calculation, 181
  interpretation, 181

Standard scores. *See* Z-scores
Standardized regression coefficient, 229–30
Statistic, 3
Statistical inference, 268, 295, 298. *See also* Confidence intervals; Hypothesis testing; Significance
Statistical significance. *See* Significance
Statistics
  defined, 3
  descriptive, 6, 10
  individuals and, 65–66
  inductive, 10
  inferential, 10
  lying with, 61
  nonparametric, 354–55
Stem and leaf displays, 56–58
Stochastic independence, 127, 184–86
Strata, sample, 255–56
Stratification, sampling, 255–56
Stratum, sampling, 255–56
Strength of association. *See* Measures of association
Student's *t*-distribution, 307–9
  table of, 308, 366
Substantive hypothesis, 326
Subtables, 196
Success, 103
Sum of squares, 90
Suppressor variables, 209–13
Symmetry
  of distributions, 93, 280
  of measures of association, 137, 138, 143, 149
Systematic sample, 253–54

T

Tables, 31
  subtables, 196
Tails, 93
  of normal distribution, 285
Tau, Goodman and Kruskal's. *See* Goodman and Kruskal's tau
Tau, Kendall's. *See* Kendall's tau
t-distribution. *See* Student's *t*-distribution
Tied pairs, 153–54
Time order of variables, 194
Trends, 52–54
  measuring, 82–85
Trials, 102
Type I error, 328
Type II error, 328

U

Unbiased estimators, 296
Unconditional probability, 106, 108
Unexplained variance. *See* Variance

Unstandardized regression coefficient, 229–30
Unstandardized scores, 287
 translating to standard scores, 290–92

V

Validity, 17–19
 construct, 17–18
 predictive, 19
Value, expected, 118–19
Variables, 14. *See also* Measurement
 antecedent, 208–9
 component, 206–8
 continuous, 22
 dependent, 125
 discrete, 22
 distorter, 213–14
 exogenous, 239–40
 extraneous, 200
 independent, 125
 intervening, 200–206
 qualitative and quantitative, 22
 suppressor, 209–13
Variance. *See also* Covariance
 analysis of, 361
 calculation of, 86–90, 91
 explained
  in linear regression analysis, 177–80
  in multiple regression analysis, 227–28
  and phi-squared, 144–45, 179
 interpretation of, 90–91
 and PQ, 311
 and relative size of slopes and correlations, 183
 of sampling distributions. *See* Standard error
 unexplained 177, 180–81
Venn diagrams, 106, 111, 117

W

Wald-Wolfowitz runs test, 354, 379
Weights, sampling, 259–60
Wilcoxon test, 354, 376

Y

Y-intercept. *See* Regression constant
Yule's Q, 138–43
 calculation, 138–41
 interpretation, 142–43
  compared with phi and phi-squared, 145–46

Z

Zero-order relationship, 193
Z-scores, 287–89
 translating to unstandardized scores, 290–92